New ICMI Study Series
VOLUME

For further volumes:
http://www.springer.com/series/6351

Information on the ICMI Study programme and on the resulting publications can be obtained at the ICMI website http://www.mathunion.org/ICMI/ or by contacting the ICMI Secretary-General, whose email address is available on that website.

Richard Barwell • Philip Clarkson
Anjum Halai • Mercy Kazima
Judit Moschkovich • Núria Planas
Mamokgethi Setati Phakeng • Paola Valero
Martha Villavicencio Ubillús
Editors

Mathematics Education and Language Diversity

The 21st ICMI Study

Springer

Editors
Richard Barwell
Faculty of Education
University of Ottawa
Ottawa, ON, Canada

Anjum Halai
Institute for Educational Development
East Africa
Aga Khan University
Dar es Salaam, Tanzania

Judit Moschkovich
University of California, Santa Cruz
Santa Cruz, CA, USA

Mamokgethi Setati Phakeng
University of South Africa
Pretoria, South Africa

Martha Villavicencio Ubillús
Ministerio de Educación
Lima, Peru

Philip Clarkson
Australian Catholic University
Fitzroy, VIC, Australia

Mercy Kazima
University of Malawi
Zomba, Malawi

Núria Planas
Universitat Autònoma de Barcelona
Barcelona, Catalonia-Spain

University of South Africa
Pretoria, South Africa

Paola Valero
Aalborg University
Aalborg, Denmark

ISSN 1387-6872 ISSN 2215-1745 (electronic)
New ICMI Study Series
ISBN 978-3-319-14510-5 ISBN 978-3-319-14511-2 (eBook)
DOI 10.1007/978-3-319-14511-2

Library of Congress Control Number: 2015944108

Acknowledgments

The editors are grateful to the following colleagues who reviewed chapters: Cris Edmonds-Wathen, Anthony Essien, Susan Gerofsky, Mercy Kazima, Michael Meyer, Frank Monaghan, Candia Morgan, David Pimm, Ricardo Scucuglia, Denisse Thompson, David Wagner, and Lynn Webb.

Richard Barwell thanks the Faculty of Education, University of Ottawa, for financial support and Farzaneh Salehi for editorial assistance during the preparation of the manuscript.

Contents

Contributors

Richard Barwell Faculty of Education, University of Ottawa, Ottawa, ON, Canada

Marcelo Borba Universidade Estadual Paulista (UNESP), Sao Jose Rio Preto, SP, Brazil

Arindam Bose Homi Bhabha Centre for Science Education, TIFR, Mumbai, India

Margarida César Associated Researcher, University of Neuchâtel, Neuchâtel, Switzerland

Lim Chapsam Universiti Sains Malaysia, Penang, Malaysia

Faïza Chellougui Faculté des Sciences de Bizerte, Bizerte, Tunisia

Nancy Chitera University of Malawi, Blantyre, Malawi

Philip Clarkson Australian Catholic University, Fitzroy, VIC, Australia

Viviane Durand-Guerrier University of Montpellier, Montpellier, France

Cris Edmonds-Wathen Charles Darwin University, Darwin, NT, Australia

Anthony A. Essien University of the Witwatersrand, Johannesburg, South Africa

Solange Hassan Ahmad Ali Fernandes Anhanguera University of São Paulo, São Paulo, SP, Brazil

George Gadanidis Faculty of Education, University of Western Ontario, London, ON, Canada

Leila Goosen University of South Africa, Johannesburg, South Africa

Anjum Halai Institute for Educational Development, East Africa, Aga Khan University, Dar es Salaam, Tanzania

Lulu Healy Anhanguera University of São Paulo, São Paulo, SP, Brazil

Mercy Kazima University of Malawi, Zomba, Malawi

Gladis Kersaint University of South Florida, Tampa, FL, USA

Jennifer M. Langer-Osuna Stanford University, Stanford, CA, USA

Paul Libbrecht Informatics, Weingarten University of Education Kirchplatz 2, Weingarten, Germany

Jackeline Rodrigues Mendes Universidade Estadual de Campinas, São Paulo, Brazil

Michael Meyer University of Cologne, Cologne, Germany

Judit Moschkovich University of California, Santa Cruz, Santa Cruz, CA, USA

Irfan Muzaffar Education and Social Research Collective, Pinner, UK

Judith Njomgang Ngansop Université de Yaoundé 1, Yaoundé, Cameroon

Thulisile Nkambule University of South Africa, Pretoria, South Africa

Eva Norén Stockholm University, Stockholm, Sweden

Aldo Parra Aalborg University, Aalborg, Denmark

Jurema Lindote Botelho Peixoto State University of Santa Cruz, Illheus, Bahia, Brazil

Núria Planas Universitat Autònoma de Barcelona, Barcelona, Catalonia-Spain

University of South Africa, Pretoria, South Africa

Arthur B. Powell Rutgers University, Newark, NJ, USA

Susanne Prediger Institute for Development and Research in Mathematics Education, TU Dortmund University, Dortmund, Germany

Elizabeth Becerra Ramos Cinvestav-IPN, Mexico City, Mexico

Leila Salekhova Kazan Federal University, Kazan, Republic of Tatarstan, Russia

Ricardo Scucuglia Universidade Estadual Paulista (UNESP), Sao Jose Rio Preto, SP, Brazil

Mamokgethi Setati Phakeng University of South Africa, Pretoria, South Africa

Denisse R. Thompson University of South Florida, Tampa, FL, USA

Hien Nguyen Thi Thu Catholic University of Louvain, Louvain, Belgium

Tony Trinick The University of Auckland, Auckland, New Zealand

Nail Tuktamyshov Kazan University of Architecture and Engineering, Kazan, Republic of Tatarstan, Russia

Paola Valero Aalborg University, Aalborg, Denmark

Sumaia Vazquez Universidade Federal de Vitoria, Vitória, Brazil

Martha Villavicencio Ubillús Ministerio de Educación, Lima, Peru

Hannatjie Vorster North-West University, Potchefstroom, South Africa

Marthie S. Van der Walt North-West University, Potchefstroom, South Africa

Lyn Webb Nelson Mandela Metropolitan University, Port Elizabeth, South Africa

Carl Winsløw Institut for Naturfagenes Didaktik, University of Copenhagen, Copenhagen, Denmark

List of Figures

List of Tables

Chapter 1
Introduction: An ICMI Study on Language Diversity in Mathematics Education

Richard Barwell, Philip Clarkson, Anjum Halai, Mercy Kazima, Judit Moschkovich, Núria Planas, Mamokgethi Setati Phakeng, Paola Valero, and Martha Villavicencio Ubillús

1.1 History of the Study

ICMI's concern about the development of research on mathematics education and language diversity dates back to 1972, and the second International Congress on Mathematical Education (ICME-2) in Exeter, United Kingdom. As explained by Howson (1973) in the introduction of the congress proceedings, the inclusion of a working group on language was an important novelty. Moreover, during the ICMI general assembly at ICME-2 a decision was made to host an international symposium on 'Interactions between Linguistics and Mathematical Education' as a response to the need for fundamental research on the relationship between the learning of basic mathematical structures and the language through which they are learnt. The Symposium was held in 1974 in Nairobi, Kenya, with sponsorship by UNESCO in cooperation with ICMI and the Centre for Educational Development Overseas. There had not been international conferences focusing exclusively on the relationship between mathematics and language before. This early event actually focused

R. Barwell (✉)
Faculty of Education, University of Ottawa, 145 J-J-Lussier, Ottawa, ON, Canada K1N 6N5
e-mail: richard.barwell@uottawa.ca

P. Clarkson
Australian Catholic University, Fitzroy, VIC, Australia
e-mail: Philip.Clarkson@acu.edu.au

A. Halai
Institute for Educational Development, East Africa, Aga Khan University, Dar es Salaam, Tanzania
e-mail: anjum.halai@aku.edu

M. Kazima
University of Malawi, Zomba, Malawi
e-mail: mkazima@cc.ac.mw

R. Barwell et al. (eds.), *Mathematics Education and Language Diversity*, New ICMI Study Series, https://doi.org/10.1007/978-3-319-14511-2_1

on the relationship between *linguistics* and mathematics education, which is different from a focus on *language* and mathematics education. Linguistics and mathematics education are two disciplines while the role, use and effect of language on mathematics education implies all happening in one discipline.

In the final report of the symposium (UNESCO, 1974), the lack of research on the relationship between language and mathematics was highlighted. The report concluded that 'difficulties in mathematics learning depend on the language of learning' (p. 8). It was further affirmed that all languages include linguistic features of benefit for the acquisition of mathematical concepts and thus can be used for mathematics teaching and learning. It was claimed that the problems of learning mathematics in an additional or foreign language are not peculiar to learning in a world language such as English or French because there are many countries such as Tanzania and India, where many learners have to learn mathematics in a national language (e.g. Kiswahili, Hindi) which is not their home language. This situation still continues. Meanwhile, European countries like Sweden or Greece experience the pressure to ensure that their learners are fluent in at least one of the world languages.

Both ICME-2 and the Nairobi Symposium gave the impetus for the first paper to appear in a mathematics education journal focusing on mathematics and language diversity; it was authored by Austin and Howson (1979) and published in *Educational Studies of Mathematics*. Austin and Howson conclude that the challenge of language and mathematics learning and teaching is not just an issue for developing countries but for the whole world. In most developing countries, the challenge is that of learners learning mathematics in a language that is not their mother tongue; in developed countries such as Belgium and Canada there are communities with well-established 'minority' languages; and in some countries learners and teachers have to face the non-standard nature of the local vernacular (e.g. Jamaica, England

J. Moschkovich
University of California, Santa Cruz, Santa Cruz, CA, USA
e-mail: jmoschko@ucsc.edu

N. Planas
Universitat Autònoma de Barcelona, Barcelona, Catalonia-Spain

University of South Africa, Pretoria, South Africa
e-mail: Nuria.Planas@uab.cat

M. Setati Phakeng
University of South Africa, Pretoria, South Africa
e-mail: phakerm@unisa.ac.za

P. Valero
Aalborg University, Aalborg, Denmark
e-mail: paola@learning.aau.dk

M. Villavicencio Ubillús
Ministerio de Educación, Lima, Peru
e-mail: MVILLAVICENCIO@minedu.gob.pe

and the United States). Austin and Howson acknowledged the fact that bilingualism is a political matter and thus change in society may lead to policy change. Indeed much has changed since 1979: the world has become more multilingual and some countries have changed their language policies and practices, which makes this ICMI Study timely and relevant.

1.1.1 What Do We Mean by Language Diversity and Why Does It Matter for Mathematics Education?

A first challenge for the International Program Committee (IPC) for this study was to decide on the title of the study. We elected to use the expression 'language diversity' rather than specify particular situations (e.g. multilingual, bilingual, etc.). In making our choice, we drew on ideas from recent work in sociolinguistics (e.g. Pennycook & Makoni, 2005; Makoni & Pennycook, 2007). This work challenges three sets of ideas about language in contemporary society in the context of increasing diversity (Blommaert & Rampton, 2011).

First, *languages* should not be understood as monolithic, static, rule-governed forms of communication. While the idea of languages is socially useful, in research terms it is more difficult to sustain in this form. In practice, for example, it is difficult to objectively identify boundaries between many languages. Indeed, the ideology of distinct languages arguably has its roots in nineteenth century Europe where a rationality of classification and creation of distinctions was not only part of the development of social sciences such as linguistics, but also underpinned the ideological formation of nation states. This view of discrete languages belonging to separate peoples (German for the Germans, French for the French) was subsequently exported to much of the rest of the world through European colonisation. In parts of Africa, for example, colonisation resulted in a language spectrum being carved up into discrete languages, which then defined the people who spoke them. Where before there had existed a language continuum, with colonisation there were distinct, named languages (Makoni & Pennycook, 2007). This ideology of languages informs how classrooms are understood today. For example, students who do not speak according to a pre-conceived standard form of a language may be seen as less educated.

Second, in the same way that languages are not monolithic, so *language speakers* are not monolithic either. This point leads to a critique of notions like 'native speaker', 'fully bilingual speaker' or designations like 'first language', 'second language', etc. From this perspective, terms like 'English as a second language' or 'English as an additional language' are open to question, since they imply a neat sequence and a separation between 'English' and the 'second language'. In many settings, however, learners of English as a second language may in fact draw on several languages for different aspects of their lives and combine aspects of these languages for different purposes. It then becomes difficult to identify what is

'English', what is 'second language' and what is 'other languages'. Similarly, a 'bilingual' student may also speak other languages, or different varieties of the official languages.

These terms are problematic politically because they privilege particular idealised or preferred forms of language use and hence privilege learners who conform to these preferences. For example, to describe a student as a learner of English as a second language in a context in which English is the official language of schooling is to erase their first language (or possibly first language*s*) and to privilege English. Similarly, to describe a classroom as bilingual can privilege two languages and a particular ideal of the bilingual speaker. The more one looks carefully at almost any mathematics classroom, the more difficult it becomes to adequately characterise language ecology in a few simple words. Each participant in a mathematics classroom, including the teacher, brings their own particular combination of languages, varieties and ways of talking.

Finally, the nature of *communication* itself has been increasingly rethought. In the context of contemporary super-diversity (Vertovec, 2007), it is apparent that speakers draw on multiple semiotic means in order to communicate. These semiotic means include parts of recognisable languages (e.g. some Hindi combined with some English). But they also draw on additional semiotic means, such as accent, style, register or genre, as well as different varieties within any particular language. In mathematics, for example, these means include gesture, diagrams, and symbols.

Rather than specify particular forms of classroom or setting, then, we elected to frame this study in terms of language diversity. The final title for the study, 'Mathematics Education and Language Diversity', ensured that the study was broad enough to encompass any conceivable language setting. There is always diversity of some kind. Moreover, the form of language diversity may vary from one classroom to another. This variety starts with the different languages and other resources on which students may draw. These resources may not be common to all students. It also includes the different languages and resources on which the teacher may draw. These languages and resources may not be the same as those used by the teacher's students. There is also variety in the different policy environments in which mathematics education takes place, for example, in terms of which languages are recognised, mandated, required or ignored. These policies themselves reflect the long-standing assumptions about the nature of language and of speakers that are questioned in this section. In this volume, we have succeeded in ensuring that a wide variety of forms of language diversity are represented.

These ideas matter for mathematics education for the simple reason that mathematics education is always happening in the context of language diversity. In our superdiverse societies, shaped by migration, mobile technologies, social media and global trade, learners and teachers of mathematics operate in a complex linguistic landscape. This landscape forms the backdrop to their participation in mathematics. This book is testament to this reality. In every specific example of mathematics classroom interaction in this book, participants are drawing on multiple resources of some kind, including multiple languages, gestures, registers, or genres.

1.1.2 *Outline of the Chapter*

The rest of this chapter begins with a brief discussion of the background and scope of the study. We then outline the process followed, including the discussions of the appointed IPC, the preparations of the discussion document and the organisation of the study conference, which not only aimed to deliberate on what relevant research had already been undertaken but envisaged what was needed to be done to push the boundaries further. The chapter concludes with some reflections on crucial themes within this volume for both the wider communities of curriculum developers, teachers, politicians and members of non-dominant communities, as well as pointers to some underlying issues that should inform continuing research in the nexus of language and mathematics education.

1.2 Changing Perspectives on Mathematics Education and Language Diversity

The beginnings of research on language and learning started with a focus on bilingualism and bilingual learners. This location of the problem in the learner was based on an underlying assumption that there is something wrong with the bilingual learner. A great majority of studies undertaken before 1979, when the paper by Austin and Howson was published, concluded that capacity in more than one language had negative effects on learners' linguistic, cognitive and educational development (Reynold, 1928 and Saer, 1923, both cited in Grosjean, 1982). Bilingualism was seen as unnatural and it was argued that a bilingual child hardly learns either of the two languages as perfectly as would be the case if only one language were learnt. There was also a widespread view that the effort required by the brain to master two languages instead of one diminished the child's power of learning other content. Weisgerber (1933, cited in Saunders, 1988) argued that bilingualism could impair the intelligence of a whole ethnic group, while Reynold (1928, cited in Saunders, 1988) was concerned about the fact that bilingualism could lead to language mixing and language confusion which in turn resulted in a reduction of the ability to think. From his study of Welsh–English bilingual children in rural areas, Saer (1923, cited in Saunders, 1988) concluded that bilingual learners had lower IQ scores than monolingual children. Commenting on this result, Saunders (1988) warned that caution must be exercised when comparing monolinguals and bilinguals on tests of intelligence, particularly on tests of verbal intelligence, and especially if, as so often happens, the bilinguals are tested in only one of their languages, often their second language.

Another stream of research, emerging in the 1960s, began to argue for a more nuanced understanding of this literature. These authors argued that under certain conditions competence in more than one language can have positive effects on the learning process (Ianco-Worrall, 1973; Ben-Zeef, 1977; Bialystok, 1987; Pearl &

Lambert, 1962, all cited in De Klerk, 1995). In 1962, Peal and Lambert conducted a seminal study that suggested that bilingualism could be an asset to the learning of the child. They studied the effects of bilingualism on the intellectual functioning of 10-year-old children from six Montreal schools. They found that instead of suffering from 'mental confusion', bilinguals were benefiting from a language asset that monolingual children did not have. While these results were criticised on the grounds that only the intellectually brighter children were chosen for the bilingual group (e.g. by Macnamara, 1966, cited in Saunders, 1988), the empirical studies that followed reinforced the idea that bilingualism can indeed be an asset (e.g. Ianco-Worrall, 1973; Been-Zeef, 1977, both cited in De Klerk, 1995). Knowing what we know now, it is strange that these arguments based on a deficit-model were accepted as plausible for so many years, and despite the fact that asset-based arguments were being documented. Perhaps this shows the power of simple, even naive, ways of understanding how the brain works and the implications this has for learning.

In 1979, Swain and Cummins compared the deficit-based arguments and the asset-based arguments in the context of the different studies and concluded that the asset-based findings were usually associated with majority language groups in immersion programmes. In other words, for these children the second language is added at no cost to the first. However they also noted that the parents of the children in these immersion programs were of relatively high socio-economic status and the parents placed a high value on knowing two languages. Deficit-based findings, on the other hand, were found with immersion students who are surrounded by negative attitudes. They were forced to learn the majority language and were not encouraged to retain their first language. Further, they did not live in a social environment that was conducive to learning. Nevertheless, Swain and Cummins (1979) argued that while there were a variety of factors impacting children's intellectual development, bilingualism was one of the significant factors that could have a positive impact, and was not automatically negative for children's learning. Cummins's (1979, 1981) theory of the relationship between language and cognition elaborates on the conceptualisation of the learner as someone who brings all the acquired strengths from the first language to use them in the process of learning a second language. In this regard, the learner's first language leads to the acquisition of fluency in the second language so that bilingualism becomes a 'cognitive advantage' rather than a factor that impedes learning. While research in this area of study at this stage did not foreground the role of the social dimension of language (although Cummins did anticipate this in some of his earlier work), it is clear that there was an acceptance that it is possible that bilingualism per se might not necessarily have any effects (either negative or positive) on the cognitive and intellectual development of children. That is, what may account for the contradictory results reported in the literature during this period are the psychosocial differences between bilinguals and monolinguals, and not bilingualism per se.

While the research described above did not specifically focus on mathematics education, its findings influenced the research that followed in mathematics education. Interestingly, the studies of language and mathematics education reviewed by

Austin and Howson (1979) were largely dominated by a deficit perspective of the learner. In the bibliography that they presented at the end of their chapter, it can be observed that the majority of the works focused either on linguistics or on issues relating to the linguistic aspects of mathematics, rather than issues relating to the challenges of teaching and learning mathematics in the context of language diversity. Nowadays the deficit perspective is being more and more overcome, particularly due to the impact of results from several studies in different research contexts all around the world. It cannot be said, however, that this perspective has been totally overcome, either in research or in educational policy, where compensatory responses and remedial approaches are common in the interpretation of the needs of the learners whose home languages are different from the language of instruction.

Using Cummins's work as a basis, in the early 1980s, Clarkson working in Papua New Guinea and, separately, Dawe working with immigrant children in England showed that Cummins's threshold language theory (Cummins, 1979) had good explanatory power for mathematics performance (Clarkson, 1991, 1992; Dawe, 1983). Their results suggested that bilingual students who were competent in their home language and the language of teaching outperformed other students in mathematics, even when other factors such as socioeconomic status and parental education were accounted for. Later, working together, similar results concerning performance on mathematics were obtained for bilingual immigrant students in Australia (Clarkson, 1996; Clarkson & Dawe, 1997). These studies, although suffering from some of the pitfalls outlined above, showed that the notion of bilingualism had to be far more nuanced than had been the case in earlier research, and global applications of deficit models were just not appropriate. The impact of bilingualism on children's learning of mathematics was neither simple nor unitary. They also showed some of the ranges of individual and social factors impacting children's learning of mathematics, including language, and showed that many of these factors, such as the socio-economic status of families, could not be influenced by schooling. But these studies also suggested that the way children used their languages was a learnt behaviour and hence could perhaps be influenced by schooling. Hence, the ways teachers and schools dealt with students' multiple languages in relation to mathematics learning was for the most part an unrecognised but critical issue in both research and practice.

More than a decade after Austin and Howson's paper, Secada (1992) provided an extensive overview of research on bilingual education and mathematics achievement (with a North-American-orientation), and pointed to findings of a significant relationship between the development of language and achievement in mathematics. This publication is not a mere update of the former review by Austin and Howson. It recognises that the social dimension of language needs to be integrated in theory and research, in order to understand and explain differences in the mathematics achievement of bilingual learners from different racial, ethnic and social class contexts. In particular, he concluded that the studies he reviewed indicated that oral proficiency in English in the absence of mother tongue instruction is negatively related to achievement in mathematics. Nevertheless, he found that correlations between language proficiency and mathematics achievement were highly variable and stated that there was much variance to be explained. He pointed to this finding from his review along with another important consistent finding: that middle and upper social class learners entered school with higher achievement levels in mathematics than lower social class learners. Social class was thus highlighted as a relevant variable in his analysis of the learning of mathematics by bilingual learners.

While questions relating to the relationship between language and mathematics had been the focus of study for over 30 years, it was only in the mid-1990s that the focus shifted to multilingual mathematics classrooms. This was mainly due to Adler's work, which was carried out in multilingual South Africa, exploring mathematics teaching and learning in classrooms where the teacher and learners did not share a home language or one in which there was an imposed language of learning and teaching which may not have been the home language for either the teacher or learners (Adler, 1995, 1997, 1998). Through her exploration, Adler (2001) captured three interrelated dilemmas that lie at the heart of teaching mathematics in multilingual classrooms: code-switching, mediation, and transparency. She provided a sharp analysis and strong theoretical grounding, pulling together research related to the relationship between language and mathematics, communicating mathematics, and mathematics in multilingual settings and offered a direct challenge to dominant research on communication in mathematics classrooms in which the normalised setting was, and sadly still is in many studies, taken to be the monolingual mathematics classroom. The shift from a focus on bilingualism to multilingualism occurred during the same period as the shift from a conceptualisation of language as a problem in mathematics teaching and learning, to language as a resource (Adler, 1995, 1997, 2001). It was through this ground-breaking work that the multilingual mathematics classroom has become the 'new normal' in research on mathematics and language diversity published after 2000 (see, for example, Setati, 2005; Moschkovich, 2008; Barwell, 2009; Clarkson, 2009).

1.3 Establishing the Scope of the Study: Preparing the Discussion Document

ICMI Study 21 was announced in July 2008 during the 11th International Congress on Mathematical Education (ICME-11) in Monterrey, Mexico, with Mamokgethi Setati Phakeng (South Africa) and Mario do Carmo Domite (Brazil) as co-chairs, and an IPC of nine further academics from around the world (Richard Barwell, Philip Clarkson, Anjum Halai, Mercy Kazima, Sinfree Makoni, Judit Moschkovich, Nuria Planas, Paola Valero and Martha Villavicencio). The preparation of the discussion document started immediately thereafter with Mamokgethi Setati Phakeng preparing the first draft, initiating and leading the interaction between members of the IPC via email and incorporating their ideas into the draft document. This early discussion, mostly via email, was robust and engaging. Key strands in the discussion concerned: how to describe and conceptualise multilingualism and language diversity; the possible effects of our choices about how to describe and conceptualise multilingualism and language diversity; and the implications of these conceptualisations for the study and for mathematics education.

The debates were not surprising given our diverse theoretical backgrounds. After 5 months of interaction on the draft discussion document, the IPC held a successful meeting on 17–21 February 2009 in Pretoria, South Africa, organised and hosted by Co-Chair Mamokgethi Setati Phakeng. She also organised travel funding for four of the IPC members, as well as accommodation for all members of the IPC. The programme for the meeting included inputs by Bernard Hodgson, ICMI Secretary General, Jill Adler, ICMI Vice President, as well as sessions for IPC members to work on aspects of the discussion document in groups. IPC members also visited schools in the west of Johannesburg to have some shared experience of mathematics education and language diversity in South Africa.

At the end of the meeting it was agreed that a small team comprising of Mamokgethi Setati Phakeng, Maria do Carmo Domite, Judit Moschkovich, Nuria Planas and Richard Barwell should work on finalising the discussion document. Thereafter the final draft of the discussion document would be distributed to all the IPC members for final comment. The completed discussion document was released in September 2009 after a year of rigorous interaction between IPC members and became the basis for the study conference and the development of this volume.

The discussion document (this volume, pp. 297–308) set out five themes, which served to organise preparations for the study conference:

1. *Teaching mathematics in diverse language contexts*. This theme focused on language issues in the teaching of mathematics in different language contexts. The assumption here is that language issues that emerge in different contexts are not only shaped by the complexities of the language of mathematics, but are also shaped by the linguistic contexts in which mathematics is taught and learned.
2. *Teacher education for diverse language contexts*. The focus here was on issues in and for teacher education in diverse language contexts. An assumption is that teacher education principles and practices are rooted in the real world of the classrooms, and therefore must take into account the different language contexts in which mathematics is taught.
3. *Researching mathematics teaching and learning in multilingual contexts*. This theme focused on the theories and methods for doing research in multilingual contexts and includes theoretical, ethical, methodological and philosophical issues.
4. *Mathematics, language diversity and society*. Diverse language settings reflect broader social, cultural and political issues. When considering classroom dynamics, an assumption is that they are shaped by and go on to shape the broader social and political settings. Language diversity issues intersect with multicultural, policy, and wider social issues.
5. *Student mathematics learning and experiences in multilingual classrooms*. This theme focused on students' learning and students' experiences of learning mathematics in different language contexts. An underlying assumption is that it is important to focus on learners in order to support them to learn mathematics

effectively. This theme includes work on classroom discussion, in different mathematical domains, and in different age groups.

Some issues were not explicitly mentioned in these themes. For example, code-switching is one of the most widespread topics of research in this area but is not referred to directly in the themes. The IPC felt that given the extent to which this issue had been written about in the mathematics education literature, contributors would be well aware of its importance and implications for our work and would address it in the context of several of the themes. This assumption proved to be the case, as can be seen from several of the chapters in this volume.

Similarly, some crucial sociological factors such as gender and poverty are not explicitly referred to. The IPC were aware that such factors have a profound impact on the way teachers teach and how students learn mathematics whether in a multilingual context or not. The IPC also understood that these factors are intertwined with issues of language learning and language use and with how language and learning impact mathematics learning in school settings. Again, the IPC was confident that contributors would foreground these issues in relation to all of the themes.

On the other hand, the IPC wanted to foreground ideas from applied linguistics. Research on language diversity in mathematics education must be informed by theory, methods and research in relevant disciplines such as applied linguistics and linguistic anthropology. There are many empirical findings and theoretical ideas in these disciplines that should be broadly disseminated within the mathematics education research and practitioner communities. To this end, Sinfree Makoni, an internationally known applied linguist was included in the IPC. In addition, two applied linguists were invited to address the study conference (see below).

1.4 The Study Conference

The study conference was held 16–20 September 2011 in Águas de Lindóia, São Paulo State, Brazil with Co-Chair Maria Do Carmo Domite as the lead organiser. The conference was planned as a working event and had two main goals: to enable discussion of the latest research of relevance to the study theme; and to generate writing teams and proposals for the study volume. To meet these goals, and following a call for papers, a total of 54 papers, authored or co-authored by 91 individuals from 27 countries, were accepted for presentation at the conference. The papers covered all of the five themes of the study as follows:

- Theme 1: Teaching mathematics in diverse language contexts: 18 papers.
- Theme 2: Teacher education for diverse language contexts: 14 papers.
- Theme 3: Researching mathematics teaching and learning in multilingual contexts: 8 papers.
- Theme 4: Mathematics, language diversity and society: 4 papers.
- Theme 5: Student mathematics learning and experiences in multilingual classrooms: 10 papers.

Table 1.1 Number of papers given per country of authors

Country	No. of papers	Country	No. of papers
Australia	2	Mexico	1
Belgium	1	Mozambique	1
Brazil	5	New Zealand	1
Cameroon	1	Pacific Region	1
Canada	5	Pakistan	3
China	1	Papua New Guinea	1
Denmark	1	Peru	1
Germany	3	South Africa	11
India	1	Spain	2
Iran	1	Swaziland	1
Kenya	1	Sweden	1
Malawi	2	Russia	1
Malaysia	1	USA	6
		Vietnam	1

Table 1.1 shows the number of papers per country. These themes were a starting point for the Conference.

That South Africa had the highest number of papers is not surprising given the nature of work focusing on teaching and learning mathematics in multilingual classrooms conducted in South Africa. Of interest are the countries that were not represented at the study conference despite how multilingual the world (and thus mathematics classrooms) has become in the last 20 years. The papers focused on different levels of schooling (primary, secondary and tertiary). A large majority of papers used a qualitative case study methodology.

The conference proceedings (Setati, Nkambule, & Goosen, 2011) containing all the accepted papers were published well ahead of the conference and delegates were asked to read papers of interest beforehand, since conference time would be devoted to discussing the papers and presentations would only be summaries of what had been written. Given the number of papers received, the program was organised into three groups focusing on the following themes:

GROUP 1: Theme 3 and 5: Focus on students' mathematics learning and experiences in multilingual classroom and on researching mathematics teaching and learning in multilingual contexts.

GROUP 2: Theme 1: Focus on teaching.

GROUP 3: Theme 2 and 4: Focus on teacher education and on mathematics, language diversity and society.

During the conference delegates were asked to join one of the three groups and to continue in that group for all sessions until the end of the conference. The activities of the groups included short presentations and follow-up discussions of important developments in research and/or practice related to the theme under consideration. Each group was co-ordinated by a team of four theme leaders drawn from the IPC.[1] Two members were appointed co-chairs for each session with the other two acting as rapporteurs for the session, with these roles changing for each session. The purpose of the groups was to provide an opportunity for presentations and discussions in order to share key findings and ideas, moving to a point when collaborations between group members could form, and possibilities for book chapters of the study volume could be discussed. This outcome was discussed at the first meeting of the groups and time was deliberately set aside each day for these extra discussions.

Two linguists (Marilyn Martin-Jones, UK, and Marco Barone, Brazil) were invited to present plenary lectures and workshops. The plenary lecture by Marilyn Martin-Jones was entitled 'New times, new dimensions of linguistic diversity: Rethinking research and practice' and the plenary lecture by Marco Barone was entitled, 'An overview of native languages in Brazil, the scientific importance of preserving linguistic diversity and how mathematics can contribute in this regard'. Marilyn Martin-Jones presented a workshop with the same title as her plenary lecture, while Marco Barone's workshop was entitled 'Mathematical methods for linguistics: How to build an Atlas'. Both speakers achieved the fine art in their presentations and workshops of drawing issues from the participants' papers and reflecting back potentially important nuances in their thinking.

At the beginning of each day, two members of the IPC working with two different thematic groups were tasked with presenting a short 'provocation' to the conference. They were asked to reflect on the preceding day's activities and raise ideas but mainly questions: in particular, questions that they found had challenged their own thinking. In this way, there was a running commentary of what was happening across the conference and an air of deep questioning and inquiry was foregrounded for all participants.

1.5 Preparation of This Volume

Although the conference and the published conference proceedings were important in their own right, the IPC had always regarded the conference as a step towards the development of the study volume. The development of chapters for the study volume had an organic and self-organising nature so that contributors from different regions who attended the conference coalesced around common topics of interest. By the second day of the conference, with the encouragement of the IPC members leading the groups, a number of delegates coalesced into writing teams. The IPC encouraged chapter proposals to include multiple authors and geographic settings. By the end of the conference most of these teams had decided they were prepared to

[1] All members of the IPC attended the conference, with the exception of Sinfree Makoni, who was not able to participate and who took no further part in the work of the IPC.

continue their thinking together and had developed written proposals for chapters for the study volume. The IPC established some criteria for reviewing the proposals:

- Quality of content and clarity of proposal
- Relevance to the study themes and questions
- Not longer than 8,000 words
- At least two authors from different settings or countries
- Representation of different mathematical/linguistic settings

In reviewing the proposals received, the IPC also sought to ensure that the study volume would reflect diversity of geography, mathematics, educational level, languages and theoretical perspectives.

Most of the 91 delegates were included in at least one chapter writing team. As had been planned, the teams of authors were composed of colleagues from different countries thus bringing to their writing a confluence of different multilingual contexts, and in many cases different traditions of educational research and very different experiences of educational praxis. Of the submitted proposals, 19 were accepted by the IPC, with one subsequently withdrawn. Of course, not every proposal developed into a completed chapter.

The 18 proposals were organised into five subgroups, to be guided and edited by one or two members of the IPC. The editors worked with their authors to prepare their manuscripts and organised blind peer reviews. Each chapter was reviewed by at least one other member of the IPC, as well as by an expert external to the IPC.

During September 2013, a meeting of most of the IPC[2] was hosted in Lima by Martha Villavicencio, sponsored by the Ministry of Education of Peru. Other members of the IPC participated by email from time-to-time during this 2-day meeting. During this time, each chapter was reviewed again by at least two IPC members who had not previously seen the chapter. After this review, a number of chapters were accepted as ready for publication, while others were accepted for inclusion in the study volume subject to additional revisions or, in one or two cases, acceptance after completion of another round of peer review. Hence each chapter that has been included in this study volume has undergone a number of reviews.

Additional meetings of multiple IPC members were also held at the ICME-12 conference in Seoul, South Korea, in July 2012, and at the PME conference in Vancouver, Canada, in July 2014. In 2012, Mamokgethi Setati Phakeng was not able to continue in her role of co-chair, although she continued to participate as a member of the IPC. At the meeting in Seoul, Richard Barwell agreed to co-ordinate the final stages of the preparation of the volume, including the preparation of the book proposal, collection and preparation of the chapter manuscripts, and preparation and final editing of the completed manuscript. The production of the volume has been in a very real sense the product of a collective process involving the nine members of the IPC whose names are listed as editors.

[2] After the study conference, Maria do Carmo Domite withdrew from the IPC.

1.6 Summary of Research Ideas in This Volume

The chapters in the volume as a whole present a variety of results from existing research and, at the same time, pose a number of challenges and questions for the future of research in this area, as well as having implications for school practices in multilingual settings (these latter two points are dealt with more extensively in the following sections). Rather than summarising all chapters, we present a synthesis of some key issues that are addressed across different chapters, and which constitute important points of insight in current research.

What is mathematics? Authors in this volume recurrently come back to the ongoing discussion in mathematics education about how mathematics is conceived and thought of in educational settings. Beyond the philosophical discussions about the nature of mathematics as a discipline or as a field of practice for professional mathematicians, a debate that rightly continues among our mathematics colleagues, the issue of what is 'mathematics' when configured in educational practices is also far from resolved. In this volume, there is clearly not one unified answer to this question. More importantly, however, the chapters point to the fact that commonly held assumptions about the nature of mathematics in the realm of educational practice are challenged when the language diversity of learners and teachers alike result in tensions in generating unified meanings. The mediation of natural languages in forming what is mathematics and mathematical in a particular educational situation cannot be ignored. As a consequence, there emerges the recognition of the multiplicity of mathematics that may be present simultaneously in one multilingual, often also multicultural, educational setting. Mathematics then becomes the sum of all the varied 'mathematics', as well as the possible relations and points of divergence among them.

How are mathematical and natural languages related? Some chapters implicitly or explicitly discuss the notion that in educational practice, as well as in existing mathematics education research, there are strong assumptions about the universality of mathematical language. Such universality is implicitly assumed in the apparent transparency of meaning of numerical and symbolic language in mathematics registers. The dominance of such assumptions overshadows the fact that the natural language(s) of teachers and learners alike are fundamental in teaching and learning, not only as mediators but also as constitutive elements of culturally bounded forms of thinking. In other words, there still seems to be an understanding that the more 'mathematical language' (of symbolic type) learners can acquire, the less learners and teachers are dependent on and hence can rely less on the resources of their natural languages. This assumption runs through the different levels of education, and takes different forms at each level. Such an assumption would mean that the problems of learners' natural language diversity are pressing and evident as learners are young and in the primary levels of schooling. Authors in this volume clearly agree with this. But as learners grow and develop more symbolic mathematical language, the diversity of possibilities of meaning created by the diversity of natural languages may be thought to decrease. From this perspective, the problems of multilingualism in mathematics education would not seem to be so relevant or pressing at the higher levels of mathematics education. Although some authors seem to agree with this, nevertheless others challenge the second part of this assumption and suggest that students are still accessing their natural languages as resources in their learning of mathematics through secondary schooling and into university level. However, the impact of natural language is also canvassed beyond the teaching and learning context by some authors. The impact of researchers' natural languages on the way *they* formulate ideas and the hegemony of English in the research literature is challenged. This seems to have a narrowing effect on what in the end is acceptable to the research community at large. Hence evidence is presented in this volume that the impact of students' natural languages on learning and teaching mathematics needs to be an ongoing research theme for mathematics education. As well, there also needs to be more detailed investigation of the impact of researchers' own natural languages on their work, and whether ideas and notions not easily expressed in English are being lost to the world-wide community of researchers.

Which are the multiple languages in multilingualism? In most chapters in this volume, language diversity refers to the variety of natural languages spoken by the learners and teachers. However, some chapters raise interesting points that move the discussion further. The growing presence of different media in teaching and learning situations makes it evident that other languages, registers and modes contribute to the linguistic complexity of today. Furthermore, international trends for the inclusion of blind and deaf students in schools also raise the issue of how other types and forms of language are dealt with. Various contributors have written about these challenges arising in such (to many) novel contexts of language diversity and about how these issues affect teachers and students and their meeting with mathematics.

Is mathematics education in contexts of language diversity setting political? Culturally dominant groups, their languages and worldviews establish the norms to follow, strive for and classify people accordingly. In any society, the mathematics in educational curricula embeds the values, worldviews and languages of the culturally dominant group(s) in society. This is a condition of the historical and social organisation of schooling and education in all societies. Thus, learners who do not share or comply with the norm are seen as deficient. This deficiency perspective is part of the form of operation of schooling as an institution in society. In settings of language diversity, many learners who are not part of the 'norm' will be deemed as deficient in relation to both the language of instruction, and in relation to many of the values and forms of being that school promotes. This means that even though some mathematics education research has tried to evidence and criticise the pervasiveness of deficit perspectives of learners (see above for a very short summary), it is quite difficult to eliminate. This terrain is highly political, as highlighted in several chapters and the mathematics education community needs to find ways to build constructive dialogue with those wielding political power. As well, new options for pedagogies and forms of understanding educational practices continue to be desirable. A number of the chapters in this volume suggest interesting paths to follow.

1.7 Implications for Policy and Practice

Many of the questions that were originally included in the discussion document remain open to further research, but even what has been accomplished so far can offer some guidance to policy-makers, curriculum developers and teachers in classrooms or lecture halls. In this section, we first offer some points of guidance. These points are a brief distillation of what we regard as important issues raised in various chapters of this volume. Following these points, we then note a number of questions that impinge on practice, which authors have suggested still need investigation.

We begin each point with a question, since the whole tenor of this volume is to open up multiple discussions and dialogue rather than promote 'one size fits all' results, and then offer some comments in note form.

What is our basic advice?

- Accept and acknowledge the reality and diversity of multilinguistic contexts.
- Language diversity is a reality for mathematics learners and their teachers and is a complex phenomenon. Learners in mathematics classrooms live in communities that use one or more languages besides the language of instruction. Language diversity also exists in language varieties, including dialects as well as regional, social and economic variation and sign languages.

How can policy and practitioners deal with this diversity and complexity?

- Know the local context, setting, and details of language history, policies, and especially the experiences of disenfranchised language groups.
- Shift from focusing on deficiencies to noticing and building on competencies.
- Although deficit models of mathematics learners or their communities are pervasive, they do not work: they do not provide an accurate picture of learners' potential for progress, they do not describe any resources, strengths, or competencies that instruction can build on, and they may condemn students to endless cycles of remedial instruction that simply do not work.

How can policy and practitioners avoid deficit models?

- Balance a focus on challenges that learners may face with an equal focus on deep and detailed knowledge of the resources, competencies, and strengths that these learners bring to the classroom. Again, know the students and their communities.
- Accept and acknowledge the complexity of language issues in mathematics classrooms.
- Language issues in mathematics classrooms are complex. It is not possible to say that using home languages is always the right thing to do. The question to ask is: when and how should home language(s) and the language of instruction be explicitly used and encouraged by teachers, depending on the goals of mathematics instruction, and in the full knowledge that multilingual students will, for a variety of reasons, code-switch whether the teacher encourages, or even

recognises this practice, or not. As well it should be acknowledged that language and mathematical discourse is much more than vocabulary, number names, or the logical structure of sentences. One important aspect of mathematical discourse is that it is multimodal and involves a variety of modes such as listening, talking, writing, reading, drawing, graphing, etc. Another aspect is that it is multisemiotic; it involves a variety of sign systems such as mathematical symbols, written text, etc. This adds up to acknowledging that language and culture are closely and intimately related and cannot be separated.

How can policy and practitioners avoid simplifying issues of language?

- Avoid reducing language issues to vocabulary, provide opportunities for students to participate in multiple modes (oral, written, concrete objects, drawings, etc.) and use multiple sign systems.
- Consider the systemic and political nature of language issues in classrooms.
- Language diversity issues function in a system that includes not only teachers and classrooms but also schools, families, communities, and language attitudes as well as classism, racism, and other systems of institutionalised oppression. Language diversity is a political issue: the hegemony of the language of dominant cultural groups shows up in the classroom, in policies, curriculum, and language choices, and learner identities with regard to mathematical competence. Language policies reflect who counts and who does not count in society. This is so even in mathematics classrooms.

How can policy and practitioners address systemic issues?

- Include research that addresses the systemic and political nature of language issues in education in policy and practice discussions.

We now list a series of questions that have emerged for us in participating in this study, and many of which have been asked by authors in this volume. They are organised according to the themes in the discussion document. It is a deliberate choice to leave them as questions, rather than giving summary statements as to what research says about these questions, since we regard the field as one that still is emerging and diverging in many constructive paths. This is not to say that research does not articulate some specific notions regarding some of these questions, as can be seen in this volume and elsewhere. Nevertheless the articulation of these questions taken together suggests our thinking is deepening, since we are no longer asking the type of questions that were foregrounded by researchers some 30 years ago. We also invite our colleagues in policy and curriculum development, as well as teachers, to think along with us, since they too are agents, along with representatives from non-dominant communities, who need to be involved in discussions of future research.

Teaching mathematics in diverse language contexts:

- What strategies and resources can educators in multilingual classrooms use to teach mathematics effectively to learners who are in the process of learning the language of instruction?

- How do assessment and curriculum systems relate to mathematics teaching and language policies in diverse language contexts?
- Which current teaching practices are sensitive to the relationship between multilingualism and mathematics learning?
- What are the relationships between teaching language and teaching mathematics?

Teacher education for diverse language contexts:

- Which current practices in teacher education are sensitive to the relationships between multilingualism and mathematics teaching and learning?
- What can be done to prepare teachers to teach mathematics effectively in multilingual classrooms?
- What kind of data from multilingual classrooms would be useful in designing teacher education programs?
- What knowledge and skills do teachers need to teach in multilingual classrooms and what are teachers' perspectives on this question?

Researching mathematics teaching and learning in multilingual contexts:

- What types of theories and methods enable the development of research in this area?
- What ethical issues arise in pursuing this kind of research and how can researchers address them?
- On what basis can researchers interpret the mathematical worlds of students who come from linguistic backgrounds with which they are not familiar?
- To what extent is mathematics education research sensitive to linguistic diversity?

Mathematics, language diversity and society:

- To what extent and for what purposes do research in multilingual contexts need to address multicultural issues?
- What role does teaching mathematics in diverse language settings play in reproducing or challenging prevailing social patterns?
- How can researchers engage productively with policy-makers involved in mathematics education to address language diversity?
- What is the relationship between the teaching and learning of mathematics in multilingual settings, and wider social discourses?

Student mathematics learning and experiences in multilingual classrooms:

- What are the characteristics of students' mathematical discussions and explanations in different languages, in multiple classroom contexts, and in multiple mathematical domains?
- What are the demands on multilingual students learning mathematics in different mathematical domains (i.e. algebra, geometry, etc.) and at different ages?
- How do students themselves see and describe their experiences in multilingual mathematics classrooms?

- What are students' strengths and resources, what can we learn from successful students, and how can instruction build on these resources, strengths, and successes in linguistically diverse settings?

1.8 Some Issues for Future Research

In this study, research is reviewed that investigates the teaching and learning of mathematics at different levels including in schools, in university undergraduate courses, and in teacher education. The work has also covered teaching and learning using ICT and digital media. Furthermore, the concept of the language of teaching and learning has gone beyond the spoken language to that of teaching and learning mathematics to deaf and blind learners. Thus this volume draws on a wide research and knowledge base. However our research needs to be taken further.

Mathematics education researchers, mathematics teacher educators, graduate students in mathematics education, mathematics teachers and student teachers, among others, can progress the work of this book further with future research. Within various chapters and indeed in this introduction, there are explicit calls for future research in specific areas. For example the call for more research in undergraduate mathematics, where often language is assumed not to be an issue of concern, and the call for more research in European classrooms, where there are many languages that are not shared, are two that are quite explicit. There are also less explicit calls, such as the work on blind and deaf learners that brings an awareness that language diversity includes non-verbal communication.

Equally, while some situations of language diversity have been extensively examined, other aspects have not. For example, most research to date has focused on what can approximately be termed bilingual and multilingual contexts. There is little that examines the specificity of trilingual contexts where learners are exposed to a home language, a national language and an official language of instruction. The study of such contexts in mathematics teaching and learning remains a gap in our field. Its specificity lies in the fact that unlike in multilingual contexts where there are multiple languages, but only two languages (home language and language of instruction) that are in competition, learners in trilingual contexts have to deal with three languages, each of which has its own power and influence: one as a home language, the second as a national language and the third as a world language.

Research in mathematics education in general, and the linkage between mathematics education and language specifically, can appear to be somewhat inward looking, in the sense that it is often mainly concerned with the issues of teaching, learning and curriculum in mathematics classrooms. These foci may, in turn, be reduced to strategies and methods to promote teaching and learning. However, the broader issue of what is termed education is not necessarily considered as an issue that needs to be deconstructed. Assumptions and values that underpin the current system of formal education appear to be taken as normative and are often not questioned. This is an issue particularly pertinent to this volume, since mainly European/

Western perspectives and values inform current education systems in many countries. As a result, approaches to education that might be beyond the 'normative' frame of reference are considered as lacking and deficient. These certainly impinge on our more focused territory in mathematics education.

A related issue is the understanding of 'what counts as mathematics?' While historical development shows mathematics as fallible, as a work in progress, and undertaken from a particular cultural and epistemological tradition (normally Eurocentric-Western), it is almost always taught in schools as a culture-free subject that involves learning supposedly universally accepted facts, concepts and contents.

In addition, mathematics enjoys a position of prestige, such that proficiency in the subject can potentially open the gates to opportunity and is, therefore, often uncritically assumed to be a worthwhile goal for teachers and learners. However, it is not any mathematics that performs this function of gate-keeping: it is only the 'valorised' mathematics of the few that performs this function. Hence, it is important to understand the nature of the subject as an educational goal.

Several issues in mathematics education go beyond the disciplinary boundaries, as does the issue of language diversity and mathematics education. There are many cases of marginalisation due to language and culture, which raise issues and concerns that require instruments of analysis from linguistics and sociology. Therefore, research in mathematics education needs to shift its 'inward' stance and look across disciplines and undertake cross-disciplinary research to understand these issues from a more multidisciplinary and nuanced perspective.

Finally, much research in mathematics education, including studies on the interplay of language and mathematics education, focuses exclusively on praxis; how can teachers teach more effectively and how do students learn what is taught. We have no problem with these outcomes but question the exclusive focus. We call for more studies that run parallel to these but also incorporate and take seriously the role of theory so that the theoretical basis of our work can be enriched and in turn give deeper insight into our praxis.

1.9 Summary Statement

This volume brings together the combined thinking of over 5 years of work by colleagues from many parts of the world working in contexts of language diversity. There is huge diversity within these contexts, and yet there is also commonality for us. There are chapters that not only summarise what research has been done, but use that work as a basis for envisaging what can be researched for the benefit of future learners, teachers, educators, policy-makers and others. Other contributions, some tentatively and others more robustly, declare ways forward that suggest options for our colleagues to re-envisage their praxis. We believe there is much here for colleagues to consider that will help push forward this area of research and practice.

Nevertheless this volume also repeatedly highlights that after more than 40 years of research that suggests that for multilingual students, well-thought out use of all their languages can be beneficial in their learning, including mathematical learning, the default setting for many politicians and bureaucrats is that students must only use the official language of teaching. In many countries the official language of teaching is a world language deemed to be important for the economy of the country as a whole. Of course there is some truth in this kind of assertion. However educational research suggests there are better options that need to be considered rather than this simple default position. Hence one of the outcomes of this volume should be that we as researchers should find ways to engage with politicians and bureaucrats productively, so that they are aware of what research does say, and work with them to ensure such dialogue continues.

References

Adler, J. (1995). Dilemmas and a paradox: Secondary mathematics teachers' knowledge of their teaching in multilingual classroom. *Teaching and Teacher Education, 11*(3), 263–274.

Adler, J. (1997). A participatory-inquiry approach and the mediation of mathematical knowledge in a multilingual classroom. *Educational Studies in Mathematics, 33*, 235–258.

Adler, J. (1998). A language of teaching dilemmas: Unlocking the complex multilingual secondary mathematics classroom. *For the Learning of Mathematics, 18*(1), 24–33.

Adler, J. (2001). *Teaching mathematics in multilingual classrooms*. Dordrecht, The Netherlands: Kluwer.

Austin, J. L., & Howson, A. G. (1979). Language and mathematics education. *Educational Studies in Mathematics, 10*, 161–197.

Barwell, R. (Ed.). (2009). *Multilingualism in mathematics classrooms: Global perspectives*. Bristol, England: Multilingual Matters.

Blommaert, J., & Rampton, B. (2011). Language and superdiversity. *Diversities, 13*(2), 1–21.

Clarkson, P. C. (1991). *Bilinguals and mathematics learning*. Geelong, Victoria, Australia: Deakin University Press.

Clarkson, P. C. (1992). Language and mathematics: A comparison of bi and monolingual students of mathematics. *Educational Studies in Mathematics, 23*, 417–429.

Clarkson, P. C. (1996). NESB migrant students studying mathematics: Vietnamese and Italian students in Melbourne. In L. Puig & A. Gutierrez (Eds.), *Proceedings of the 20th conference of the International Group for the Psychology of Mathematics Education* (pp. 225–232). Valencia, Spain: PME

Clarkson, P. C. (2009). Potential lessons for teaching in multilingual mathematics classrooms in Australia and Southeast Asia. *Journal of Science and Mathematics Education in Southeast Asia, 32*(1), 1–17.

Clarkson, P. C., & Dawe, L. (1997). NESB migrant students studying mathematics: Vietnamese students in Melbourne and Sydney. In E. Pehkonen (Ed.), *Proceedings of the 21st Conference of the International Group for the Psychology of Mathematics Education* (pp. 153–160). Lahti, Finland: PME.

Cummins, J. (1979). Linguistic interdependence and the educational development of bilingual children. *Review of Educational Research, 49*, 222–251.

Cummins, J. (1981). *Bilingualism and minority language children*. Toronto, Ontario, Canada: Ontario Institute for Studies in Education.

Dawe, L. (1983). Bilingualism and mathematical reasoning in English as a second language. *Educational Studies in Mathematics, 14*, 325–353.

De Klerk, G. (1995). Bilingualism, the devil and the big wide world. In K. Heugh, A. Siegrühn, & P. Plüddemann (Eds.), *Multilingual education for South Africa* (pp. 53–62). Johannesburg, South Africa: Heinemann.

Grosjean, F. (1982). *Life with two languages: An introduction to bilingualism*. Cambridge, MA: Harvard University Press.

Howson, A. G. (Ed.). (1973). *Developments in mathematical education: Proceedings of the Second International Congress on Mathematical Education*. New York: Cambridge University Press.

Makoni, S. B., & Pennycook, A. (2007). Disinventing and reconstituting languages. In S. B. Makoni & A. Pennycook (Eds.), *Disinventing and reconstituting languages* (pp. 1–41). Clevedon, England: Multilingual Matters.

Moschkovich, J. N. (2008). "I went by twos, he went by one:" Multiple interpretations of inscriptions as resources for mathematical discussions. *The Journal of the Learning Sciences, 17*(4), 551–587.

Pennycook, A., & Makoni, S. (2005). The modern mission: The language effects of Christianity. *Journal of Language, Identity and Education, 4*(2), 137–155.

Saunders, G. (1988). *Bilingual children: From birth to teens*. Philadelphia: Multilingual Matters.

Secada, W. G. (1992). Race, ethnicity, social class, language, and achievement in mathematics. In D. A. Grouws (Ed.), *Handbook of research on mathematics teaching and learning* (pp. 623–660). New York: Macmillan.

Setati, M. (2005). Teaching mathematics in a primary multilingual classroom. *Journal for Research in Mathematics Education, 36*(5), 447–466.

Setati, M., Nkambule, T., & Goosen, L. (Eds.) (2011). *Proceedings of the ICMI Study 21 conference: Mathematics education and language diversity*. São Paulo, Brazil: ICMI Study 21

Swain, M., & Cummins, J. (1979). Bilingualism, cognitive functioning and education. *Language Teaching, 12*(1), 4–18.

UNESCO. (1974). *Final report of the symposium 'interactions between linguistics and mathematical education'*. Paris, France: UNESCO.

Vertovec, S. (2007). Super-diversity and its implications. *Ethnic and Racial Studies, 30*(6), 1024–1054.

Chapter 2
Impact of Differing Grammatical Structures in Mathematics Teaching and Learning

Cris Edmonds-Wathen, Tony Trinick, and Viviane Durand-Guerrier

2.1 Introduction

Mathematics is taught in many different languages around the world. In some countries mathematics is taught in only one language, for example in the medium of English in England, but in other countries in two or more languages. In countries such as Tunisia and Papua New Guinea, the language of instruction is dependent on the year level. Multilingualism is increasingly becoming the norm for many communities, and for others has been a way of life for thousands of years. However the fact that a community is multilingual does not imply that every member of the community can speak all the represented languages (Trinick, 2015). Therefore a constant issue in multilingual communities is the choice of what language or languages to use in schools, a choice which is 'closely bound up with issues of access, power and dominance' (Barwell, 2003, p. 37; see also Setati, 2008). For example, a widely held view by early policy makers, generally representing the colonising power, was that linguistic diversity, that is, multilingualism, presented obstacles for national development, while linguistic homogeneity was associated with modernisation and Westernisation (Ricento, 2000). As a result there has been a worldwide trend for

C. Edmonds-Wathen (✉)
Charles Darwin University, Darwin, NT, Australia
e-mail: cris.edmonds-wathen@cdu.edu.au

T. Trinick
The University of Auckland, Auckland, New Zealand
e-mail: t.trinick@auckland.ac.nz

V. Durand-Guerrier
University of Montpellier, Montpellier, France
e-mail: vdurand@math.univ-montp2.fr

R. Barwell et al. (eds.), *Mathematics Education and Language Diversity*, New ICMI Study Series, https://doi.org/10.1007/978-3-319-14511-2_2

schooling to be in English or in another of the world's dominant languages such as French, Russian, Spanish or Chinese, both as a direct colonial legacy, and in order to access the dominant discourse.

In more recent times, where populations have attempted to throw off the colonial yoke, in several countries there have been attempts to modernise the indigenous language or elaborate the national language (other than the colonising language) to enable Western school mathematics to be taught (Trinick, 2015). Therefore, mathematics is being taught in a linguistic spectrum from those with a recently created mathematics register such as Māori, to those who have had a formal tradition of teaching mathematics for several centuries, such as French or English. Not only are there differences in linguistic traditions, but there are also contexts in which children from these various linguistic traditions are all learning in the same mathematics classroom in a language which is not their mother tongue.

This chapter concerns itself with issues regarding the different grammatical structures of the languages used in mathematics learning: both languages of instruction and other languages that are used in classrooms. The linguistic term that refers to the particular kind of language used in mathematics is the mathematics register (see Halliday, 1978). This chapter examines features of the mathematics register and the evolution of the mathematics register for a few selected languages. Mathematics is understood to operate in and through language, and different languages offer different resources with which to do this. Building on the earlier work of Barton (2009), this chapter also considers the nature of the mathematics that is made possible by the linguistic expressions of different languages. It discusses linguistic differences that occur in the key mathematical areas of logic, reasoning, space and number, in a general manner and with specific examples from languages from different parts of the world. In most of the cases discussed, the impact of grammatical structures on mathematical teaching and learning has not been empirically investigated. In this chapter, we point out potential impacts and make suggestions for educators and researchers. There remains much scope for future research both on the cognitive impacts of differing grammatical structures and their implications for mathematics education.

The analysis in this chapter is influenced by Whorf's (1956) *linguistic relativity* hypothesis, that is, the idea that the structure of a language can affect the thought processes of speakers of that language. While there are many similarities in how diverse languages have developed their grammars, there are also many remarkable differences. If the forms and constructions of one language do not always have exact counterparts in other languages, this may suggest that the thinking processes of the speakers of one language will differ from those of a speaker of any other language. While there are few modern proponents of 'linguistic determinism' in its strongest form, many linguists have accepted a more moderate linguistic relativity, namely that the ways in which we see the world may be influenced by the kind of language we use (Chandler, 2004). For example, there are certain areas, such as perception of space, where some Whorfian effects have been demonstrated by empirical investigation. Research shows various indigenous Australian language perceptions of space are incongruent with spatial descriptions in European languages (see Levinson, 2003; Levinson & Wilkins, 2006; Edmonds-Wathen, 2011).

Culture, language and cognition are intertwined in a complex manner. More general indications of the effect of culture on mathematical learning can be found in Gay and Cole's (1967) a much quoted study of the Kpelle of Liberia. According to Gay and Cole, the Kpelle are proficient in types of mathematical reasoning for which they have cultural uses, such as estimation, but less proficient in other areas for which they do not have cultural needs (see Austin & Howson, 1979). However, in this chapter we are more concerned with the impact of language features on mathematical thinking than with the cultural practices that have led to the development of these language features.

Lucy (1992) points out that linguistic relativity effects relate to habitual thought, rather than potential thought. There is evidence that people can perceive and reason in ways that their languages do not facilitate, such as enumerative capacity amongst people who do not have number words (Butterworth, Reeve, Reynolds, & Lloyd, 2008). Not having number words does not mean that they do not have the potential to enumerate. However, it does mean that they are likely to apply different strategies to solve certain problems than those who do have number words (Butterworth, Reeve, & Reynolds, 2011).

The issue for mathematics with which we are concerned is not just whether a certain concept can be expressed in a certain language, but the ease of expression of the concept: that is, how the grammatical structure facilitates or impedes this expression. Barton (2012) suggests that 'we bring mathematics into existence by talking about it, and the way we talk about it changes the questions we can ask' (p. 227). Becoming aware of differences between languages can help teachers and learners avoid confusion as well as enrich the learning environment. Although the chapter describes both limitations and facilitations of individual language features, there are more limitations discussed because it is when difficulties occur that language differences are investigated as a possible factor. However, there are also pedagogical opportunities to be exploited in the relations between language and mathematics, particularly when there are language-derived alternative ways of approaching aspects of conventional mathematics (Barton, 2009, 2012). These linguistic issues are also important for curriculum development particularly when it is occurring in a language that has not previously had a formal mathematics register.

All of the authors of this chapter have worked in multilingual environments, encountering, and addressing these issues at first hand. Our own language backgrounds and our histories influence how we approach this topic, including preconceptions which we may not be aware of ourselves. Cris is an English-speaking Australian who has taught and researched in remote schools in northern Australia where Australian Indigenous-language-speaking students are taught in English. Tony is a bilingual Māori/English speaker, a lecturer in Māori-medium mathematics initial teacher education and provides professional learning support to Māori-medium schools. His main area of research is the complex relationship between language and learning mathematics including the mathematics register. Viviane is a French-speaking teacher and researcher who has worked for many years with PhD students from the francophone area of Africa, such as Tunisia and Cameroon.

2.2 The Mathematics Register

A significant body of research examining language issues in the learning and teaching of mathematics in schools has recognised that language use in school differs in some important general ways from language use outside of school and, moreover, those subjects such as mathematics are characterised by specific registers (see Halliday, 1978; Halliday & Hasan, 1985). Discussion on the features and definitions of the *mathematics register* can be traced back to studies on register theory and the much broader field of Systematic Functional Linguistics (SFL), sometimes known as Hallidayan linguistics (Schleppegrell, 2004). Essentially, the four theoretical claims of SFL are (1) that language is functional, (2) two of the functions are to make meaning, and to develop and maintain relationships, (3) these functions are influenced by the social and cultural contexts in which the interactions occur and (4) the process of using language is a semiotic process, the process of making meanings by choice (Eggins, 2004). According to Halliday (1996), the term *register* refers to specific lexical and grammatical choices made by speakers, with varying degrees of consciousness, depending on the situational context, the participants in the conversation and the function of the language in the discourse.

Although researchers have long recognised the vital role that language plays in learning and teaching (Aiken, 1972), it was not until at least the 1970s that they began to highlight its importance in the process of acquiring mathematical knowledge and skills (e.g. Cocking & Mestre, 1988; Mousley & Marks, 1991; Pimm, 1987). Similarly, interest in the problems of mathematics learners whose first language differs from the language of instruction was also brought to the fore in the early 1970s, particularly by the work of Halliday (1975). He addressed language difference and distance as instructional obstacles and described a *register of mathematics*, which to this day is considered definitive in discussions about language and mathematics (Schleppegrell, 2007). In a subsequent publication, Halliday (1978) extended his description of the mathematics register, highlighting that the kind of mathematics that students need to develop through schooling uses language in new ways to serve new functions. This is not just a question of learning new words, but also new 'styles of meaning and modes of argument [...] and of combining existing elements into new combinations' (p. 196). Halliday defined the mathematics register as:

> a set of meanings that is appropriate to a particular function of language, together with the words and structures that express these meanings. We can refer to 'mathematics register', in the sense of the meanings that belong to the language of mathematics (the mathematical use of natural language, that is: not mathematics itself), and that a language must express if it is being used for mathematical purposes. (p. 195)

The primary motivation to consider the features of the mathematical register has its roots in research considering issues to do with the language of the learner, aspects of the register that are challenging for learners, and the relationship between thought and language (Trinick, 2015). Interest in the relationship between language and the development of thinking is not new and has been studied by many psychologists, for example Bruner (1966) and Vygotsky (1978). Although not mathematically oriented, the work of theorists such as the linguist Whorf (1956), who suggested that language affects habitual thought, also influenced Halliday and his theories on the register.

Collectively this work influenced researchers examining mathematics learning in a second language, for example, the research work of Cuevas (1984) and Spanos, Rhodes, Dale, and Crandall (1988). Concerned with the considerable underachievement of Hispanic second language learners in the United States, a group of language educators including Spanos et al. (1988) categorised the linguistic features of mathematical problem-solving. To support the development of a framework to examine the language of mathematical problems, they resurrected and mathematised a model first proposed by Morris (1955) in his seminal work in semiotics and adopted by Carnap (1955) to categorise the linguistic features of particular scientific domains (Spanos et al., 1988). The Morris (1938, 1955) model distinguished between the following three linguistic categories:

1. Syntactics, the study of how linguistic signs, or symbols, behave in relation to each other.
2. Semantics, the study of how linguistic signs behave in relation to the objects or concepts they refer to or their senses or how 'meaning' is conveyed through signs and language (Halliday, 1978).
3. Pragmatics, the study of how linguistic signs are used and interpreted by speakers (Spanos et al., 1988) and the study of how context affects meaning (Leech, 1983).

It is important to note that the terms such as semantics and pragmatics have contested definitions and meanings and are a study in their own right (see Levinson, 1993). According to Da Costa (1997), it is necessary to take account of these three aspects for a proper understanding of mathematical logic.

2.2.1 Register Development: Modern European Languages

The language of modern mathematics is part of the continuum of technical language development that had began in Europe by at least the seventeenth century. Modern mathematics has also drawn heavily on the ancient language stocks in Europe, Asia Minor and North Africa, and may be said to represent the cumulative technical language development of diverse peoples over thousands of years (Closs, 1977). The evolution of mathematics was/is also the evolution of the grammatical resources of the natural languages by which Western mathematics came to be constructed (Halliday & Martin, 1993).

Inevitably, this involved the introduction of ways of referring to new objects or new properties, processes, functions and relations. Halliday (1978) suggested that the most typical procedure in contemporary European languages for the creation of new technical terms was to create new words out of non-native stock, and that these terms are not normally used in everyday situations. For example, some mathematical terms such as *quadrilateral* and *parallelogram* are made up out of Latin and Greek elements, even if the actual word did not exist in the original languages. Using these terms also requires using specific grammatical patterns (Schleppegrell, 2007). The mathematics register also imbued existing everyday words with specific mathematical meanings, such as *constant*. Sometimes this reinterpretation of existing words changed their grammatical category and function; for example numbers in ordinary English function somewhat as adjectives, but in mathematics discourse can serve as nouns (Pimm, 1987).

A particularly notable feature of the mathematics register is *nominalisation*, in which processes, originally verbs, become packed into noun phrases (Halliday, 2004). These dense noun phrases are then used in complex relational clauses (Schleppegrell, 2007). One of the consequences of this is that it reconstructs processes as objects. Nominalisation both concretises, though turning processes into things, and is part of what is considered in mathematics education to be abstraction. A *concept* in mathematics is an abstract noun. This is worth noting because different languages place different emphases on the roles of nouns and verbs.

2.2.2 Register Development: Multilingual Contexts

More recently, the discussion on the mathematics register has moved from a focus on issues confronting monolingual and bilingual students to multilingual contexts as migrants from different countries move around the world or within countries that have different regional languages seeking employment and or educational opportunities (see Barwell, Barton, & Setati, 2007). Additionally, there has been the rise of minority and indigenous peoples' movements, movements that usually incorporate a strong educational focus as the means to political and economic emancipation (Smith, 1999). These political and educational movements frequently involve the teaching and learning of mathematics in the indigenous language (Barton, Fairhall, & Trinick, 1998; Meaney, 2002).

Indigenous groups attempting to modernise their indigenous language have been confronted by a range of challenges, including linguistic ones. Literature has highlighted the limitations of the lexicons of indigenous languages to express modern Western mathematics. For example, some languages such as Igbo and Yoruba do not have simple word equivalents to 'zero', a concept which plays a central role in most mathematics (Austin & Howson, 1979; Verran, 2001). Yet 'zero' was a late arrival upon the European mathematical scene (Austin & Howson, 1979). It was an invention in the Hindu–Arabic number system to mark empty places in graphical representations of numerals, before evolving into a number. Verran (2001) claims it is not surprising that exclusively oral number systems would not have developed 'zero', a number that developed in response to a need in written mathematics.

While many modernising indigenous languages did not initially have the range of terminology necessary to teach Western mathematics, the deficiencies in mathematics vocabulary can be applied at some point in time to all languages, including the English language:

> History repeats itself. Where people are now wondering whether mathematics can ever be adequately taught through the medium of Australian Aboriginal languages, and many are stating the opinion that these languages are "too primitive" and mathematics can only be effectively taught through the medium of English, it is worthwhile to remember that back in the 1500s in England people had to fight hard to be allowed to teach mathematics and other subjects in English. The language of instruction was Latin, and English, which is a creole, was considered inadequate to convey the higher forms of learning. However, it was argued that if the common people could learn in their mother tongue (English) they would learn better and more of them would be able to take advantage of the education offered (Harris, 1980, p. 2).

Over time, English has borrowed words such as 'cosine', 'sine' and the symbol for zero from other languages (Pimm, 1995). However, while the mathematics register in English has developed incrementally over hundreds of years, from time-to-time borrowing words from other languages, this developmental process has not been similarly accorded to indigenous languages, particularly those that are endangered. From a language planning perspective, Kaplan and Baldauf (1997) contend that all languages have some mechanism for elaboration. Finlayson and Madiba (2002) noted that there is a substantial body of research work that shows that languages will develop through use. According to Cooper (1989), form will always follow function. The post-colonial experience in various African countries has shown that successful language planning and development, while eminently possible, needs to be supported at all levels from government to grassroots (Bamgbose, 1999; Bokamba, 1995). Cumulatively this research suggests languages have the ability to develop a mathematics register to meet the demand of modern mathematics given the functional need.

2.3 Grammatical Systems

While attention to register development in mathematics often focusses on creating or borrowing mathematical terminology, the ease with which this is done depends in part on the grammatical system of the developing language. Languages tend to have both closed and open classes of words. Open classes can be easily added to as needs arise, and in many (although not all) languages include nouns and verbs. Grammatical functions tend to be performed by words in closed classes, which in English and French include pronouns and prepositions. It is much more difficult for languages to add to these classes. Where a language has a mathematical process or function grammatically encoded such additions can be performed with relative ease by a speaker of the language. Where a mathematical function is not encoded in the grammar, the articulation of the function can be more difficult. It can also be difficult for speakers to accept the addition of the function to a language if it involves adding to a closed class.

As well differing in the mathematical functionality in these closed grammatical systems, languages classify mathematical ideas and functions in different grammatical categories. For example, in this section, we look at whether a language classifies numbers as nouns, adjectives, verbs or something of all three.

2.4 Number

This section discusses two aspects of number that differ in the grammatical systems of language. The first is the matter of into which syntactic category a language puts its numbers. The second is the transparency and regularity of the number system, including how well it articulates with the written system. This section does not propose to cover the many different counting systems that exist in the world (e.g. see Zaslavsky, 1979).

First however, is the controversial question of what constitutes a number system. The Amazonian language Pirahã has been described as not even having a word for 'exactly one', but only for 'approximately one' (Gordon, 2004; Frank, Everett, Fedorenko, & Gibson, 2008). There are whole groups of languages which have been characterised as having very few numbers. Writers such as Von Brandenstein (1970) and Blake (1981) stated that no Australian language has a word for a number higher than four, and Dixon (1980) described a typical Australian number system as having the numbers 'one', 'two', possibly 'three' and 'many' (Dixon, 1980). However, there are Australian languages with extensive number systems, such as Anindilyakwa (Stokes, 1982) and Tiwi (McRoberts, 1990). Harris (1982, 1987) warned that characterising Australian languages as non-counting may seriously misrepresent the mathematical systems and abilities of indigenous Australians. Additionally Meaney, Trinick, and Fairhall (2012) maintained that in some cases, indigenous mathematics may have been done in ways that were different to Western cultural norms and thus remain unrecognised by the researchers, who have been predominately European and very few in number, making this research contentious.

Researchers with a positive view of the capacity of indigenous languages to be modernised, such as Bender and Heller (2006), critiqued the earlier work and showed that traditional indigenous mathematics systems have been more than adequate to cope with their traditional cultural demands (Trinick, 2015). In their review of literature of the mathematics concepts of the Native Americans, Schindler and Davison (1985) noted that amongst the different groups there was little functional use for large numbers. Some cultures have not developed extensive number systems because they have not had the need for them, not because they could not do so (Harris, 1987; Lancy, 1983). In those Australian languages which do have small number systems, such as Warlpiri and Iwaidja, the numbers are additive and larger numbers can be created if necessary (Hale, 1975). For example 'four' in Iwaidja is *ngarrkarrk lda ngarrkarrk* 'two and two'. However, this is clearly unwieldy and if a business or educational need for larger numbers arises, it is a matter of either importing the numbers from another language, or of creating new numbers. Mendes (2011) describes the creation of numbers in Kaibi, a language of Brazil, due to the desire of the Kaibi people for their own numbers, rather than the continued use of Portuguese numbers.

We are thus less concerned with whether an individual language has an extensive number system, but are interested in the grammatical features associated with the number system.

2.4.1 Syntactic Category

When we talk about syntactic category with number, we are talking about whether numbers operate as nouns, verbs or adjectives. The term *natural number* implies a relation between numbers and nature (Verran, 2001). This is likely to make one think that there is something natural about how one's own languages express and use numbers. In fact, numbers can operate syntactically in very different ways in different languages. In Yoruba, numbers are nominalised verb phrases that function modally. Verran (2001) says that the Yoruba *Ó rí ajá méta* 'He saw three dogs' would be better translated 'He saw dogmatter in the mode of a group in the mode of three' (p. 69).

Barton (2009) describes some of the variety of roles that number can take in different languages. In English, numbers can not only act as nouns (in much mathematical discourse) or as adjectives (in much everyday discourse) but also form their own grammatical class. In Kankana-ey, a language spoken in the northern Philippines, numbers can act as adjectives. Numbers in Polynesian languages such as Māori are in their own grammatical class, but have more of the nature of verbs. Barton (2009) gives the example of how the Māori request *Homai kia rima nga pene* 'Give me five pens' would be more literally translated as 'Give me, let them be fiv-ing, the pens' (p. 43). Verbal numbers have also been described for North American languages such as Mi'kmaq, where they must be conjugated according to what is being counted, as well as distinguished for animacy or inanimacy (Lunney Borden, 2010).

The educational implication of the varied syntactic roles of numbers in different languages is that some languages find certain mathematical expressions, or uses of numbers, far more easy to deal with than others (Barton, 2009). Because numbers in English can function in varied ways, but also because they are strongly noun-like, they concord with the way numbers are used in mathematics. In contrast, in the Māori language, numbers are verb-like, and thus the grammar of numerical quantification (as opposed to geometry) are treated as verbal sentences, rather than as nouns, like in English.

The way numbers operate can also vary in *numeral classifier languages* (Allan, 1977), where the way of counting depends on the type of thing to be counted, which might include shape, as in Chinese, or animacy, as in Yucatec Mayan. Things classified differently may not be able to be added together easily, as Lancy (1983) notes of some languages of Papua New Guinea, such as Loboda. This may need to be taken into account in teaching mathematics in these languages.

2.4.2 Transparency and Regularity

Number systems can also vary in the concordance of their names with the base system used, and whether this in turn concords with the symbolic written notation. The base-ten written notation is dominant in the world today, but even some languages that predominantly use a base-ten system have irregularities. For example, the numbers between 11 and 19 are irregular in English, with 'eleven' and 'twelve' hiding their 'one' (and ten) and 'two' (and ten) origins. While French has a regular *dix-neuf* 'ten nine' for 19, 80 is *quatre-vingts* 'four twenties'.

Research that links the transparency and regularity of the number system to better performance in arithmetic calculation can be taken as working within a framework of linguistic relativity. Chinese and Vietnamese both have base-ten number systems that are regular and transparent, such that the spoken number in these languages explicitly corresponds to the base-ten composition of the number, so for example, 14 is said *ten-four*, and 44 as *four-ten(s)-four* (Miura, Kim, Chang, & Okamoto, 1988; Nguyen & Grégoire, 2011). This transparency has been linked to the ease of acquisition of counting and place value understanding (Geary, Bow-Thomas, Fan, & Siegler, 1993; Nguyen & Grégoire, 2011).

Some languages have complex multibase systems. Yoruba, for instance, uses a primary base of 20, with secondary bases of 10 and five. It also uses subtraction more than addition, so that 47 can be decomposed as $(-3-10+(20\times3))$ (Verran, 2001). There are multiple ways of deriving large numbers; Verran lists seven ways of deriving 19,669. While this system would be very complicated to write, and particularly to take account of multiple representations of large numbers, it facilitates mental computation. Verran claims that

> a Yoruba numerator, with a well-honed memory of factorial relations, would scorn the cumbersome graphic processes that must be adopted to remember where you are when calculating with a base-ten system. For a reckoner skilled in the Yoruba system, writing things down would constitute a significant interference in working the system. (p. 64)

As seen from the examples above, how well a number system fits with the requirements of mathematics education can depend on syntactic category and the regularity of the system. While these things can be and are at times modified for educational or other goals, attention must also be paid to what could be lost, such as the ease of mental calculation in Yoruba, or a dynamic world view that prioritises process (numbers as verbs) over objectivity (numbers are nouns).

2.5 Logic and Reasoning

According to Hunter (1990), the issue of whether or not logic, an underpinning of mathematical behaviour, is governed by language, was first raised by Whorf (1956). Logical connectives are one of the resources that a language uses to link and sequence ideas. As well as 'if' and 'then,' additional connectors include 'because, for example, but, either, or'. When students read problems they must be able to recognise logical connectors and what situation they signal (Dale & Cuevas, 1987). These situations include similarity, contradiction, cause and effect, and logical sequence. Dawe (1983) found that the knowledge of logical connectives in the language of instruction was the most important variable on a test of deductive reasoning for bilinguals from four different countries. Logical connectives tend to be a closed grammatical class in a language. Hence the presence or absence of particular logical connectives in a language can facilitate or impede reasoning.

Gay and Cole's (1967) famous study of mathematical reasoning among the Kpelle people of Liberia found that the Kpelle performed better on tests of logical disjunction than English-speaking US college students, but performed less proficiently on tests of implication. Gale and Cole attributed these differences between groups to differences in the class of logical operators in the Kpelle and English languages: 'the precision of the Kpelle language with respect to disjunction aids them with this task' (p. 82). Kpelle has words for both 'inclusive or' and 'exclusive or', whereas everyday English has one word that includes both concepts. On the other hand, there is no easy way to express a condition such as 'if and only if' in Kpelle.

On the other hand, Iwaidja speakers of North West Arnhem Land in northern Australia seem to be adopting common conjunctions such as 'but' and 'or' from English into Iwaidja. Traditionally, Iwaidja had only a single conjunction *lda*, which fulfilled roles such as 'and', 'but' and 'or'. A construction such as 'A or B' was rendered 'maybe A and/or/but maybe B'. It now appears that *bad* 'but' and *u* 'or' tend to be quite common in Iwaidja speech.

The syntax of mathematics is often seen as the language that describes relationships (Carrasquillo & Rodriguez, 1996). Traditional *te reo Māori* (Māori language) already had a great quantity of logical connectives that could be used in mathematical discussion. For example, numbers are related to other numbers by such relations as 'greater than' (*nui ake*), 'less than' (*iti iho*) and 'equal to' (*ōrite ki*). While the Māori language has an abundance of logical connectives, how they are used in mathematics classrooms has implications not only for the learning of mathematics but also for the cultural teaching in the classroom. For example, the word 'relation' can be translated as either *whanaunga* or *pānga*. However, both these words are context specific. *Whanaunga* is a generic term applied to kin of both sexes related by marriage, adoption and/or descent. This word implies some human kinship relation. Thus *whanau* terms are inappropriate to use when describing 'relationships between mathematical objects' and it is more appropriate to use terms like *pānga* (a connection) or *tūhono* (join), for non-kinship/human relations (Trinick, 1999).

2.5.1 Negation

How a language expresses negation can also affect mathematical reasoning. Kazima's (2007) study of Chichewa students' understandings of the language of probability in English found that their attribution of meanings to words such as 'likely' and 'unlikely' was influenced by the meanings of these words in Chichewa. In Chichewa, 'unlikely' is *zokayikitsa*. 'Likely' is *zosakayikitsa*, the negative of 'unlikely' and so means 'not unlikely'. Modifying 'likely' thus can create double or complex negatives where 'not very likely' translates as 'not very not unlikely'. Kazima suggests that teachers need to be aware of preconceptions about mathematical meanings that students bring to the mathematics classrooms from their home languages. She also points out that students need opportunities to construct for themselves the meanings they need in mathematics classrooms, as opposed to being just presented with definitions, and that this needs to be done through multiple examples of how the words are used in their mathematics lessons.

In the Māori language, negation is a complex phenomenon. For example, mathematical practices, such as quantification and location in time and space, are treated like verbal sentences, hence the term *kāore* 'not' to negate is used. Non-verbal sentences are negated by terms such as *ehara* 'not'. Therefore, a sentence such as 'there are <u>not</u> three in the group' is translated as *<u>kāore</u> e toru kei roto i te rōpū*. A sentence such as 'the group is <u>not</u> big' is translated *<u>ehara</u> te rōpū i te nui*. The teacher needs to be aware that this makes the presentation and discussion of examples of negation more complex than in a language where there is a single construction to negate a statement.

In French, negation presents unexpected mathematical challenges, both for students whose French is not the preferred language and for native speakers, due to the relationship between syntax and semantics. In singular sentences, negation is applied to the verb using 'ne… pas', such that *7 divise 27* '7 divides 27' is negated *7 ne divise pas 27* '7 does not divide 27'. When sentences involve an existential quantifier, applying the negation on the verb does not provide a logical negation. Both the statements *certains nombres entiers sont pairs* 'some integers are even' and *certains nombres entiers ne sont pas pairs* 'some integers are not even' are true, so that while the second sentence is negative (syntax) the truth values are not exchanged (semantics). This is not specific to French; it occurred in ancient Greek, as Aristotle noted in *On Interpretation* (Organon, Book 2). Another problem, more specific to French, appears with sentences involving a universal quantifier. Accordingly, using the French linguistic norm, applying the negation on the verb provides the negation of the sentence: *tous les entiers sont pairs* 'all integers are even' and *tous les entiers ne sont pas pairs* 'all integers are not even' exchange their truth values. However, the substitution of *sont impairs* 'are odd' for *ne sont pas pairs* 'are not even' modifies the meaning of the sentence. *Tous les entiers sont impairs* 'all integers are odd' is a false statement, while using the norm *tous les entiers ne sont pas pairs* 'all integers are not even' is true.

Even for French native speakers, such sentences are ambiguous (Durand-Guerrier & Njomgang-Ngansop, 2009).

In Tunisia, mathematics is taught in Arabic until the end of the *Ecole de base* (Grade 8) and then in French at secondary school. A study by Ben Kilani (2005) showed clearly that these ambiguities were reinforced by the specific linguistic context (Durand-Guerrier & Ben Kilani, 2004). In the Arabic language, when the negation is inside the sentence, its scope is not the sentence but the verb or the predicate, so that in a word-to-word translation, the meaning is changed; for example the statement *tous les entiers ne sont pas pairs* 'all the integers are not even', that according to the norm means 'not all integers are even', will be interpreted as *tous les entiers sont non-pairs* 'all integers are odd'. Ben Kilani's (2005) study showed that for most students, the universal sentences with an internal negation were not interpreted as the negation of the corresponding universal affirmative sentence, but as its *contrary* in Aristotle's sense.

The ongoing research of Njomgang-Ngansop in Cameroon shows that the grammatical structure of negation in Ewondo also differs from French, leading to ambiguities or inadequate interpretation of negative sentences (See Njomgang-Ngansop & Durand-Guerrier, 2011; this volume, Chap. 5).

Hence for formal language in a variety of languages, the logical formalisation of such statements can point out the grammatical difference, and help teachers and advanced students to become aware of such phenomena, and be less susceptible to misinterpretations of the intended meaning.

2.5.2 Formal Semantics

A relationship between syntax and semantics is an ancient discussion, posited by Aristotle, in terms of opposition in *On interpretation* (Organon, Book 2), and concerning the relationship between truth and validity in *Prior Analysis* (Organon, Book 3; see Durand-Guerrier, 2008). The modern introduction of semantics into logic was undertaken by Frege (1984), further developed by Wittgenstein (1921) and Tarski (1944), and then influenced Morris (1938, 1955).

While Tarski (1944) had considered that his semantic definition of truth did not apply to languages allowing self-reference, including natural languages, Montague (1974) argued that there were no theoretical differences between natural and formal languages. He applied Tarski's theoretical model to natural language, and introduced a unifying mathematical theory known as *Montague Grammar* encompassing the syntax and semantics of both kinds of languages (Montague, 1974). This originated formal methods in linguistics such as the Discourse Representative Theory (Kamp, 1981; Kamp & Reyle, 1993). These formal methods can be useful in mathematics education, where natural language and formal language are used concomitantly. Logical analysis can offer conceptual clarification (Quine, 1997), and in multilingual contexts offers a common reference for comparison.

As an example, we discuss briefly here the well-known *Donkey sentences* problem concerning quantification and anaphora. The Donkey sentences owe their name to a famous example in Kamp (1981) of how to represent in predicate calculus, the sentence:

$$\text{Every farmer who owns a donkey beats it.} \tag{1}$$

To formalise this sentence, we must determine what type of quantifier should be used to formalise 'a donkey', existential or universal. A common choice would be to represent the 'a' in 'a donkey' existentially, and that the bounded quantifier 'every farmer' introduces an implication, so that (1) is paraphrased by:

For all x, if x is a farmer and there exists y such that y is a donkey and x owns y, then x beats y,

and is formalised as:

$$\forall x\,[[\text{Farmer}\,(x)\ \&\ \exists y\,[\text{Donkey}\,(y)\ \&\ \text{Owns}\ (x,y)]] \Rightarrow \text{Beats}\,(x,y)] \tag{2}$$

However, from a formal point of view, there is a problem, since in (2) 'y' is a free variable, while it should be within the scope of the existential quantifier.

An alternative is to represent 'a' with a universal quantifier and to formalise (1) as

$$\forall x\,\forall y\,[[\text{Farmer}\,(x)\ \&\ \text{Donkey}\,(y)\ \&\ \text{Owns}\,(x,y)] \Rightarrow \text{Beats}\,(x,y)] \tag{3}$$

The linguistic question is why 'a' should be sometimes represented by an existential quantifier, and sometimes by a universal quantifier. Kamp (1981) assumes that it is necessary to modify the language use for representing such sentences. The Discourse Representative Theory has been designed for this purpose, and other theories have since been elaborated (i.e. Abbott, 1999). Discourse Representative Theory is a non-quantificational approach aiming to extend the narrow conception of meaning as truth conditions to a more dynamic notion of meaning relative to context. In particular, it assumes that indefinites (syntax) introduce discourse referents (semantics) remaining in mental representation (pragmatics), which are accessible to anaphoric elements, such as pronouns.

Mathematical discourse at all levels includes natural language, where phenomena such as anaphora are common. In anaphora words take meaning from relationships with other parts of a statement. Because the meaning is not directly contained in the word, anaphora offers the potential for confusion to students. Durand-Guerrier (1996) reports an experiment with students beginning with the university in France given a questionnaire on implication. One of the questions was:

(u_n) is the name of a sequence of real numbers determined by a recursive law of type '$u_{n+1}=f(u_n)$', where f is a continuous function on the set of real numbers.
One has then the following theorem:

$$\text{If the sequence } (u_n) \text{ converges to the real number } L, \text{ then } L \text{ is a solution for equation } (\mathrm{E}): \text{'} f(x)=x \text{'} \quad (4)$$

Questions:
What can be said about the convergence of the sequence (u_n) if:

(a) The equation (E) has no solution?
(b) The equation (E) has at least one solution?

What can be said about eventual solutions for the equation (E) if:

(c) The sequence (u_n) converges?
(d) The sequence (u_n) does not converge?

The structure of sentence (4) is more complex than the Donkey sentence previously mentioned. Three variables are needed (sequence, function, limit) and there are various relationships between these objects. Unlike sentence (1) above, the implication is already introduced and the universal quantifier is implicit. The answer to both (b) and (d) is 'one cannot say anything': it is possible that the sequence converges/that the equation has at least a solution, and it is also possible that the sequence does not converge/that the equation has no solution. For (a), we can deduce that the sequence does not converge; and for (c) that the equation has at least a solution.

The first appearance of L in the theorem introduces an anaphoric discourse referent that remains in the mental representation. With (b) and (d), although there is no more referent for L, many students seem to consider that L is a given element as in the following answers to those questions:

(b) 'If equation (E) has at least a solution, this solution might be either L, or not be L. We can't conclude about convergence of this sequence.'
(d) 'If sequence u doesn't converge, then it is possible that there exists one or several solutions to (E), but none of them is L.'

For others, the uniqueness of the limit is moved to uniqueness for the equation:

(b) 'u may converge to only one limit or not at all. Hence there exists a unique limit for u if (E) has a unique solution.'
(b) 'If the equation (E) has a solution, then u converges; if the equation (E) has more than one solution, then u diverges.'

Anaphora contributes to the misunderstanding of the sentence and to difficulties in using it in inferences. Teachers should become aware of such didactical phenomena. Moreover, formalising the sentence (4) with students could open discussion on features which provide clues to both the concept and to the logical connectives and quantifiers involved.

2.6 Space and Geometry

Space is a fundamental part of many areas of mathematics including graphing, geometry, calculus and mechanics, among others (Lean & Clements, 1981). Spatial skills are also used to manage information on the page or in the mind when performing complex computations (Booth & Thomas, 1999; Wheatley, 1998). Spatial visualisation can be used to solve number and fraction problems (Lean & Clements, 1981). Spatial language has also been historically assumed to reflect a natural order of perception of the world (Miller & Johnson-Laird, 1976) and thus to vary less than some other parts of language. Children's development of spatial thought has also been taken to follow a regular trajectory (Piaget and Inhelder 1948/1956). However, linguistic research into spatial frames of reference has revealed more variation in spatial language than was previously thought (Levinson, 2003; Levinson & Wilkins, 2006; Pederson et al., 1998). Piaget and Inhelder (1956) described three main stages in conceptual spatial development: topological, projective and Euclidean. These stages parallel the acquisition of spatial frames of reference in European languages, but the same order of acquisition does not apply to the Mayan languages Tzeltal (Brown & Levinson, 2000) or Tzotzil (De León, 1994). The stages are thus in part language-dependent rather than universal as has been often assumed. Variations in spatial language have consequences for mathematics beyond the clearly spatial fields. A great deal of the operational language of mathematics uses metaphorical extensions of spatial language, so variations in spatial language will affect how mathematical processes can be described. The topics of topological language and spatial frames of references are discussed below.

2.6.1 *Spatial Frame of Reference*

European languages among others favour the use in small-scale space of an egocentric or *relative* spatial frame of reference, which uses left and right, and front and back that are projected from the speaker's viewpoint (Levinson, 2003). In written mathematics, we constantly use left and right to place and order things on the page or screen. Many Australian languages such as Warlpiri (Laughren, 1978) and Guugu Yimithirr (Levinson, 1997) favour *absolute* frames of reference, using terms for north, south, east and west constantly, including most importantly for mathematics learning, in small-scale space. Speakers of these languages tend not to use the relative left and right. They might approach the written organisation of mathematics differently to speakers of strongly relative languages. An absolute-preferring speech community might consider developing a convention of assigning absolute axes to the page/screen/workspace. If you were to do that in English, north would be assigned to the top of a page, but there are many speech communities with a sun orientation (see O'Grady, 1998) who might prefer to use east. Aymara speakers who construe the past as in front of them, use a word meaning 'front' for the east (Núñez & Cornejo, 2012). This connection of the past with the east is also seen in the Kuuk Thaayorre speakers of Pormpuraaw in northern Australia (Boroditsky & Gaby, 2010), suggesting that a preference for absolute conceptions of space sometimes leads to absolute-oriented conceptions of time.

The Māori spatial concepts are multilayered and are derived from a range of traditions all now merged into one. For example, some direction terms are derived from the concept of the North Island of Aotearoa (New Zealand) being a fish, therefore the head of the fish is 'up' (south) and the tail of the fish 'down' (north). The sky is also referred to as 'up' (north) and the land as 'down' (south). Traditionally many spatial terms were very localised, but one of the consequences of standardising the Māori-medium mathematics language has been to decontextualise spatial terms.

People who speak languages that favour an *intrinsic* frame of reference, like the Australian language Iwaidja, and talk about things in terms of their relation to each other, but not to the speaker or external referent (Edmonds-Wathen, 2011), might organise objects mentally or on the written page differently again. Discussing the implications of the favouring of the intrinsic frame of reference in Mopan, a Mayan language, Danziger (1996) discussed mental rotation activities that are often given as spatial mathematics problems in schools, such as deciding whether pictured dice are the same or not. She suggested that such problems could be solved using the intrinsic frame of reference without mentally rotating.

There is certainly scope for further research on spatial frames of reference in mathematics education, particularly in terms of children's development of spatial language from a cross-linguistic perspective.

2.6.2 Topological Language

Topological information in language includes concepts such as closure, proximity, separation and continuity (Piaget and Inhelder 1948/1956). In English and some other European languages, this is provided predominantly with prepositions such as 'in', 'on', 'at', 'by', 'under', 'behind' and 'in front of'. Doing mathematics, the primary spatial meaning of many of these terms is extended metaphorically. The prepositions are used to indicate the roles of numbers or other operands (Barton, 2009). Some languages do not have this range of prepositions. Iwaidja has one general locative preposition *wuka*, which means 'in', 'on', 'at' or 'by', or 'in the vicinity of'. Iwaidja speakers using English may have trouble differentiating the prepositions, and use the English word 'where' in the general way that *wuka* is used in Iwaidja, that is, to mean 'in', 'on' or 'at'.

Other languages encode topological information using case-marking rather than separate words. In these languages, the roles are marked on the dependent nouns. Turkish is an example of a case-marking language where topological relationships are marked with a suffix. For example 'A in B' is expressed 'A B-*in*', where *–in* is not a separate word. Johnston and Slobin (1979) studied order of acquisition of certain spatial terms in English, Italian, Turkish and Serbo-Croatian, finding a similar order for each language but different ages of acquisition which they attributed to linguistic factors such as morphological complexity and, in the case of Turkish, to the fact that these terms are 'postpositions' in Turkish, rather than prepositions. There is definitely scope for mathematics education research comparing prepositional and 'postpositional' or case-marking languages.

2.7 Suggested Directions for Teachers and Researchers

So how can teachers and researchers in multilingual contexts become aware of grammatical differences among the languages in their context, and what can they do with this awareness? Throughout this chapter we have pointed out potential avenues for future research, particularly in logical and spatial areas. We have pointed out that mathematical terminology may fall into different syntactic categories in different languages. Teachers might consider the general linguistic features of the languages of their students: are relationships between words shown by a fixed word order, or by affixes on keywords? Are the languages verb-rich and productive, or are verbs a closed class? How are logical relations expressed? Are features such as evidentiality (such as whether the speaker has personal evidence for what is being said or not) grammaticalised or optional? These aspects of language can all potentially affect how mathematical ideas are processed and expressed and manipulated in the language. As educators, we want to give our students access to the richness of mathematical discourse. In many of the world's dominant languages, this discourse is especially marked by the process of nominalisation. Halliday (2004) suggests that the language of science, including, we might imagine, the language of mathematics will 'back off from its present extremes of nominalisation and grammatical metaphor and go back to being more preoccupied with processes and more tolerant of indeterminacy and flux' (p. 224). While he states that this would be unlikely to be able to be achieved by design, teachers are in a position to influence the relative status of noun phrases and verb groups in their mathematics classroom. If they continue to accord status to complex nominalisations as evidence of abstraction and higher-order thinking, then these types of language use will continue to be privileged in mathematical discourse.

Teaching mathematics in the medium of Māori has supported the maintenance of cultural knowledge as a functional system that has applicability to everyday functional use. For example, Māori has multiple quantitative pronouns in comparison to

English or French. These terms take into account the hierarchical relationship between the speaker and the listener, for example *tōku hoa* (my friend-higher status) versus *taku hoa* (my friend-equal status), but also the relationship between the listener and other people. In addition, Māori has a special set of plural pronouns that refer to two persons only. *Tāua* (us two) includes the speaker and listener, while *māua* (us two) includes the speaker, another person but excludes the listener. Similarly, *tātau* includes the listener, the speaker, and others, but *mātau* includes the speaker and others, but not the listener. In English, both of these would be translated like *us*. Also, in Māori there is no distinction of case or gender.

Verran (2001) described the disconcertment that she experienced in Nigerian classrooms while observing the practices with number of Yoruba teachers of mathematics and science, who were her students in a teacher education programme. Rather than trying to explain away this disconcertment, she set out to compare and contrast the different 'generalizing logics' of English and Yoruba. Eventually, she says:

> I learned to trust my students' classes and to trust them as teachers and their pupils as learners. Encouraging my students to do science and mathematics lessons in practical ways, bringing to the fore the actual doing of the little rituals of the quantifying with hands, eyes, water, string, and rulers as well as with utterances turned out to be a useful and generative way to deal with the generative tensions between English and Yoruba logics of numbering. (pp. 235–236)

Similarly, in this chapter we have not sought to explain away grammatical differences in languages that influence mathematical learning, but to see how this diversity of expression might be comprehended and utilised by teachers and researchers. Finally, mathematicians might like to investigate more deeply this diversity, which can point to ways to create new mathematics (Barton, 2009).

2.8 Conclusion

The impacts of features of grammatical structures on mathematical thinking are still underresearched. We have shown that languages express mathematical ideas in diverse ways. These different ways of exploring mathematical ideas provide an opportunity to enrich the mathematical experiences of learners in multilingual contexts. They can also introduce ambiguities or misunderstanding between teachers and students and impede the process of mathematical learning. While at times, multilingualism and/or teaching mathematics in the medium of indigenous languages has been considered from deficit perspectives, this chapter considers these challenges as more enabling and enriching. Teaching mathematics in indigenous languages supports the revitalisation and maintenance of the languages, particularly those that are endangered. This revitalisation may or may not involve grammatical changes.

We have also shown that languages such as English and French, with long traditions of developing a mathematics register, nevertheless contain some grammatical features which are not always ideal mathematically. If opportunities arise for linguistic–mathematical innovations in these world languages, language planners might like to consider innovating for features which research shows facilitate mathematics learning.

References

Abbott, B. (1999). The formal approach to meaning: Formal semantics and its recent developments. *Journal of Foreign Languages, 119*(1), 2–20.

Aiken, L. R. (1972). Language factors in learning mathematics. *Review of Educational Research, 42*, 359–385.

Allan, K. (1977). Classifiers. *Language, 53*(2), 285–311.

Austin, J. L., & Howson, A. G. (1979). Language and mathematical education. *Educational Studies in Mathematics, 10*(2), 161–197.

Bamgbose, A. (1999). African language development and language planning. *Social Dynamics, 25*(1), 13–30.

Barton, B. (2009). *The language of mathematics: Telling mathematical tales*. New York: Springer.

Barton, B. (2012). Preface to "Ethnomathematics and philosophy". In H. Forgasz & F. D. Rivera (Eds.), *Towards equity in mathematics education: Gender, culture and diversity* (pp. 227–229). Berlin, Germany: Springer.

Barton, B., Fairhall, U., & Trinick, T. (1998). Tikanga Reo Tātai: Issues in the development of a Māori mathematics register. *For the Learning of Mathematics, 18*(1), 3–9.

Barwell, R. (2003). Linguistic discrimination: An issue for research in mathematics education. *For the Learning of Mathematics, 23*(2), 37–43.

Barwell, R., Barton, B., & Setati, M. (2007). Multilingual issues in mathematics education: Introduction. *Educational Studies in Mathematics, 64*(2), 113–119.

Ben Kilani, I. (2005). *Les effets didactiques des différences de fonctionnement de la négation dans la langue arabe, la langue française et le langage mathématique.* Unpublished doctoral thesis, University of Tunisia and Claude Bernard University, Lyon 1, Tunisia and France.

Bender, A., & Heller, S. (2006). Numeral classifiers and counting systems in Polynesian and Micronesian languages: Common roots and cultural adaptations. *Oceanic Linguistics, 45*(2), 380–403.

Blake, B. J. (1981). *Australian Aboriginal languages*. Sydney, New South Wales, Australia: Angus and Robertson.

Bokamba, E. G. (1995). The politics of language planning in Africa: Critical choices for the 21st century. In M. Putz (Ed.), *Discrimination through language in Africa: Perspectives on the Namibian experience* (pp. 11–27). Berlin, Germany: Walter de Gruyter.

Booth, R. D. L., & Thomas, M. O. J. (1999). Visualization in mathematics learning: Arithmetic problem-solving and student difficulties. *The Journal of Mathematical Behavior, 18*(2), 169–190.

Boroditsky, L., & Gaby, A. (2010). Remembrances of times east: Absolute spatial representations of time in an Australian Aboriginal community. *Psychological Science, 21*(11), 1635–1639.

Brown, P., & Levinson, S. C. (2000). Frames of spatial reference and their acquisition in Tenejapan Tzeltal. In L. Nucci, G. Saxe, & E. Turiel (Eds.), *Culture, thought, and development* (pp. 167–197). Mahwah, NJ: Lawrence Erlbaum.

Bruner, J. (1966). *Studies in cognitive growth: A collaboration at the center for cognitive studies.* New York: Wiley & Sons.

Butterworth, B., Reeve, R., Reynolds, F., & Lloyd, D. (2008). Numerical thought with and without words: Evidence from Indigenous Australian children. *Proceedings of the National Academy of Sciences of the United States of America, 105*(35), 13179–13184.

Butterworth, B., Reeve, R., & Reynolds, F. (2011). Using mental representations of space when words are unavailable: Studies of enumeration and arithmetic in Indigenous Australia. *Journal of Cross-Cultural Psychology, 42*(4), 630–638.
Carnap, R. (1955). On some concepts of pragmatics. *Philosophical Studies, 6*, 89–91.
Carrasquillo, A., & Rodriguez, R. (1996). *Language minority students in the mainstream classroom*. Clevedon, England: Multilingual Matters.
Chandler, D. (2004). *Semiotics: The basics* (6th ed.). New York: Routledge.
Closs, M. (1977). *A survey of mathematics development in the New World* (Report No. 410-77-0222). Ottawa, Ontario, Canada: University of Ottawa.
Cocking, R., & Mestre, J. P. (Eds.). (1988). *Linguistic and cultural influences on learning mathematics*. Hillsdale, NJ: Lawrence Erlbaum.
Cooper, R. L. (1989). *Language planning and social change*. Cambridge, England: Cambridge University Press.
Cuevas, G. (1984). Mathematics learning in English as a second language. *Journal for Research in Mathematics Education, 14*(2), 133–144.
Da Costa, N. C. A. (1997). *Logiques classiques et non classiques: Essai sur les fondements de la logique*. Paris, France: Masson.
Dale, T., & Cuevas, G. (1987). Integrating language and mathematics learning. In J. Crandall (Ed.), *ESL through content-area instruction: Mathematics, science, social studies* (pp. 9–54). Englewood Cliffs, NJ: Prentice Hall.
Danziger, E. (1996). Parts and their counterparts: Spatial and social relationships in Mopan Maya. *Journal of the Royal Anthropological Institute, 2*(1), 67–82.
Dawe, L. (1983). Bilingualism and mathematical reasoning in English as a second language. *Educational Studies in Mathematics, 14*(4), 325–353.
De León, L. (1994). Exploration in the acquisition of geocentric location by Tzotzil children. *Linguistics, 32*(4–5), 857–884.
Dixon, R. M. W. (1980). *The languages of Australia*. Cambridge, England: Cambridge University Press.
Durand-Guerrier, V. (1996). *Logique et raisonnement mathématique: Défense et illustration de la pertinence du calcul des prédicats pour une approche didactique des difficultés liées à l'implication*. Unpublished doctoral thesis, Claude Bernard University, Lyon 1, France.
Durand-Guerrier, V. (2008). Truth versus validity in mathematical proof. *ZDM: Mathematics Education, 40*(3), 373–384.
Durand-Guerrier, V., & Ben Kilani, I. (2004). Négation grammaticale versus négation logique dans l'apprentissage des mathématiques: Exemple dans l'enseignement secondaire Tunisien. *Les cahiers du Français contemporain, 9*, 29–55.
Durand-Guerrier, V., & Njomgang-Ngansop, J. (2009, April). *Questions de logique et de langage à la transition secondaire–supérieur: L'exemple de la négation, à paraitre dans les actes du colloque*. Paper presented at Espace Mathématique Francophone, Dakar, Senegal.
Edmonds-Wathen, C. (2011). What comes before? Understanding spatial reference in Iwaidja. In M. Setati, T. Nkambule, & L. Goosen (Eds.), *Proceedings of the ICMI Study 21 Conference: Mathematics Education and Language Diversity* (pp. 89–97). São Paulo, Brazil: ICMI Study 21.
Eggins, S. (2004). *Introduction to systemic functional linguistics*. New York: Continuum.
Finlayson, R., & Madiba, M. (2002). The intellectualisation of the indigenous languages of South Africa: Challenges and prospects. *Current Issues in Language Planning, 3*(1), 40–61.
Frank, M. C., Everett, D. L., Fedorenko, E., & Gibson, E. (2008). Number as a cognitive technology: Evidence from Pirahã language and cognition. *Cognition, 108*, 819–824.
Frege, G. (1984). *Collected papers on mathematics, logic and philosophy* (B. McGuiness, Ed.; M. Black, V. Dudman, H. Geach, E. Kaal, & W. Kluge, Trans.). New York: Basil Blackwell.
Gay, J. H., & Cole, M. (1967). *The new mathematics and an old culture: A study of learning among the Kpelle of Liberia*. New York: Holt Rinehart and Winston.
Geary, D. C., Bow-Thomas, C. C., Fan, L., & Siegler, R. S. (1993). Even before formal instruction, Chinese children outperform American children in mental addition. *Cognitive Development, 8*(4), 517–529.

Gordon, P. (2004). Numerical cognition without words: Evidence from Amazonia. *Science, 306*, 496–499.
Hale, K. (1975). Gaps in grammar and culture. In C. F. Voegelin, M. D. Kinkade, K. Hale, & O. Werner (Eds.), *Linguistics and anthropology: In honor of C. F. Voegelin* (pp. 295–316). Lisse, The Netherlands: Peter de Ridder.
Halliday, M. A. K. (1975). *Some aspects of sociolinguistic interactions between linguistics and mathematics education: Final report of the symposium sponsored by UNESCO, CEDO and ICMI, Nairobi Kenya, 1-11 September, 1974*. Paris, France: UNESCO.
Halliday, M. (1978). Sociolinguistic aspects of mathematics education. In M. A. K. Halliday (Ed.), *Language as social semiotic: The social interpretation of language and meaning* (pp. 194–204). London: Edward Arnold.
Halliday, M. A. K. (1996). Systemic functional grammar. In K. Brown & J. Miller (Eds.), *Concise encyclopedia of syntactic theories* (pp. 321–325). New York: Elsevier Science.
Halliday, M. A. K. (2004). *The language of science* (J. Webster, Ed.). New York: Continuum.
Halliday, M. A. K., & Hasan, R. (1985). *Language, context and text: A social semiotic perspective*. Geelong, Victoria, Australia: Deakin University Press.
Halliday, M. A. K., & Martin, J. (1993). *Writing science: Literacy and discursive power*. London: Falmer.
Harris, P. J. (1980). *Mathematics in Aboriginal schools project: British Council program research notes*. Darwin, Northern Territory, Australia: Department of Education, Professional Services Branch.
Harris, J. W. (1982). Facts and fallacies of Aboriginal number systems. In S. Hargrave (Ed.), *Work papers of SIL–AAB* (Series B, Volume 8: Language and Culture, pp. 153–81). Darwin, Northern Territory, Australia: Summer Institute of Linguistics.
Harris, J. W. (1987). Australian Aboriginal and Islander mathematics. *Australian Aboriginal Studies, 2*, 29–37.
Hunter, L. (1990). *The ESL student in the mathematics classroom: Student questions as a mode of access to knowledge*. Unpublished master's thesis, The University of British Columbia, Canada.
Johnston, J. R., & Slobin, D. I. (1979). The development of locative expressions in English, Italian, Serbo-Croatian and Turkish. *Journal of Child Language, 6*, 529–545.
Kamp, H. (1981). A theory of truth and semantic representation. In J. A. Groenendjik, T. M. Janssen, & M. B. Stokhof (Eds.), *Formal methods in the study of language* (pp. 277–322). Amsterdam, The Netherlands: Mathematical Centre Tracks.
Kamp, H., & Reyle, U. (1993). *From discourse to logic: Introduction to model theoretic semantics of natural language, formal logic and discourse representation theory*. Dordrecht, The Netherlands: Kluwer.
Kaplan, R., & Baldauf, R. (1997). *Language planning from practice to theory*. Clevedon, England: Multilingual Matters.
Kazima, M. (2007). Malawian students' meanings for probability vocabulary. *Educational Studies in Mathematics, 64*(2), 169–189.
Lancy, D. F. (1983). *Cross-cultural studies in cognition and mathematics*. New York: Academic Press.
Laughren, M. N. (1978). Directional terminology in Warlpiri. *Working Papers in Language & Linguistics* (Vol. 8, pp. 1–16). Launceston, Australia: Tasmanian College of Advanced Education.
Lean, G., & Clements, M. A. K. (1981). Spatial ability, visual imagery, and mathematical performance. *Educational Studies in Mathematics, 12*(3), 267–299.
Leech, G. (1983). *Principles of pragmatics*. London: Longman.
Levinson, S. C. (1993). *Pragmatics*. Cambridge, England: Cambridge University Press.
Levinson, S. C. (1997). Language and cognition: The cognitive consequences of spatial description in Guugu Yimithirr. *Journal of Linguistic Anthropology, 7*(1), 98–131.
Levinson, S. C. (2003). *Space in language and cognition: Explorations in cognitive diversity*. Cambridge, England: Cambridge University Press.

Levinson, S. C., & Wilkins, D. (2006). *Grammars of space: Explorations in cognitive diversity*. Cambridge, England: Cambridge University Press.
Lucy, J. (1992). *Language diversity and thought: A reformulation of the linguistic relativity hypothesis*. New York: Cambridge University Press.
Lunney Borden, L. (2010). *Transforming mathematics education for Mi'kmaw students through mawikinutimatitimk*. Unpublished doctoral thesis, University of New Brunswick, Canada.
McRoberts, R. W. (1990). *Counting at Pularumpi: A survey of a traditional mathematics and its implications for modern learning*. Canberra, Australia: Library of the Australian Institute of Aboriginal and Torres Strait Islanders Studies.
Meaney, T. (2002). Symbiosis or cultural clash? Indigenous students learning mathematics. *Journal of intercultural studies, 2*(23), 167–187.
Meaney, T., Trinick, T., & Fairhall, U. (2012). *Collaborating to meet languages challenges in indigenous mathematics classrooms*. Dordrecht, The Netherlands: Springer.
Mendes, J. R. (2011). The meaning of 'number' in Kaiabi: Identity and language in the context of indigenous teacher education. In M. Setati, T. Nkambule, & L. Goosen (Eds.), *Proceedings of the ICMI Study 21 Conference: Mathematics Education and Language Diversity* (pp. 218–223). São Paulo, Brazil.
Miller, G. A., & Johnson-Laird, P. N. (1976). *Language and perception*. Cambridge, MA: Belknap.
Miura, I. T., Kim, C. C., Chang, C.-M., & Okamoto, Y. (1988). Effects of language characteristics on children's cognitive representation of number: Cross-national comparisons. *Child Development, 59*(6), 1445–1450.
Montague, R. (1974). *Formal philosophy: Selected papers of Richard Montague* (R. Thomason, Ed.). New Haven, CT: Yale University Press.
Morris, C. (1938). *Foundation of the theory of signs*. Chicago, IL: Chicago University Press.
Morris, C. (1955). *Signs, language, and behavior*. New York: Braziller.
Mousley, J., & Marks, G. (1991). *Discourses in mathematics*. Geelong, Victoria, Australia: Deakin University Press.
Nguyen, T. T. H., & Grégoire, J. (2011). A special case of natural numbers denomination: A comparison between French and Vietnamese languages. In M. Setati, T. Nkambule, & L. Goosen (Eds.), *Proceedings of the ICMI Study 21 Conference: Mathematics Education and Language Diversity* (pp. 259–266). São Paulo, Brazil.
Njomgang-Ngansop, J. & Durand-Guerrier, V. (2011). Negation of mathematical statements in French in multilingual contexts: An example in Cameroon. In M. Setati, T. Nkambule, & L. Goosen (Eds.), *Proceedings of the ICMI Study 21 Conference: Mathematics Education and Language Diversity* (pp. 268–275). São Paulo, Brazil.
Núñez, R. E., & Cornejo, C. (2012). Facing the sunrise: Cultural worldview underlying intrinsic-based encoding of absolute frames of reference in Aymara. *Cognitive Science, 36*(6), 965–991.
O'Grady, G. N. (1998). Toward a proto-Pama-Nyungan stem list, part I: Sets J1-J25. *Oceanic Linguistics, 37*(2), 209–233.
Pederson, E., Danziger, E., Wilkins, D., Levinson, S., Kita, S., & Senft, G. (1998). Semantic typology and spatial conceptualization. *Language, 74*(3), 557–589.
Piaget, J., & Inhelder, B. (1948/1956). *The child's conception of space* (F. J. Langdon & J. L. Lunzer, Trans.). London: Routledge & Kegan Paul.
Pimm, D. (1987). *Speaking mathematically: Communication in mathematics classrooms*. London: Routledge Keegan Paul.
Pimm, D. (1995). *Symbols and meanings in school mathematics*. London: Routledge.
Quine, W. V. O. (1997). *Quiddities: An intermittently philosophical dictionary*. Cambridge, MA: Harvard University Press.
Ricento, T. (2000). Historical and theoretical perspectives in language policy planning. In T. Ricento (Ed.), *Ideology, politics and language policies: Focus on English* (pp. 9–24). Amsterdam, The Netherlands: John Benjamins.
Schindler, D. E., & Davison, D. M. (1985). Language, culture and the mathematics concepts of American Indian learners. *Journal of American Indian education, 24*(3), 27–34.

Schleppegrell, M. (2004). *The language of schooling: A functional linguistics perspective.* Mahwah, NJ: Lawrence Erlbaum.

Schleppegrell, M. J. (2007). The linguistic challenges of mathematics teaching and learning: A research review. *Reading & Writing Quarterly, 23*(2), 139–159.

Setati, M. (2008). Access to mathematics versus access to the language of power: The struggle in multilingual mathematics classrooms. *South African Journal of Education, 28*(1), 103–116.

Smith, L. T. (1999). *Decolonizing methodologies: Research and indigenous peoples.* London: Zed Books.

Spanos, G. N., Rhodes, N., Dale, T. C., & Crandall, J. (1988). Linguistic features of mathematical problem solving: Insights and applications. In R. M. Cocking & J. Mestre (Eds.), *Linguistic and cultural influences on learning mathematics* (pp. 221–240). Hillsdale, NJ: Lawrence Erlbaum.

Stokes, J. (1982). *A description of the mathematical concepts of Groote Eylandt Aborigines.* Canberra, Australia: Library of the Australian Institute of Aboriginal and Torres Strait Islanders Studies.

Tarski, A. (1944). The semantic conception of truth. *Philosophy and Phenomenological Research, 4*, 13–47.

Trinick, T. (1999). *The relationships between Maori culture and Māori mathematical language.* Unpublished master's thesis, The University of Auckland, New Zealand.

Trinick, T. (2015) *Te Reo Tātai: The development of a mathematics register for Māori-medium schooling*. Unpublished doctoral dissertation, Waikato University, New Zealand.

Verran, H. (2001). *Science and an African logic.* Chicago, IL: University of Chicago Press.

Von Brandenstein, C. G. (1970). What next in Aboriginal Australian linguistics? *The Etruscan, 19*(2), 11–15.

Vygotsky, L. S. (1978). *Mind in society: The development of higher psychological processes.* Cambridge, MA: Harvard University Press.

Wheatley, G. H. (1998). Imagery and mathematics learning. *Focus on Learning Problems in Mathematics, 20*(2/3), 65–77.

Whorf, B. L. (1956). *Language, thought, and reality: Selected writings.* Cambridge, MA: MIT Press.

Wittgenstein, L. (1921). *Tractatus logico-philosophicus* (C. K. Ogden, Trans.). London: Routledge.

Zaslavsky, C. (1979). *Africa counts: Number and pattern in African culture.* Chicago, IL: Lawrence Hill.

Chapter 3
Making Use of Multiple (Non-shared) First Languages: State of and Need for Research and Development in the European Language Context

Michael Meyer, Susanne Prediger, Margarida César, and Eva Norén

3.1 Introduction

As Durkin and Shire (1991) underline, "Mathematics begins and proceeds in language, it advances and stumbles because of language, and its outcomes are often assessed in language" (p. 3). The importance of language in mathematics learning illuminates the need for speaking a language that everyone is able to understand. Historically, the development of the European nations in the eighteenth and nineteenth centuries went along with the development of a monolingual self-concept of many of them, although some experienced a multilingual way of living, such as, for example, Switzerland and Belgium. But the monolinguistic *habitus* was strengthened during the Second World War and since then it has been one reason for the development of social and cultural communities. Today the role of language has changed. Emigration and immigration movements in Europe have produced a mixture of languages and cultural backgrounds (Gogolin, 2010). Vertovec (2007) uses the designation "Super-diversity" to describe the actual situation in the majority of European countries: a complete mixture of (first) languages and cultures.

M. Meyer (✉)
University of Cologne, Cologne, Germany
e-mail: michael.meyer@uni-koeln.de

S. Prediger
Institute for Development and Research in Mathematics Education,
TU Dortmund University, Vogelpothsweg 87, Dortmund 44227, Germany
e-mail: prediger@math.uni-dortmund.de

M. César
Associated Researcher, University of Neuchâtel, Neuchâtel, Switzerland
e-mail: macesar@fc.ul.pt

E. Norén
Stockholm University, Stockholm, Sweden
e-mail: eva.noren@mnd.su.se

R. Barwell et al. (eds.), *Mathematics Education and Language Diversity*,
New ICMI Study Series, https://doi.org/10.1007/978-3-319-14511-2_3

Although Europe is a multicultural and multilingual region, discussion and practices regarding multiple languages as resources for mathematics learning still seem to be underdeveloped. If the multilingual dimension of mathematics classrooms is neglected, this can under-privilege mathematics learning for those students whose first language does not correspond to the language of instruction. That is why making use of the multiple first languages[1] is often claimed as necessary. In her survey on different, mostly non-European, bilingual studies (not only in mathematics education), Gogolin (2011) emphasizes the relevance of asking about the effect of teaching in the first language on mastery of the second language. However, in this chapter we do not want to address this question. Instead, we handle the issue of whether (the use of) a plurality of languages can be useful in order to learn mathematics.

This chapter explores the *status quo*, actual tendencies and desiderata in research into the use of first languages for the European language contexts, which are often shaped by a large diversity of first languages in the same classroom. Thus, the chapter reviews these aspects and the development of classroom practices in the following steps: we briefly describe worldwide discussion and research on benefits of first language use, then consider European language contexts with their specificities, and finally report on current European developments, practices, and research on first language use in mathematics classrooms.

3.2 First Language as a Widely Accepted Resource for Giving Access to Mathematics

Many studies all over the world have highlighted the relevance of the first language for giving access to mathematics (e.g., Baker, 1996; Barwell, 2009; Clarkson, 1992). Various case studies have shown how the first language can provide wider options to participate in classroom interactions. Often this first language use is naturally interrupted by moments of code-switching, considered as a social practice of flexible use of languages (Moschkovich, 2007; Setati & Duma, 2009).

Other studies have emphasized the cognitive and meta-cognitive benefit of the first language while making sense of mathematical texts, for example, for mathematically successful bilinguals (Clarkson, 2007). Kern (1994) has specified the cognitive benefit of facilitating semantic processing, relieving the short-term memory and, especially, allowing concepts to become alive more easily, since first languages offer richer connections to students' networks of associations. This last aspect seems particularly important for conceptual understanding (Ellerton & Clarkson, 1996).

[1] In this chapter, we use the term "first language" for the language that students usually speak at home, in which they think and feel comfortable using. Although acknowledging subtle differences, we use it as synonymous with "mother language," since no single term can reflect the complexity of different individual language profiles with more than one language spoken in families. In the quoted data, different conceptualizations are used, mostly implicit, for example "language that is mostly spoken in the family" (OECD, 2007, p. 120) or "mother tongue" (Eurobarometer, 2006).

Additionally, Clarkson (2007) emphasized the meta-cognitive use of language switching, as bilinguals seem to self-correct themselves more frequently in their first language. These results are in line with general results on the relevance of individual languages for making sense of mathematical expressions and for developing conceptual understanding (Ellerton & Clarkson, 1996).

In the light of these international results, surveyed by Barwell (2009), it appears as a logical consequence that the Council of Europe nowadays advocates the inclusion of first languages into school subjects such as mathematics or chemistry, among others (cf. Little, 2010). However, these ideas spread very slowly and the majority of classroom practices do not match these ideals. This chapter explores why it seems particularly difficult in the European language context to activate first languages as a resource for mathematics learning.

Nevertheless, we have to mention other studies showing opposite results. On the basis of their meta-analysis of studies concerning bilingual education in the USA, Rossell and Kuder (2005) came to the result that a monolingual English education program has to be preferred to a mixture of English and Spanish education. Although their examinations did not focus on mathematics education, we have to be aware of an obstacle: in order to establish knowledge which can be taken as shared, children have to communicate their results to the other children. In a multilingual classroom this can often not be done by the use of the first language, if these children do not have the same first language.

As teaching and learning practices cannot simply be transferred from one language context to another, we felt the need to explore the opportunities and limits of using first languages in the specific language context of different countries and their mathematics classrooms. For this, we make a rough differentiation between those countries where the first languages are shared or non-shared between the different persons participating in the classroom interaction.

3.3 Cultural and Political Dimension: Presenting the European Language Context

3.3.1 Multilingualism in Europe and European Schools: Demographic Facts and Emerging Consciousness

In 2012, about 500 million people lived in the 27 countries of the European Union and there were 23 languages officially accepted in the European Union (Eurobarometer, 2006). Six countries had more than one official language, the majority of them regionally distributed. The main first language in Europe was German (with 90 million native speakers, 18 % of all Europeans), then English, French, and Italian (each about 60 million speakers, about 13 %) (European Commission, 2004; Eurobarometer, 2006). Although Europe as a whole is multilingual, however, many countries conceptualize themselves as monolingual societies.

Since the European Union was constructed, dealing with multiple languages has been a political aim. However, it is interesting to see how the European conceptualization of multilingualism has slowly changed from dealing with multiple foreign languages as a *learning goal* (for improving international relations between countries) to the awareness of multilingualism within each country as a *learning condition*, due to different first languages of minorities and immigrants: even as recently as 2004, the European Commission published a brochure titled *Many Tongues, One Family: Languages in the European Union* (European Commission, 2004) where the multilingualism within each country is marginalized to a brief reference to some regional languages such as Welsh or Sardinian, while immigrant languages are not even mentioned. Instead, it promotes the European campaign "mother tongue plus two" with its focus on the aim that each European citizen should speak two foreign languages (which is currently reached by about 30 % of all Europeans, most of them speaking English, German, and/or French). The rate of people mastering two foreign languages varies from the small countries (92 % in Luxembourg, 75 % in the Netherlands) to larger countries (Germany 27 %, UK 18 %) (European Commission, 2004). This indicates that speaking a second or third language is less common when your first language is one of the most spoken languages in Europe.

In contrast, a similar booklet published by the European Commission 2 years later (Eurobarometer, 2006) documents and acknowledges the variety of first languages, including not only regional languages but also immigrant languages (which have higher percentages in the majority of Western European countries). Typical immigrant languages are Arabic in Sweden, Turkish in Germany, and African languages in Portugal. Another important issue is that many people with immigrant backgrounds are born in European countries. Thus, they belong to a second (or even a later) generation. Table 3.1 shows an overview of the given percentages of people with a first language other than each country's official languages. Note that the data comes from selective inquiries (Eurobarometer, 2006), since the variable "language backgrounds" has still not been captured in the countries' official statistics (only nationality and migration status, see e.g., Mikrozensus, 2009). Other sources give higher rates of nonofficial first languages for the younger generation. For example, in Germany and Sweden, rates of 20 %[2] are estimated for students in compulsory education with first language other than German (Chlosta & Ostermann, 2008) or Swedish (Swedish National Agency of Education [Skolverket], 2013). The increasing rates of children with other first languages can be traced back to increasing immigration to European countries and to demographic factors (i.e., higher birth rates in some language communities) (Chlosta & Ostermann, 2008).

The development in the European Commission's conceptualization of multilingualism (from a learning goal for European ideals to a multicultural learning condition within each country) reflects the slowly increasing awareness in some of the countries about their immigrant *status*. For example, Germany acknowledged only

[2] In North Rhine-Westphalia (the most populous federal state of Germany) exactly 19.50 % of the primary school students have another first language (IT.NRW, 2012, p. 2).

Table 3.1 Multiple first languages in European countries: "What is your mother tongue?"

	Percentages of indications of first language being ... (Multiple answers possible)		
Country	... a state language or other official language (%)	... another EU language (%)	... another language (%)
Austria	96	3	2
Belgium	Dutch 56, French 38, German 0.4	5	3
Bulgaria	90	0.4	11
Cyprus	98	2	1
Czech Republic	98	2	0.7
Denmark	97	2	2
Estland [Estonia]	82	1	18
Finland	Finnish 94, Swedish 5	0.8	0.4
France	93	6	3
Germany	90	3	8
Greece	98	1	0.8
Hungary	100	0.8	0.6
Ireland	English 94, Irish 11	2	0.2
Italy	95	5	1
Latvia	73	1	27
Lithuania	88	5	7
Luxembourg	Luxembourgish 77, French 6, German 4	14	0.8
Malta	Maltese 97, English 2	0.6	–
Netherlands	96	3	3
Poland	98	1	1
Portugal	100	0.6	0.1
Romania	95	6	0.7
Slovakia	88	12	2
Slovenia	95	1	5
Spain	Spanish 89, Catalan 9, Galician 5, Basque 1	1	2
Sweden	95	5	2
United Kingdom	92	3	5

Source: Eurobarometer (2006, p. 9)

in 2000 that it was an immigration country. These trends are also reflected in the political context of the schools.

These developments have been supported by the *Charter of Fundamental Rights of the European Union*, proclaimed by the European Parliament, Council, and Commission in 2000. The charter guarantees the rights of cultural, religious, and language diversity, including a non-discrimination law for languages (European Parliament, 2000, §21). These new perspectives on multilingualism are also reflected in changing language policies for schools.

3.3.2 Multiple Languages in European Schools

Notwithstanding the non-discrimination law for languages, nearly all European countries mainly organize their schools as monolingual institutions in which the official regional or state language is the only accepted language of instruction (except for some special private schools or Content and Language Integrated Learning (CLIL) classrooms with English or French as a foreign language of instruction). Within these similar policies, the school systems attain significantly different rates of success in allowing students with other first languages to achieve mathematics performance comparable to their native speaking classmates. As Table 3.2 shows, some countries (Germany, Denmark, Switzerland, Austria) particularly fail in mathematics achievement for students with other first languages.[3]

The comparative results of some less adequate school systems have raised discussions in these countries on how to give better access to mathematics for learners with different first languages. Among all national strategies for a better inclusion of second (or third or fourth language) speakers, most emphasis is given to language appropriation in the language of instruction (e.g., der Bundesrepublik Deutschland, 2010; Thürmann, Vollmer, & Pieper, 2010). In many countries, immigrant students' low performance in mathematics is attributed to their lack of competence in the language of instruction and a lack of cultural inclusion, such as, for example the so-called lack of "Swedishness" (Haglund, 2005; Runfors, 2003), and similarly for Denmark (Holmen, 2008). That is why, besides this focus on the language of schooling, many countries increasingly discuss a new language policy allowing the use of first languages, as reflected in the Council of Europe's goal of *plurilingualism starting from the individual resources*, which comprise also first languages (Beacco et al., 2010). Although these European first languages policies mainly refer to the early years of schooling (e.g., learning how to read), some documents also refer to learning in secondary schools and in subjects such as science and mathematics (e.g., Thürmann et al., 2010).

The use of first languages in education is not only supported by political demands, but also by educational reasons, for example, "Language is a tool for acquiring knowledge, one aspect of the development of the person, as both individual and social actor, a means of and factor in understanding and making sense of reality, and a vehicle for imaginative creativity" (Coste, Cavalli, Crisan, & van den Ven, 2009, p. 5). In spite of all the claims and reasons for first language use in European schools, its use in European mathematics classrooms is still quite rare. In order to understand this reluctance, it is important to consider that these types of practices are not equally easy to follow in all language contexts, since the existence of multiple languages among students in European schools has at least two significantly different facets.

[3] Although this chapter focuses on language issues, we emphasize that not only the language background, but also other factors are crucial for school success of minority students, for example, their socioeconomic status, parents' literacy and educational background, and other individual circumstances (cf. Alrø et al., 2003; César, 2009; Heinze, Reiss, Rudolph-Albert, Herwartz-Emden, & Braun, 2009).

Table 3.2 Different equity success of school systems—comparing European and non-European countries

Country	Students with first language ≠ language of instruction		Students with first language = language of instruction		Difference between both groups' mean scores
	Mean score	(S.E.)	Mean score	(S.E.)	
Austria	442	(12.8)	515	(3.5)	73
Belgium	515	(6.7)	530	(2.6)	15
Denmark	440	(7.0)	519	(2.6)	79
France	441	(10.1)	502	(3.2)	61
Germany	438	(8.4)	519	(3.3)	81
Greece	408	(9.7)	462	(2.7)	54
Luxembourg	494	(1.3)	525	(4.7)	30
Netherlands	472	(10.3)	536	(2.2)	64
United Kingdom	458	(9.8)	499	(2.0)	41
Norway	448	(7.5)	495	(2.4)	47
Portugal	445	(15.3)	468	(3.0)	23
Switzerland	473	(5.5)	546	(2.9)	73
Australia	523	(7.7)	521	(2.0)	–2
Canada	522	(4.3)	531	(1.8)	9
New Zealand	522	(6.7)	526	(2.3)	4
USA	440	(5.9)	480	(4.2)	40
OECD Average	467	(2.2)	504	(0.5)	37

Data from PISA, 2006. *Source*: OECD (2007, p. 120, Table 4.3b)

Some regions such as Catalonia (Catalan and Castilian, see Planas & Setati, 2009) or parts of Ireland (Irish and English, see Table 3.1) experience a more or less *shared bilingualism* where most people share the two main languages, at least if they are not immigrants (e.g., Catalonia has many immigrants from Africa). Here, the use of the other language is easy since usually many teachers and students understand each other. However, in those countries commonly designated as immigration countries in Europe, there are five to seven, and sometimes even more, different first languages spoken in multilingual classrooms and the only intersection is the language of instruction. In this language context, we talk about multiple (non-shared) languages. The teachers of these classes are usually speakers of the language of instruction. They might even speak other languages, but they normally do not speak the languages of (all of) the students (César & Oliveira, 2005; Gogolin, 1994).

Speaking another language having immigrated to another country often comes with participating in a group of lower socioeconomic status (Nusche, 2009). Concerning Germany, Prenzel and his associates (2005) analyzed the PISA results and found that a child from the second lowest socioeconomic background is four times less likely to join the highest track of secondary education compared with a child from the top quartile. Comparable results have been reported for black Caribbean people in the UK (Strand, 2007). Eckhardt (2008) points out that a migration background is not the reason for systematic disadvantage of some children. Moreover, the analyses indicate that problems of appropriating the language of instruction can primarily be ascribed to the social backgrounds of the families and their communicative practices. Moreover, the disadvantages do not go along with the first languages of the children (comparable results in Gogolin, 2006). In her survey on research on mathematics teaching and learning of immigrant students in Europe, Civil (2010) points out that language issues are only a part of a culturally complex reality that needs further research. Of course, the policy and practices of allowing or forbidding first languages reflect the multicultural sensitivities of a country.

3.4 Practical Dimension: Teachers' Options for Including First Languages in European Classrooms

3.4.1 Options for First Language Use Under Different Language Conditions

In this section, we present a variety of ways in which first languages are used for increasing access to mathematics in European mathematics classrooms. We do not discuss bilingual or CLIL classrooms in which the official language of instruction is enriched or substituted by one of the major foreign languages, such as English, French, or German. Also we do not want to discuss those models of bilingual schools in immigrant minority languages, which are paid for and work under the supervision of the "home countries" and are not supported by mathematics education research. Instead, we refer to minority or immigrant first languages and focus on practical matters in the mathematics classroom.

The most far-reaching model for first language use refers to all language domains, namely oral and written language production and reception concerning all moments of mathematical lessons (see first column of Table 3.3). In such an "ideal maximum" model, the textbooks, the mathematical tasks, and the presented mathematical knowledge are translated into all languages spoken by the students in the specific classroom. Every utterance can be translated (by bilingual teaching assistants), thus guaranteeing full participation in the classroom communication. In such a classroom, students can switch between languages in all language domains (reading, writing, listening, speaking; see Table 3.3) and choose in each situation which language to use. A model close to this idealistic one has been tested in a Swedish project with students of different language and migration language backgrounds (Norén, 2007). Although the textbooks have not been translated, many teaching materials and blackboard writings were provided in Swedish and Somali/Arabic. The teachers were bilingual with migration backgrounds, and teachers and students could use these languages.

However, the language conditions can be more difficult when the teacher does not speak the students' first languages or when five to seven (or even more) first

Table 3.3 Necessary language conditions for different options of first language use

	... not possible, if	... possible or... allowed, if	... encouraged, if	... obliged, if
Written language reception (READING)	Materials all in L2	Materials in L1 and L2	Materials in L1 and L2 and translation supported	Materials only in L1
Written language production (WRITING)	Readers understand only L2	One reader understands L1 and L2 and writing allowed	One reader understands L1 and L2 and writing in L1 valued	All readers understands only L1
Oral language reception (LISTENING)	Teacher and students only speak L2	Some other students speak L1	Some other students speak L1 and communication valued in L1	Spoken language input only in L1
Oral language production (SPEAKING)	Teacher and students only understand L2	Some other students understand L1 and speaking allowed	Some other students speak L1 and communication valued in L1	Some partners only understand L1

languages are in the classroom. Thus, in this case, as well as in many others, it is even harder to provide translated materials. For these language conditions, we can still find options for first language use, mainly in the informal communication between students of the same first language community (César & Kumpulainen, 2009; Elbers & de Haan, 2004, 2005; Favilli, César, & Oliveras, 2004).

Table 3.3 gives an overview of different options for first language use. All language domains (reading, writing, listening, speaking) are *obliged* in the first language if some materials or some communication partners only use the first language, for example, the teaching assistant or parent with migration background (last column in Table 3.3). The *obligation* to use the first language in selected moments helps to initiate first language use even when students are not accustomed to it (Meyer & Prediger, 2011). However, this mode cannot be applied constantly since L2 (being the official language of instruction) evidently must be appropriated too.

The usual model is to make first language use *possible* or *encourage* it (second and third column of the table), without *obligation*. For example, the UK's National Association for Language Development in the Curriculum (NALDIC) promotes (among other things) the method of defining words for all key concepts in first languages (see www.naldic.org.uk). This method not only allows students to speak and write in their first language but also encourages them to really do so, even if the teacher does not speak these languages. The inclusion of parents as emphasized in the UK project "Our Languages" also encourages students to speak their first languages (see Melhuish, Sammons, Siraj-Blatchford, & Taggart, 2004).

Of course, providing texts in the first language cannot guarantee successful *encouragement* of first language use, as Meyer and Prediger (2011) show. In their design experiments in Germany, the Turkish-speaking sixth graders often hesitated to use the tasks written in Turkish L1 because they did not change usual patterns of language that were established over years in the monolingual culture of mathematics classrooms. In contrast, a Belgian longitudinal study (van Avermaet, Slembrouck, and Verhelst, in Little, 2010) indicates that if students are confronted with texts in two languages from the beginning of their schooling, language production in L1 can be successfully encouraged. The part-time *obligation* to read in L1 might even catalyze this process towards part-time L1 language production. But one must not forget that the use of language(s) is also shaped by the number of official languages in a country and in central exams, as they are becoming more and more usual in many European countries.

A comparable situation is given in Greenland, an autonomous country within Denmark. According to Pedersen (2010), Greenlandic is spoken by less than 56,000 people. It has been used as a language of instruction for over a century, but from the 1950s there has been a growing tendency to use Danish as the language of instruction (Patrick & Shearwood, 1999). Today, Greenland has its own educational curricula, its own textbooks, including mathematics, and there are two languages used in schools: Greenlandic and Danish.

The first language option that is easiest to implement *allows* or *encourages* students to use their first language in oral situations, namely in oral communication. Especially for group-work settings, a multilingual teacher may not be needed. Sometimes it is sufficient if some other students understand the same first languages. This option happens in many classrooms, even when not intended by the teacher (Clarkson, 2009; Elbers & de Haan, 2004, 2005; Planas & Setati, 2009). Hence, especially for classrooms with many non-shared first languages, interesting options for enhancing first language production include building language-homogeneous small groups (Gorgorió & Planas, 2001) and consequently enhancing mathematical communication. However, some authors also stress possible risks of internal segregation by language-homogeneous small groups, which might attract lower teachers' expectations (and support), promoting a different type of segregation and discrimination (see César, 2009, 2013a, 2013b; Favilli et al., 2004).

To sum up, Table 3.3 offers a wide variety of options, their systematization according to the language modes (reading, writing, listening, speaking) and according to degrees of obligation helping to adapt suitable options purposefully, according to different local language conditions. However, so far, only a few empirical studies have investigated the effects and conditions of these different options of first language use under different European language conditions. In Sect. 3.5, we report on these studies and formulate further research needs.

3.4.2 Cultural Dimensions Beyond Language: Bridging Cultural Gaps

Many researchers and practitioners have emphasized that first language use cannot be isolated from its social and cultural dimensions (see the survey in Civil, 2010). The switch from one language to another (e.g., from the first language to the language of instruction) is not only a matter of one-to-one translation, but also of an interplay between different cultures which needs interpretation, because every attribution of meaning is culturally bound and embedded in a complex network of symbols, norms, sociocultural practices, and identities (de Abreu, Bishop, & Presmeg, 2002). As a consequence, many projects do not restrict themselves to languages in the narrow sense, but consider them as parts of different minority cultures that have to be included into school life in order to bridge the gap between school and everyday life (César, 2009, 2013a, 2013b; de Abreu et al., 2002). This happens, for example, by including parents in home-school partnerships, by collaboration between mainstream and complementary schools (e.g., Melhuish et al., 2004), by including handicrafts or other cultural artifacts from minority cultures to promote ethnomathematics in classrooms (e.g., Favilli et al., 2004; Jannok-Nutti, 2011), and by many other aspects that value minority cultures and help to bridge the sociocultural gap between schools and homes, particularly using regulatory dynamics between schools and families (César, 2013b) or other ways of empowering those who participate in vulnerable minorities which are socially undervalued (see César, 2013a; César & Kumpulainen, 2009). These considerations of the cultural embeddedness of languages make clear why including first languages can be difficult: it is not only a matter of concrete pedagogy, but also of multicultural attitudes that value and support cultural diversity (Alrø, Skovsmose, & Valero, 2003), something that can be particularly seen in sign languages used as first languages—an issue that is even more lacking in research and a robust and sustained discussion (César, 2010; Melro & César, 2010).

The cultural dimension also refers to some risks that have been stressed in the context of promoting first language use (César, 2009, 2013a, 2013b; Favilli et al., 2004; Melro & César, 2010). The aim of preparing students for universities or professional careers might be in conflict with promoting linguistic competencies in both languages. From this perspective, encouraging first languages might also include a certain risk of creating barriers to their access to the most socially valued universities, jobs, or social positions, such as representatives of the community. In short, research should include this dimension for understanding the conditions and challenges of first language use in different settings, scenarios, and situations.

3.5 Research Dimension: Three Examples of Empirical Studies on Effects on Students' Learning, Obstacles, and Conditions

The evaluations of many practical development projects in Europe have provided empirical evidence that including first languages and other aspects of students' out-of-school cultures facilitate their access to mathematics achievement, even under more complicated language conditions of multiple non-shared languages (see, for instance, Elbers & de Haan, 2004, 2005; Favilli et al., 2004; Melhuish et al., 2004; César & Oliveira, 2005; Norén, 2010; Ventura, 2012; César, 2013a, 2013b; Machado, 2014).

Beyond these evaluations that illuminate the effects for mathematics learning, research studies aim to understand why and under what conditions teaching and learning practices have effects on students' mathematics achievement. For this deeper understanding, theoretical frameworks are needed to explain the connection between language, social interactions, and mathematics learning. The following three examples of studies show that not only the theoretical framework but also the concrete aims of the research shape its methods and results.

3.5.1 Language and Agency

Norén (2007, 2010) investigated practices in a model project in multilingual mathematics classrooms by means of the theoretical constructs discourse and agency, taking into account wider societal issues (Foucault, 2005). Two languages were used for teaching and learning mathematics: Swedish and Arabic, or Swedish and Somali. The findings indicate that bilingual communication in mathematics classrooms enhances students' identity construction as engaged mathematics learners. Language- and content-based mathematics instruction seemed to do the same, though monolingual instruction may jeopardize students' identities as bilinguals while the discourse may normalize Swedish and Swedishness exclusively. The focus on linguistic dimensions in mathematics, and students' first languages valued as resources for learning mathematics, allowed building up a communicative reform-oriented school mathematics discourse. The competing and intersecting discourses available in the multilingual mathematics classroom affected students' agency, foreground,[4] and identity formation as engaged mathematics learners. As a consequence, Norén concludes that each student's first language, in various ways, has to be acknowledged in the mathematics classroom, as it is crucial for a student's possibilities to act agentively, to become a participant in the classroom, and to engage in learning mathematics. This acknowledgment may vary from bilingual mathematics instruction (Norén, 2010) to peer-group work in the classroom, where students who use the same first language can work together (Planas & Setati, 2009).

A similar role for first languages in students' identity and agency was found in a study on Portuguese students' dialogical self and learning when they migrated from Portugal to the UK (de Abreu & Hale, 2009). Although not focusing on mathematics learning, this study illuminates the role of language in adapting into another country, society, and school system.

[4] A foreground is formed through a person's interpreteation of future possibilities regarding education and a "good" life, it is rebuilt and reconstructed in contexts of social interaction and learning processes (Skovsmose, 2012).

3.5.2 Language and Cultures

César (2009, 2013a, 2013b) analyzes the use of languages in mathematics learning in a theoretical framework based on Bakhtin (1929/1981), Vygotsky (1934/1962), and Wertsch (1991). Bakhtin's (1929/1981) distinction between meaning (as a social attribution for a particular word) and sense (as an individual attribution for a particular word in a specific situation, scenario, and context) leads to conceiving language, teaching, and learning as shaped by culture, time, and space. This approach gives great importance to the symbolic systems and cultural resources used by students and those needed to solve particular mathematical tasks, as exemplified in the analysis of social interactions taped while solving mathematics tasks (César, 2009, 2013a, 2013b; César & Santos, 2006). The mathematical tasks, the working instructions, the didactic contract, and even the evaluation system, shape the way social interactions are established among the different educational agents. Moreover, they also shape students' mathematical performances and their access to school achievement, or the various forms of segregation that emerge from teachers' practices, including the way they use language in learning situations within formal educational settings. But these studies also illuminate the need to promote regulatory dynamics that allow students' families, above all those participating in minority cultures that are often socially undervalued, to participate in more active ways in schools (César, 2013b).

In this approach, language is considered to be social before being internalized and becoming individual (Vygotsky, 1934/1962) and language only exists within a network of social interactions that allow participants to give sense to their talk (Bakhtin, 1929/1981). Thus, facilitating students' access to mathematics learning and achievement means creating conditions for those students whose voices are usually silenced to become legitimate participants in the learning community. This means not only changing teachers' practices within mathematics classes, but also changing the school organization and families' participation, such as, for example, when some teachers began learning Creole with parents so that the parents would also feel tempted to learn Portuguese, and also as a way of empowering those students and their families (César, 2013b).

César considers linguistic diversity to be directly connected to cultural diversity and has undertaken a long-term research project, Interaction and Knowledge, that lasted 12 years and had a 10-year follow-up to collect an empirical corpus of differences and similarities (for more details, see César, 2009, 2013a, 2013b; Ventura, 2012). The most striking empirical evidence referred to ideographic and mainly oral languages, and the mathematical reasoning and solving strategies students tended to prefer (César, 2009, 2013a; Machado, 2014; Ventura, 2012). For instance, some students from Cape Verde have Creole as their L1, a language that is mainly learnt and used in oral practices and which is an ideographic language. These students usually prefer a global approach rather than a step-by-step approach to problems and investigative tasks (Machado, 2013). They also prefer geometrical reasoning rather than the analytical reasoning that is usually prioritized in school mathematics (César,

2009, 2013a). These students and their families tend to highly value learning from older and experienced people, directly connected to daily life practices, rather than learning through books and in schools. Thus, it is not only their L1 language that shapes their engagement in mathematical school activities but also their culture and their expectations towards schooling and their future life trajectories of participation (César, 2013a, 2013b). Collaborative practices based on challenging mathematics tasks, higher teachers' expectations, and the use of very careful criteria to form dyads and groups, proved to be effective ways to promote intercultural and inclusive mathematics education (César, 2009, 2013a; Machado, 2014). Collaborative practices, developed within and outside classrooms, allow students to achieve in school and in mathematics. This applies particularly to those students participating in minority cultures and using other first languages (also in César & Oliveira, 2005; César 2013b), or presenting special educational needs (César & Santos, 2006). As a consequence of these observations, teaching practices should be guided by Vygotsky's claim for addressing each student's zone of proximal development (Vygotsky, 1934/1962; César, 2013a). Empirical evidence shows that these teaching practices favor students' learning and their mathematics and linguistic development (César, 2009, 2013a, 2013b; César & Santos, 2006; Machado, 2014; Ventura, 2012).

3.5.3 Installing Teaching Strategies Against Established Monolingual Classroom Norms

The empirical studies reported in Sect. 3.3 show that bilingual learners profit more from culturally comprehensive projects that encompass several years of schooling, all subjects, and not only language but also other cultural issues. However, classroom reality in many countries is much more restrictive, which is why it is also important to conduct research not only on maximum models but also on very modest attempts to make use of students' first languages. In this regard, Meyer and Prediger (2011) conducted a design research project with clinical interviews in which they investigated how to implement first language use in mostly monolingual classroom cultural contexts.

Starting from an interactionist theoretical background (Blumer, 1969; Goffman, 1959; Mead, 1934), the focus is not on the larger sociocultural context but on the classroom culture in itself. In the interactionist perspective, mathematical knowledge is established by elaborating shared meanings in the interactions between learners and between learners and teachers (Meyer, 2009; Voigt, 1998). Language is crucial as a mediator that gives the opportunity to elaborate meanings, but only for those students who can participate in the interaction. Hence, first language use is conceived as a vehicle to enlarge students' access to interaction and to the construction of meanings.

However, each classroom interaction is regulated by explicit and implicit norms (Voigt, 1995). If monolingualism is one of the implicit norms that students have been socialized to, the inclusion of first languages cannot be started easily in later grades. In the interviews, different ways to promote written and oral first language production and reception were tested and some of them proved to be effective for developing conceptual understanding. For example, after observing the students' hesitations to use their first language (Turkish), the students were asked to teach the interviewing person counting in Turkish by making them believe that he/she would be going on holiday to Turkey. In this way, the Turkish language would be more valued in the situation. In a second experiment, it was found that talking to a monolingual Turkish-speaking interviewer resulted in a more frequent use of the first language. In many interviews this approach has been useful in order to elaborate mathematical concepts (e.g., fractions of fractions, Meyer & Prediger, 2011) or to identify defining characteristics of geometrical forms in the first language. Furthermore, the first language use seemed to influence social functions in the interaction, such as increasing verbal exchanges or clarifying the flow of work. This influence has not been observed before while using the language of instruction, and can be explained by moments of privacy established by the first language (detailed descriptions and examples are given in Meyer & Prediger, 2011; Krägeloh & Meyer, 2012).

Interpretative analysis of the design experiments provided some evidence that students can be invited to make use of their first languages in order to solve mathematical tasks even if they are used to a monolingual mathematics classroom and that they profit from it for constructing shared meanings of mathematical concepts and hence for developing their conceptual understanding. This way of allowing groups of students to use their first language in the phases of gaining and negotiating mathematical knowledge is not only a useful way but also an easy way to make use of the first languages of these students in mathematics. Nevertheless, interaction in the whole classroom necessitates a commonly shared language.

3.6 Final Remarks

There is a huge need for further research and development on multiple first languages in European mathematics classrooms.

In the *cultural and political dimension*, it can be shown that monolingualism has become an unrealistic fiction in many European classrooms due to increasing rates of migration and multiculturalism. However, the unequal distribution of achievement between native speakers and those students whose first language is not the language of instruction indicates that European school systems have not yet found adequate answers to this multilingual reality. Making more use of students' first languages is one of the approaches for dealing with the challenges of multilingual students, particularly in classes with few first languages among students.

In the *practical dimension*, different mathematics classroom practices are reported of how the inclusion of multiple first languages might be put into practice. The presented "ideal maximum model" and the systematization of different settings for written versus oral language and language production versus reception might offer orientations for decision-making and application in classrooms. However, an enormous need for research in the European language context must be stated.

In the *research dimension*, brief excerpts from three empirical studies on effects of first language use and on the contributions of collaborative work are presented as examples to show different facets of the state of research, all being shaped by the different theoretical perspectives that were applied. The research deals not just with effects of first language use on students' learning and achievement but also with obstacles to and conditions of its realization.

Several studies give us hope that even under the more complex language conditions in Europe, with multiple non-shared first languages in each classroom, ways can be found to make use of students' important (cultural) resources, namely their first languages and the ways they shape their mathematical performances. Some research shows that obstacles can be overcome and benefits are illuminated in the learning processes that concern socio-cognitive and emotional aspects, as well as cultural identities and students' agency. These results suggest that the inclusion of the students' first languages and/or the valuing of social interaction and of different cultures can be useful in order to enable better access to the learning of mathematics.

However, many research questions have so far only been addressed as first attempts and need further exploration. Europe is in need of much research: learning conditions vary between the countries and sometimes even between one country's federal states. The migration backgrounds and the structures and proximities of the first languages differ. We also need to consider Deaf students, as although they are not migrants they do use a different first language—a sign language. Thus, we are not only in need of investigating useful ways of using selected first languages as resources in mathematics classrooms and their sociocultural implications but we also need to assume that learning is a situated process. Thus, for example, a method which has proven useful for Turkish immigrants in Germany, may not be useful for Italian immigrants, or even for Turkish students in another country or another community.

Acknowledgments The *Interaction and Knowledge* project was partially supported by the Instituto de Inovação Educacional, medida SIQE 2 (project n°. 7/96) in 1996/1997 and 1997/1998, and by the Centro de Investigação em Educação da Faculdade de Ciências da Universidade de Lisboa since 1996. The *MuM-Project* (*Mathematics learning under conditions of multilingualism*) at TU Dortmund University was partially funded by the BMBF, grant 01JG1067. We thank the teachers, students, and colleagues who helped to conduct the different projects underlying this chapter.

References

Alrø, H., Skovsmose, O., & Valero, P. (2003). Communication, conflict and mathematics education in the multicultural classroom. In M. A. Mariotti (Ed.), *Proceedings of the Third Conference of the European Society for Research in Mathematics Education.* Bellaria, Italy: ERME. Retrieved from http://www.dm.unipi.it/~didattica/CERME3/proceedings/tableofcontents_cerme3.html

Baker, C. (1996). *Foundations of bilingual education and bilingualism* (2nd ed.). Clevedon, England: Multilingual Matters.
Bakhtin, M. (1929/1981). *The dialogical imagination* (M. Holquist, Ed.; M. Holquist & C. Emerson, Trans.). Austin: University of Texas Press. [First published 1929 in Russian.]
Barwell, R. (Ed.). (2009). *Multilingualism in mathematics classrooms: Global perspectives*. Bristol, England: Multilingual Matters.
Beacco, J.-C., Byram, M., Cavalli, M., Coste, D., Egli Cuenat, M., Goullier, F., et al. (2010). *Guide for the development and implementation of curricula for plurilingual and intercultural education*. Strasbourg, France: Council of Europe.
Blumer, H. (1969). *Symbolic interactionism: Perspective and method*. Englewood Cliffs, NJ: Prentice-Hall.
César, M. (2009). Listening to different voices: Collaborative work in multicultural maths classes. In M. César & K. Kumpulainen (Eds.), *Social interactions in multicultural settings* (pp. 203–233). Rotterdam, The Netherlands: Sense.
César, M. (2010). Comment to Paola's conference: Dialogism in action. In V. Durand-Guerrier, S. Soury-Lavergne, & F. Arzarello (Eds.), *Proceedings of CERME 6* (pp. LXXXVII–XCIII). Lyon, France: INRP.
César, M. (2013a). Collaborative work, dialogical self and inter-/intra-empowerment mechanisms: (Re)constructing life trajectories of participation. In M. B. Ligorio & M. César (Eds.), *Interplays between dialogical learning and dialogical self* (pp. 151–192). Charlotte, NC: Information Age.
César, M. (2013b). Cultural diversity and regulatory dynamics of participation between schools and families. In P. Marsico, K. Komatzu, & A. Iannaccone (Eds.), *Crossing boundaries: Intercontextual dynamics between family and school* (pp. 35–81). Charlotte, NC: Information Age.
César, M., & Kumpulainen, K. (Eds.). (2009). *Social interactions in multicultural settings*. Rotterdam, The Netherlands: Sense.
César, M., & Oliveira, I. (2005). The curriculum as a mediating tool for inclusive participation: A case study in a Portuguese multicultural school. *European Journal of Psychology of Education, XX*(1), 29–43.
César, M., & Santos, N. (2006). From exclusion into inclusion: Collaborative work contributions to more inclusive learning settings. *European Journal of Psychology of Education, XXI*(3), 333–346.
Chlosta, C., & Ostermann, T. (2008). Grunddaten zur Mehrsprachigkeit im deutschen Bildungssystem. In B. Ahrenholz (Ed.), *Deutsch als Zweitsprache* (pp. 17–30). Baltmannsweiler, Germany: Schneider.
Civil, M. (2010). A survey of research on the mathematics teaching and learning of immigrant students. In V. Durand-Guerrier, S. Soury-Lavergne, & F. Arzarello (Eds.), *Proceedings of the Sixth Congress of the European Society for Research in Mathematics Education* (pp. 1443–1452). Lyon, France: Institut National de Recherché Pédagogique.
Clarkson, P. C. (1992). Language and mathematics: A comparison of bilingual and monolingual students. *Educational Studies in Mathematics, 23*, 417–429.
Clarkson, P. C. (2007). Australian Vietnamese students learning mathematics: High ability bilinguals and their use of their languages. *Educational Studies in Mathematics, 64*, 191–215.
Clarkson, P. C. (2009). Mathematics teaching in Australian multilingual classrooms: Developing an approach to the use of classroom languages. In R. Barwell (Ed.), *Multilingualism in mathematics classrooms: Global perspectives* (pp. 145–160). Bristol, England: Multilingual Matters.
Commission, E. (2004). *Many tongues, one family: Languages in the European Union*. Luxembourg: Office for Official Publications of the European Communities.
Coste, D., Cavalli, M., Crisan, A., & van den Ven, P.-H. (2009). *Plurilingual and intercultural education as a right*. Strasbourg, France: Council of Europe.
de Abreu, G., Bishop, A., & Presmeg, N. C. (2002). *Transitions between contexts of mathematical practices*. Dordrecht, The Netherlands: Kluwer.

de Abreu, G., & Hale, H. (2009). 'Self' and 'other' imposed withdrawing in social interactions at school: Portuguese students talk about their experiences in adjusting to schooling in Britain. In M. César & K. Kumpulainen (Eds.), *Social interactions in multicultural settings* (pp. 91–115). Rotterdam, The Netherlands: Sense.

Bundesregierung der Bundesrepublik Deutschland. (2010). *Der Nationale Integrationsplan. Neue Wege: Neue Chancen*, Berlin. Retrieved January 6, 2014, from http://www.bundesregierung.de/Content/DE/Archiv16/Artikel/2007/07/Anlage/2007-07-12-nationaler-integrationsplan.pdf?__blob=publicationFile

Durkin, K., & Shire, B. (Eds.). (1991). *Language in mathematics education: Research and practise*. Milton Keynes, England: Open University Press.

Eckhardt, A. G. (2008). *Sprache als Barriere für den schulischen Bildungserfolg. Potentielle Schwierigkeiten beim Erwerb schulbezogener Sprache für Kinder mit Migrationshintergrund*. Münster, Germany: Waxmann.

Elbers, E., & de Haan, M. (2004). Dialogic learning in the multi-ethnic classroom: Cultural resources and modes of collaboration. In J. van der Linden & P. Renshaw (Eds.), *Dialogical perspectives on learning, teaching and instruction* (pp. 17–43). Dordrecht, The Netherlands: Kluwer.

Elbers, E., & de Haan, M. (2005). The construction of word meaning in a multicultural classroom: Mediational tools in peer collaboration during mathematics lessons. *European Journal of Psychology of Education, XX*(1), 45–59.

Ellerton, N., & Clarkson, P. (1996). Language factors in mathematics teaching and learning. In A. Bishop et al. (Eds.), *International handbook of mathematics education* (pp. 987–1033). Dordrecht, The Netherlands: Kluwer.

Eurobarometer. (2006). *Europeans and their languages. Special Eurobarometer [243 / Wave 64.3]*. European Commission, TNS Opinion & Social. Retrieved January 6, 2014, from http://ec.europa.eu/public_opinion/archives/ebs/ebs_243_en.pdf

European Parliament. (2000). *Charter of fundamental rights of the European Union*. Retrieved January 13, 2014, from http://www.europarl.europa.eu/charter/pdf/text_en.pdf

Favilli, F., César, M., & Oliveras, M. L. (2004). *Projecto IDMAMIM: Matemática e intercultura*. Pisa: Universidade de Pisa. [3 CdRoms: La Zampoña, Os Batiques e Las Alfombras].

Foucault, M. (2005). *The hermeneutics of the subject: Lectures at the College de France 1981–1982* [F. Gross & F. Ewald Ed.]. New York: Palgrave Macmillan.

Goffman, E. (1959). *The presentation of self in everyday life*. New York: Doubleday.

Gogolin, I. (1994). *Der monolinguale Habitus der multilingualen Schule*. Münster, Germany: Waxmann.

Gogolin, I. (2006). Chances und risiken nach PISA: über die bildungsbeteiligung von migrantenkindern und reformvorschläge. In G. Auernheimer (Ed.), *Schieflagen im bildungssystem. Die benachteiligung der migrantenkinder* (pp. 33–50). Wiesbaden, Germany: Springer VS.

Gogolin, I. (2010). Stichwort: Mehrsprachigkeit. *Zeitschrift für Erziehungswissenschaften, 13*, 529–547.

Gogolin, I. (2011). Bilingual education. In J. Simpson (Ed.), *The Routledge handbook of applied linguistics* (pp. 229–242). London: Routledge.

Gorgorió, N., & Planas, N. (2001). Teaching mathematics in multilingual classrooms. *Educational Studies in Mathematics, 47*(1), 7–33.

Haglund, C. (2005). *Social interaction and identification among adolescents in multilingual suburban Sweden: A study of institutional order and sociocultural change*. Doctoral dissertation, Stockholm University, Sweden.

Heinze, A., Reiss, K., Rudolph-Albert, F., Herwartz-Emden, L., & Braun, C. (2009). The development of mathematical competence of migrant children in German primary schools. In M. Tzekaki, M. Kaldrimidou, & H. Sakonidis (Eds.), *Proceedings of the 33rd Conference of the International Group for the Psychology of Mathematics Education* (Vol. 3, pp. 145–152). Thessaloniki, Greece: PME.

Holmen, A. (2008). Dansk som andresprog I den flerkulturelle skole. In J. Bundsgaaard (Ed.), *Positioner I dnaskfagets didaktik* (CURSIV nr 2 2008) (pp. 59–76). Institut for

Curriculumforskning, Danmarks Paedagogiske Universitetsskole ved Aarhus Universitet. Retrieved January 6, 2014, from http://www.dpu.dk/site.aspx?p=12829

IT.NRW. (Ed.) (2012). Information und Technik NRW. *Statistik Kompakt. Bildungsbeteiligung von Schülerinnen und Schülern mit Zuwanderungsgeschichte.* Düsseldorf, Germany. Retrieved January 6, 2014, from https://webshop.it.nrw.de/ssearch.php?kategorie=1005&prefix=Z25

Jannok-Nutti, Y. (2011). *Ripsteg mot spetskunskap i samisk matematik: Lärares perspektiv på transformeingsaktviteter i samisk förskola och sameskola.* Luleå, Sweden: Luleå Tekniska Universitet.

Kern, R. G. (1994). The role of mental translation in second language reading. *Studies in Second Language Acquisition, 16*, 441–461.

Krägeloh, N., & Meyer, M. (2012). "Erkläre es mir auf Türkisch"—Anknüpfen an die Ressource Erstsprache im Mathematikunterricht. *PM—Praxis der Mathematik, 45*(54), 24–27.

Little, D. (2010). *The linguistic and educational integration of children and adolescents from migration backgrounds.* Geneva, Switzerland: Council of Europe.

Machado, R. (2014). Trabalho colaborativo e matemática: Um estudo de caso sobre o instrumento de avaliação de capacidades e competências do projecto Interacção e Conhecimento. Lisboa: Associação de Professores de Matemática. [Doctoral thesis presented at Faculdade de Ciências e Tecnologia, Universidade Nova de Lisboa].

Mead, G. H. (1934). *Mind, self and society.* Chicago, IL: University of Chicago Press.

Melhuish, S. K., Sammons, P., Siraj-Blatchford, I., & Taggart, B. (2004). *Effective pre-school education. Final report.* London: Institute of Education/DfES.

Melro, J., & César, M. (2010). Educação inclusiva: Desafios à profissionalidade docente e às aprendizagens dos alunos Surdos. *Educação Inclusiva, 1*(2), 10–17.

Meyer, M. (2009). Use of words: Language-games in mathematics education. In V. Durand-Guerrier, S. Soury-Lavergne, & F. Arzarello (Eds.), *Proceedings of the Sixth Congress of the European Society for Research in Mathematics Education. Lyon 2009* (pp. 904–913). Lyon, France: Institut National de Recherché Pédagogique.

Meyer, M., & Prediger, S. (2011). The use of first language Turkish as a resource: A German case study on chances and limits for building conceptual understanding. In M. Setati, T. Nkambule, & L. Goosen (Eds.), *Proceedings of the ICMI Study 21 Conference: Mathematics Education and Language Diversity* (pp. 225–234). São Paulo, Brazil: ICMI Study 21.

Mikrozensus. (2009). *Alterspyramide 2009 nach Migrationshintergrund. Bevölkerung und Erwerbstätigkeit. Bevölkerung mit Migrationshintergrund—Ergebnisse des Mikrozensus 2009.* Wiesbaden, Germany: Bundesamt für Statistik 14.

Moschkovich, J. (2007). Using two languages when learning mathematics. *Educational Studies in Mathematics, 64*, 121–144.

Norén, E. (2007). *Det går att lära sig mer. En utvärdering av tvåspråkig matematikundervisning.* Stockholm, Sweden: Kompetensfoden.

Norén, E. (2010). *Flerspråkig matematikundervisning. Diskurser i grundskolans matematikundervisning.* Stockholm, Sweden: Stockholm University.

Nusche, D. (2009). *What works in migrant education? A review of evidence and policy options. OECD Working Papers, No. 22.* Paris, France: OECD.

OECD. (2007). *Science Competencies for Tomorrow's World: Vol. 2: Data (PISA 2006).* Paris, France: OECD.

Patrick, D., & Shearwood, P. (1999). The roots of inuktitut-language bilingual education. *The Canadian Journal of Native Studies, XIX*(2), 249–262.

Pedersen, I. S. (2010). *Rethinking bilingualism: A sociolinguistic analysis of language planning and education legislation in Greenland.* Unpublished master's thesis, University of Tromsø, Norway.

Planas, N., & Setati, M. (2009). Bilingual students using their languages in their learning of mathematics. *Mathematics Education Research Journal, 21*(3), 36–59.

Prenzel, M., Baumert, J., Blum, W., Lehmann, R., Leutner, D., Neubrand, M., et al. (2005). *PISA 2003: Ergebnisse des zweiten Ländervergleichs Zusammenfassung.* Retrieved from https://www.gew.de/Binaries/Binary34586/PISA2003_E_Zusammenfassung.pdf

Rossell, C. H. & Kuder, J. (2005). Meta-murky: A rebuttal to recent meta-analyses of bilingual education. In Arbeitsstelle Interkulturelle Konflikte und gesellschaftliche Integration (Ed.), *The effectiveness of bilingual school programs for immigrant children* (pp. 43–76). Berlin, Germany: Wissenschaftszentrum Berlin für Sozialforschung, AKI.

Runfors, A. (2003). *Mångfald, motsägelser och marginaliseringar: en studie av hur invandrarskap formas i skolan*. Doctoral dissertation, Stockholm University, Sweden.

Setati, M., & Duma, B. (2009). When language is transparent: Supporting mathematics learning multilingual contexts. In M. Tzekaki, M. Kaldrimidou, & H. Sakonidis (Eds.), *Proceedings of the 33rd Conference of the International Group for the Psychology of Mathematics Education*, (vol. 5, pp. 65–72). Thessaloniki, Greece: PME.

Skovsmose, O. (2012). Students' foregrounds: Hope, despair, uncertainty. *Pythagoras*, *33*(2), 1–8, article 162.

Strand, S. (2007). *Minority ethnic pupils in the longitudinal study of young people in England (LSYPE)*. Warwick, England: CEDAR and London: DCSF.

Swedish National Agency of Education [Skolverket]. (2013). *Schools and pupils in compulsory school [Skolor och elever i grundskolan läsår] 2012/13. Table 8A_2012_13webb*. Retrieved January 13, 2014, from http://www.skolverket.se/statistik-och-utvardering/statistik/grundskola/skolor-och-elever

Thürmann, E., Vollmer, H., & Pieper, I. (2010). *Language(s) of schooling: Focusing on vulnerable learners. The linguistic and educational integration of children and adolescents from migrant backgrounds studies and resources*. Strasbourg, France: Council of Europe.

Ventura, C. (2012). Interacção e Conhecimento: Um estudo de caso que analisa a história de um projecto. Lisboa: Associação de Professores de Matemática. [Doctoral thesis presented at Faculdade de Ciências e Tecnologia, Universidade Nova de Lisboa].

Vertovec, S. (2007). Super-diversity and its implications. *Ethnic and Racial Studies*, *30*(6), 1024–1054.

Voigt, J. (1995). Thematic patterns of interaction and sociomathematical norms. In P. Cobb & H. Bauersfeld (Eds.), *The emergence of mathematical meaning: Interaction in classroom cultures* (pp. 163–201). Hillsdale, NJ: Lawrence Erlbaum.

Voigt, J. (1998). The culture of the mathematics classroom: Negotiating the mathematical meaning of empirical phenomena. In F. Seeger, J. Voigt, & U. Waschescio (Eds.), *The culture of the mathematics classroom* (pp. 191–220). Cambridge, England: Cambridge University Press.

Vygotsky, L. S. (1962). *Thought and language*. Cambridge, England: MIT Press.

Wertsch, J. (1991). *Voices of mind: A sociocultural approach to mediated action*. Hemel Hempstead, England: Harvester Wheatsheaf.

Chapter 4
Mathematics Education in Multilingual Contexts for the Indigenous Population in Latin America

Aldo Parra, Jackeline Rodrigues Mendes, Paola Valero, and Martha Villavicencio Ubillús

4.1 Introduction

Within internationally published research in mathematics education dealing with multilingualism, little is known about the challenges faced in the education of the Indigenous peoples of Latin America. This chapter has a double contribution. On the one hand, it will present trends in the mathematics education of Indigenous people in Brazil, Colombia, and Peru. This presentation intends to inform an international mathematics education audience about the common traits as well as particular developments of Indigenous mathematics education in the continent.

On the other hand, and based on the information presented, we discuss three interrelated issues. First, the education of Indigenous populations in Latin America needs to be understood in the framework of Spanish and Portuguese colonization. Within colonization, the project of religious conversion and evangelization is a constitutive element of the models of education, multilingualism, and mathematics. Through history, the struggle between different Indigenous communities and state policies and programs have made it evident that (mathematics) education is a terrain of cultural politics for indigenous communities.

Second, in the construction, negotiation, and implementation of different models for Indigenous education, varying from monocultural monolingualism, to bilingualism, and more recently cultural and linguistic diversity, conflicting claims about

A. Parra • P. Valero
Aalborg University, Aalborg, Denmark
e-mail: aiparras@unal.edu.co; paola@learning.aau.dk

J.R. Mendes
Universidade Estadual de Campinas, São Paulo, Brazil
e-mail: rodrigues@mpc.com.br

M. Villavicencio Ubillús (✉)
Ministerio de Educación, Lima, Perú
e-mail: MVILLAVICENCIO@minedu.gob.pe

R. Barwell et al. (eds.), *Mathematics Education and Language Diversity*, New ICMI Study Series, https://doi.org/10.1007/978-3-319-14511-2_4

what counts as Indigenous languages and cultures have been present. Indigenous education in Latin America is a space of struggle for the recognition of Indigenous worldviews and knowledge; as such, it is a space where language, knowledge, and identity go hand in hand. In this context, issues of language revitalization as well as tension between oral traditions and written registers are important discussions that shape education and mathematics education.

Third, the current move towards the recognition of cultural and linguistic diversity also poses challenges as to what counts as the mathematics of the Indigenous people and the mathematics of the mainstream cultures in the context of the life of communities and also in schooling. Different state policies of Indigenous mathematics education have resulted in diverse meanings and appropriations of ethnomathematics. Here research in mathematics education has provided tools to deal with these challenges.

In the three main sections of this chapter, we address each one of these issues. Our strategy is intentionally descriptive of the historical development and current situation in three countries: Brazil, Colombia, and Peru. We use the descriptions of these national contexts to depict both commonalities and differences in the discussion about how mathematics education in the cultural and linguistic diversity of the Latin American Indigenous population is highly political.

4.2 Models of Indigenous Education in History

Latin American countries share similar histories around the treatment of their Indigenous communities. The intersection points can be traced back to the Spanish and Portuguese incursions into the continent in the sixteenth century. Colonization represented a process of extermination and domination of Indigenous communities. The natives in this period were treated as *nonhuman, wild savages*. At that time, the definition of "wild" was structured around three components: "unaware of the Castilian language, the Christian religion, and the larger society model" (Roldán & Gómez, 1994, p. 65). From the very beginning of the contact between European and native South American civilizations, language and religion appeared as a decisive factor structuring relations with Indigenous peoples. The Iberian monarchies entrusted to the Catholic Church the tutelage of Indigenous peoples, to carry out their spiritual salvation through evangelization, with the work of missions being central in such task. Religious conversion and with it, cultural assimilation, led to the loss of cultural traditions like written codices and languages. The hard and extenuating work for the economic exploitation of the natural richness of the Americas was also associated with the systematic extermination of large proportions of the native population. In some cases, religious conversion and the exoticization of the "wild" native resulted in the annihilation of Indigenous people. The most extreme case of the eradication of an Indigenous people is that of the Charrúas, the aboriginal settlers of the Uruguayan territory.

In the first quarter of the nineteenth century, independence from the European empires took place. The creation of national states and republics produced a change in philosophy towards the few native populations left. Indigenous people were not seen any more as belonging to an inferior species, but rather as *potential citizens*. The work to integrate them was once again left under the tutelage of the Catholic Church. The policy of the Church to control the education and the language of the natives remained alive in Brazil, Peru, and Colombia until the last decade of the twentieth century. Even some of the early efforts to register Indigenous languages were associated with religious movements. For example, the Summer Institute of Linguistics (SIL), an organization supported by Christian factions in the USA, created alphabets for various Indigenous languages in the continent and advanced the recording and study of native languages, for the purpose of evangelizing the natives in their indigenous language. Educational initiatives during this period correspond to a transition model (Larson, Davis, & Ballena, 1979, p. 57), in which it is hoped that indigenous students will move from their cultural and traditional practices to become part of the national society. The instrumental use of language and of religion was fundamental in taking that step.

In the course of the 1960s and 1970s alternative institutional educational programs emerged around *bilingualism* in schools. In Peru, the Universidad Nacional Mayor de San Marcos was the first to start a pilot program of research and experimentation of bilingual education with the Quechua-speaking population in the area of Quinua, Ayacucho Department. Since then and within the Peruvian government's Language Development Plan, studies on bilingualism were strengthened. Years later, that university was able to influence the idea of a National Policy for Bilingual Education in the framework of the 1972 Peruvian Education Reform (López and Küper, 1999).

In the 1970s in Brazil and Colombia a series of political processes and social movements started to discuss and promote changes in public policies for Indigenous peoples. There was a critique of the transitional bilingual model, which had prevailed for more than two centuries. In Colombia, the Regional Indigenous Council of Cauca and the Tayrona Indigenous Confederation were founded. These organizations demanded the presence of bilingual Indigenous teachers in schools to respect the history and traditions of the communities. In Brazil, the Indigenous Missionary Council was created in 1972. Two years later, this Council organized the first of a long series of "Indigenous assemblies." The first Indigenous organization in Brazil, the National Indigenous Union, was created in the 1980s and it made alliances with non-Indigenous NGOs and with the Alliance of the People of the Forest in Amazonia, which included non-Indigenous rubber gatherers in Amazonia.

The first experiences of bilingual education for Indigenous peoples were characterized by their linguistic bias, downplaying the enculturating role of education. In the early 1980s, and as a reaction to the focus on language, the concept of *interculturality* emerged strongly. It involved the search for educational models that are specific, differentiated and bilingual and that recognize the importance of Indigenous cultures and languages. Such a view is expressed in educational policy documents and programs in Peru (Bilingual Education Experimental Project of Puno and General Directorate of Bilingual Education) and Colombia (Office of Ethnic Education). It also permeated the production of bilingual teaching materials in mathematics. Elements of the sociocultural practices of the communities were incorporated. Although these initiatives were not sufficiently grounded in indigenous worldviews, such programs played a role in the discussions that took place in Brazil about Indigenous teacher education and preparation of materials for bilingual Indigenous schools.

During the 1980s, many countries entered a process of change in their political constitutions, which had a huge influence on educational development. Leaving a narrative of national unification, the new constitutions became pluralistic and acknowledged the *cultural and linguistic diversity* of the population. This move facilitated reclaiming the rights of Indigenous nations and safeguarding their worldviews, cultural practices and languages. In Peru, the Political Constitution of 1979 promoted the study and knowledge of Indigenous languages and guaranteed the right of the Quechua, Aymara and other communities to receive primary education in their own languages. It declared that Indigenous peoples have the right to preserve their ethnic and cultural identity and that the state should recognize and protect the ethnic and cultural plurality of the nation. In Brazil, the Constitution of 1988 included the rights of Indigenous peoples to the preservation and maintenance of their values, languages, and cultures. The Political Constitution of 1991 in Colombia declared a multiethnic and multicultural state. In the three countries, after the new political constitutions, several indigenous languages were recognized as official, and many initiatives were developed aimed at including the cultural specificities of each group in the educational system. This context drew more attention to the need for indigenous teachers in schools and led to the development of Indigenous teacher education programs to prepare these teachers.

The attempt to create and establish bilingual, intercultural, and culturally specific Indigenous schools resulted in curricular changes in the three countries during the 1990s. It also increased the participation of Indigenous teachers in writing bilingual materials. Since 2000, the development of Indigenous, bilingual education in each of our countries has a sociocultural orientation that makes education relevant and increases quality. This may be due to several reasons. First, there is a record of 30 years of implementation of experiences in bilingual education. Second, educational initiatives are decentralized and organized around each Indigenous people, as an attempt to take into account their views and wishes. Finally, sociocultural research in mathematics teaching and learning has advanced progressively in the region. This has motivated a search to explicitly include in local curricula the mathematical knowledge of each people.

Nevertheless, interest in Indigenous worldviews has not only produced better conditions for dealing with bilingualism and mathematics teaching but also generated new challenges and emerging tensions. We elaborate on them in the rest of the chapter.

4.3 Dynamics and Tensions Between Languages and Cultures

The historical changes in models of education of Indigenous peoples in Latin America are not unique to the continent. Meaney, Trinick, and Fairhall (2012) discuss the political tensions experienced in the (mathematics) education of Māori people in New Zealand, and also review similar discussions in Australia, Africa, and North America. Their research highlights the two main topics of this section, namely language revitalization and the creation of written registers on the basis of an oral tradition of knowledge. These two dynamics of language are of particular interest given how Indigenous language, worldviews, and identity intersect in one of the powerful tools of the colonizer, namely, mathematics.

4.3.1 Language Extinction, Revitalization, and Development

In Brazil, Colombia, and Peru there is a great diversity of cultures and languages. Notwithstanding that, sociolinguistic studies conducted in the three countries depict a critical scenario for the vitality of Indigenous languages. Table 4.1 summarizes the situation.

The study carried out by UNICEF and FUNPROEIB (2009) discussed the status of preservation of Indigenous languages in Latin America. In general, Indigenous communities with better organization and more efficient and consistent adaptive strategies have coexisted for centuries with the national society and its dominant language. Their strategies result in the vitality of Indigenous languages and their articulation with the Spanish language. Language vitality is also strong in communities living in isolation, whereas in communities with recent and increasing contact with the national community there will be a tendency to lose language use, leading to a decrease in vitality.

More open political constitutions in the countries around the 1990s promoted changes in the colonial model of a monolingual language policy. Indigenous communities were granted rights over their land, cultures, and languages. This change went together with the development of language policies decreasing the speed of language extinction and preserving and maintaining Indigenous languages. One of the results of such a trend was the formulation of educational policies for preservation, aimed at generating special professional development of Indigenous bilingual teachers, as well as the creation of alphabets and written registers for these indigenous languages.

These new settings have greatly impacted the possibilities of claiming an Indigenous cultural identity, not only from a legal and institutional point of view but also from the point of view of Indigenous peoples themselves. We are taking here the notion of identity as a discursive concept. Identity is seen as a construct that happens in and through language. According to Woodward (1997), identity construction processes are always built in relation to and maintained by history and culture and are dependent upon the nature of social relationships established over time.

The inseparability of identity, language, and culture has resulted in the politicization of research on Indigenous languages and education. From the denial of the Indigenous in colonial policies to the awareness of multilingualism and multiculturalism, many Indigenous education policies have adopted a naïve, but still colonizing, approach to the "development" of Indigenous languages and cultures.

Table 4.1 Indigenous groups, their languages and their language vitality

	Basic statistics				Vitality		Languages by number of speakers			
Country	Indigenous groups	National population (%)	Indigenous languages	Linguistic families	Endangered languages	Severely endangered	1–99	10–999	1,000–50,000	>50,000
Brazil	305	0.47	184	42	24	10	34	106	44	1
Colombia	91	3.3	66	13	19	5	25	33	33	2
Peru	60	15.9	47	19	4	17	6	20	17	4

Sources: For Brazil (IBGE. Instituto Brasileiro de Geografia e Estatística, 2010); for Colombia (Landaburu, 2004; UNICEF and FUNPROEIB, 2009); for Peru (INEI. Instituto Nacional de Estadística e Informática, 2009; Ministerio de Educación del Peru, 2013, pp. 13–16)

Simultaneously political discourses have evidenced tensions in the meeting between Indigenous identities and the dominant cultures, their languages and related forms of identity represented in schooling. For example, the development of Indigenous languages leads to their normalization, as a prerequisite to achieve a standard style of writing. In Peru in 1985, the Quechua and Aymara alphabets were normalized, and at present 21 more Indigenous languages have been normalized. However, the processes of negotiation for settling alphabets and rules of writing have not ended.

There have been integrationist policies or policies oriented to the revitalization and maintenance of the cultural and language practices of minority groups. Concerning the latter, various educational programs have been developed with a focus on language and culture. These programs have generated the production of alphabets and written registers for some Indigenous languages. The categorization of these languages as “written” or “unwritten,” or having an alphabetic system or not is a politically highly contentious issue which is part of the history of the power relationship between the colonial power and dominant groups and Indigenous communities. Indigenous world-views conceive as legitimate texts registers of various kinds, such as inscriptions, fabrics, or pictograms, which are part of their cultural artefacts. The fact that the linguistic structures of the Indigenous languages have been recognized as extremely complex (UNICEF and FUNPROEIB, 2009) is of importance when moving into the discussion of the relationship between language and mathematics (Parra, 2013). This is of particular relevance for understanding how different national histories have dealt with the “lack” of alphabetical registers in Indigenous languages.

4.3.2 Tensions Between Oral and Written Registers

Indigenous teachers and communities have argued for the need to raise awareness of the political dimension and impact of introducing schooling and writing in their communities. Criticism has been raised of the inclusion of local cultural knowledge in the curricula. In discussions about mathematics education for Indigenous people, a major point is the relationship between orality and writing, because the introduction of alphabetical writing in contexts shaped by an oral tradition creates many tensions. We can see these tensions in standardization processes, technical problems in the production of alphabets, teaching materials, loanwords, neologisms, and processes of re-semantization. Moreover, these tensions are related to the ultimate purpose of writing oral languages to affirm cultural identity. Written languages find a role in the school, and they create the necessity of exploring social uses for that writing outside the school.

When addressing the relationships between orality and writing in Indigenous contexts of mathematics education, a source of tension is the fact that Indigenous languages have strong oral traditions, while Portuguese and Spanish have a strong written tradition. With respect to Indigenous students’ development of mathematical thinking, local mathematical knowledge and Indigenous worldviews appear close to each other. This is evidenced in oral expression and graphic representations that are valid only in the limited scenario of one culture. School mathematical knowledge, by contrast, demands a different type of link with the students’ sociocultural context. It implies the use of expressions not only at the oral, concrete, and graphic level but also at a symbolic one.

The oral-written tension can also be discussed in situations where oral patterns and other forms of symbolic representation that make part of the language practices of these communities appear in written texts, such as, for instance, when aspects of orality influence the writing of mathematical problems, as discussed by Mendes (2007). The writing of these problems in Indigenous languages and in Portuguese by Indigenous teachers have a narrative form. The problems are presented without question(s), which is the expected form. They preserve the discursive characteristics that could be associated with language use to solve problems in daily life. The narrative form incorporates alphabetical and numerical writing and visual representation, i.e., drawings are part of the narrative. They transfer oral narrative characteristics of an everyday problem situation into writing and give a narrative function to drawings that appear in the problem. This type of writing of mathematical problems in a narrative form brings to light the difference between oral and written traditions. "Narrative is a strong trait in the process of production, incorporation and maintenance of social and cultural knowledge and norms among indigenous groups" (Mendes, 2007, p. 224).

The development of a bilingual and intercultural mathematics education for Indigenous students entails another challenge related with expressions in Indigenous languages, since their structure may sometimes complicate the understanding of symbolic expressions. For instance, the expression "7>4," in Spanish is read "7 es mayor que 4" [7 is greater than 4], while in the Amazonian language Shipibo Conibo is expressed as "7 riki 4 bebon iká," an expression that has a different syntactic structure. In Shipibo Conibo this expression means "7 is in front of 4," appealing to the order of natural numbers, instead of their cardinality. In the same way, there are tensions generated by the mismatch between some Indigenous concepts and those used in mathematics. In some Andean cultures, time and space form a conceptual unit, termed in Quechua as *pacha,* where there is no distinction between the two (Yáñez Cossío, 1990, p. 4). Another example is that in some Indigenous numeration systems, the terms for integer numbers in the Indigenous language could be a useful resource from a pedagogical point of view. For example, "in Quechua, Aimara and in some Amazonian languages the numbers up to ten have each one a name. From ten on, they are named "ten and one," "ten and two," "ten and three"... "two tens," "two tens and one," etc. This structure facilitates both understanding of the decimal positional numbers system and the construction of algorithms for the basic operations" (Villavicencio, 2013, p. 36). All these tensions can be seen either as obstacles to overcome or as opportunities to enrich the cultural dialogue of different mathematical knowledge.

4.4 Relevant Mathematics Education in Indigenous Contexts

The changes towards diversity in linguistic and educational policies took place at the heart of cultural, social, and political claims of a national nature. Due to this, the dynamics of multilingual education are not restricted to bilingualism and they are not separated from conceptions, policies, and projects based on interculturality in each country. With regard to mathematics education, and in order to contribute to the cultural reaffirmation of Indigenous peoples, we understand mathematics as a cultural phenomenon and thus recognize the mathematics of each Indigenous people.

Although we use distinct perspectives for the concept of *ethnomathematics*, we find common ground in our countries with respect to the importance of research on connections between language, culture, and mathematics. Drawing on D'Ambrosio (2011, pp. 111–112), ethnomathematics is not only the study of the mathematics of various ethnicities but also more than that; it is a spatially and temporally differentiated study of the various *technes* or *ticas* (ways, techniques, abilities) of *matema* (to explain, to understand, to deal, to coexist) in different *ethnos* (natural cultural and socioeconomic contexts). Since cultural encounters take place in a network of power relations, these relations set in operation a hierarchy of knowledge that declares what counts as official, valid, or even invalid (Knijnik, 1996).

The cultural forms particularly associated with mathematics are produced in the weaving of cultural understanding as interactive processes or as a web of meanings, using Geertz's (1973) idea of culture as a web of significance. If culture is produced like a web of meanings in interaction, there is a close connection between language and culture. This is because we do not understand language only in a verbal sense, as a vehicle of expression of culture, but as a symbolic system. Thus language can be seen as the stage of cultural production.

Ethnomathematics in the mathematics curriculum of basic education for Indigenous peoples in Brazil, Colombia, and Peru has the aim of contributing to cultural reaffirmation of Indigenous students, through their language and in the context of their worldviews. The inclusion of mathematical aspects of local cultures in the school curriculum is not planned to build a bridge that serves to facilitate students' learning of Western mathematics. Rather, it is a strategy for building cultural identity. This idea is aligned with the ethnomathematical aim of understanding and making visible the cultural forms in which mathematical thinking is produced. The routes that each country has taken into formulating relevant mathematics education for Indigenous are diverse.

4.4.1 Officializing Ethnomathematics in Peru

In Peru, current mathematics education in Indigenous contexts comes from the systematization of accumulated experiences. The documentation on this matter dates from the 1950s. Arithmetic texts produced by SIL were used by students and teachers in Amazonian bilingual schools (Larson et al., 1979). In the 1980s, the study "Numeration, algorithms and application of numerical and geometric relations in the rural communities of Puno" identified the mathematical knowledge of Quechua and Aymara communities (Villavicencio, 1983, pp. 135–141). The results were taken into account in the Experimental Project of Bilingual Education in

Puno. In this pilot project and other similar ones, as well as in the creation in 1989 of a section in the Ministry of Education responsible for Bilingual Education, a new attitude towards and awareness of Indigenous mathematics started.

Since 2003, educational law establishes interculturalism as a mainstream trend in the entire education system. This includes an intercultural mathematical education for all, i.e., for Spanish speakers and speakers of Indigenous language. This law also contributes to the recognition of ethnomathematics as part of the official discourse. For the purpose of implementing and developing an intercultural and bilingual education, ethnomathematics is operationally understood as the knowledge of a cultural group, identifiable as part of their worldview, manifested through the activities of counting, measuring, locating, designing, playing, and explaining.

For basic mathematics education and monolingual Spanish speakers, learning is planned through school activities that enable students to acknowledge the presence of mathematics in Indigenous cultures. In the mathematics education for students whose culture and language are Indigenous, the Peruvian government promotes and implements a pedagogical proposal in the context of the Bilingual Intercultural Education program (Dirección General de Educación Intercultural Bilingüe y Rural [DIGEIBIR] 2012). This proposal has been developed in consultation with Indigenous representatives and the participation of teachers and wise men from the involved communities. The pedagogical approach for mathematics in this educational model is problem solving, in addition to the focus on democratic, intercultural, and welfare rights.

When a school year begins, an elementary teacher in the above mentioned program identifies the students' language proficiency using a psycholinguistic diagnosis tool. There are four levels of proficiency. In classrooms where children speak mainly Indigenous language, the intercultural approach starts with learning-oriented activities of ethnomathematical knowledge, using strategies from their cultural practices, complemented later with school mathematics. While school mathematics has its own codes that can be verbalized in different languages, ethnomathematical non-Western knowledge is inseparable from the corresponding native language and culture. Hence, in multilingual Indigenous contexts, an intercultural bilingual mathematics education is necessary. With respect to the teaching and learning of school mathematics, the language of instruction is chosen, taking into account the students' proficiency in the indigenous languages and Spanish.

For several reasons, mathematics education in Indigenous Peruvian schools has not encountered adequate technical support. Since 2012, the Ministry of Education started prioritizing the education of Indigenous peoples when it was made evident that Indigenous students' achievement in mathematics was low. However, bilingual education strategies may have a positive impact on these students' achievement. López (1998) found that "children served by the PEEB-P [bilingual education program in Puno province] outperformed their peers in the control schools in terms of reading comprehension, oral proficiency of Spanish and mathematical problem solving" (p. 70). Results indicate that learning achievements are better when mathematics education for indigenous children is bilingual, in Andean languages (Quechua or Aymara) and Spanish.

In a study in rural schools in Puno within the bilingual and intercultural approach, Cueto and Secada (2003) did not find statistically significant differences between students' performance in monolingual Spanish and students in these schools. Rather than accepting these results as evidence that the introduced program does not work (or just does not "harm"), the authors argue that the program is not running adequately. They suggest that:

> Any effort to create a truly operational program should take into consideration the language teacher and his mastery of reading and writing, the values and beliefs of the students and their parents about the importance of indigenous languages, and the support given to the program inside and outside classroom (Cueto & Secada, 2003, p. 19)

Studies providing more evidence about the quality of mathematics education in Indigenous contexts are still needed.

4.4.2 Using Language to De-colonialize Indigenous Education in Colombia

During the last 20 years in Colombia, there has been a growing development of multilingualism in mathematics education, emerging from Indigenous communities and researchers. Such development corresponds to a wider context, in which a bottom-up process has now reached policy levels. This section shows important elements of that process for mathematics education as the political changes in the country since 1991 have helped Indigenous communities to legitimately claim the right to defend their cultural identity.

In the 1980s, Indigenous peoples of Cauca were working on creating school readiness booklets on arithmetic. As a result of ethnolinguistic work, Queixalos (1988) created "neonumerations" for the Sikuani language, and Cauty and Ramos (1990) for the Nasayuwe. These first attempts had the intention of creating a basic register for arithmetic and numerals in these languages. It was even intended that these new registers would enter schools. They were introduced in schools, but the dominant use of Spanish in out-of-school situations was difficult to challenge.

By tracking the evolution of curricular proposals for Indigenous education, whether governmental or initiated by Indigenous people, a vacuum on the subject of mathematics can be seen. More often than not, mathematics is considered to be an autonomous differentiated discourse that cannot be addressed using Indigenous knowledge. There has been a tendency, therefore, to deal with mathematics in Spanish. In the few cases where mathematics is connected with Indigenous cultures, it is reduced to arithmetic, with numeral translation, and implemented in school problems of trading. This situation can be related to the fact that during the 1980s, Indigenous movements lacked theoretical tools to associate elements of mathematics with cultural practices.

In the mid-1990s, Ochoa and Peláez (1995), supported by an Indigenous association, presented the mathematical knowledge of the Tule people and a mathematics that they called "Western." They explained the idea of number in the Tule worldview, topological notions, some classifiers and operations with numbers. They also exposed basic school mathematics tied to arithmetic and its operations. In their text, we find together, but separately, two conceptions of mathematics. Issues of language are made evident in the bilingual writing in both Spanish and Tule language.

Cauty and Tovar developed a project with the Wayuu people, reported in Cauty (1998). The project provided a new approach to structure the fieldwork dynamics for articulating Indigenous claims with advances in mathematics education and ethnomathematics. Cauty engaged in the difficult task of translating an algebra textbook into the Wayuunaiki language. He convened an interdisciplinary and intercultural team of experts: traditional knowledge-holders, linguists, and mathematicians. They worked on creating new knowledge, generating explanations of different concepts from cultural legacies, academic mathematical knowledge, and the structure of Wayuunaiki. It became clear that such an endeavor was not about making literal translations, or equivalences word to word, but about building networks of explanations and representations around concepts and practices. The process could not only be steered by the mathematical knowledge of the dominant society. The product was not a translated text, but a process that brought together different peoples and knowledge. They proposed new words and alternative conceptualizations in mathematics, which expanded the original fields of knowledge of each culture. One contribution of this work was to overcome the idea that arithmetic is the only content to be worked in Indigenous education. Another element was to suggest a multidirectional approach that evidenced how this type of encounter cannot be a translation from academic mathematical knowledge to words used in the native language, tacitly maintaining a relationship of hierarchy between the two cultures. It was also necessary to describe with Western mathematical representations some topics and features of the Wayuu worldview, which were conceptualized by the Wayuu as hallmarks of their rationality.

Since then, there has been an increase in research on Indigenous knowledge associated with mathematics in different Colombian communities. Such work privileges a strategy of dialogue of knowledge which emphasizes cultural and cosmological issues, while it downplays a linguistic focus (Aroca, 2007; Parra, 2003).

In 2006, some of the Caucan communities expressed interest in researching their worldview, forming an Indigenous intercultural research center, and within it a research team on Nasa people's mathematics. Such a team was formed by Indigenous teachers who had been involved in the Nasa-alphabet unification as well as in the standardization of the writing process. The team also involved experts in mathematics and language, with experience in Indigenous education. The work process reflected some elements of Cauty's proposal, but coordinated with the Nasa people's ways of producing knowledge. The work was collectively developed in different shelters, with the participation of knowledge-holders, elders, children, teachers, and educational authorities.

Previous published information about the mathematical practices of the Nasa people was refuted, complemented, and increased. The discussions during the exploration took place in the Nasayuwe language and the teachers recorded, transcribed, and summarized the elements found in them. When the community elders expressed mathematical concepts, unexpected words that had fallen into disuse or were not known to newer generations appeared. Some common words also gained a new meaning. Considering issues of localization, there emerged old expressions in Nasayuwe for movement and stillness. Words indicating specific lapses of time during the day were commonly used as time markers. Also the existence of some local measure units for weight, length, and volume led the researchers to propose a word for the concept of "measure unit." When conceptual issues appeared, e.g., velocity or continuity, common words gained a new meaning, as the constant and uninterrupted presence of the territory and its spirits has a particular word in Nasayuwe. That word was proposed to express the concept of continuity. The road, that sometimes disappears, was used as an image for "the discrete" as a mathematical concept. Research findings were published in a bilingual book, in which Indigenous teachers were the authors (Caicedo et al., 2009). The book includes several stories of the oral tradition and rescues what the community considers as mathematics.

The book also raises conceptual elements to be developed by other teachers in their process of *Educación Propia. Educación propia* (in Spanish) is an educational approach resulting from 35 years of political negotiation between Indigenous organizations and the Colombian government. This approach encapsulates three meanings of the word *propia*: propia as *their own* or belonging to them, propia as *adequate* for their needs, and propia as other forms of knowledge and education that get *appropriated* by the community for their political struggles (Parra, 2011). In the same way, attuned to government regulations on Indigenous education, Viluche and Yujo (2006) and Tamayo (2012) have initiated studies on mathematics education in other communities, discussing the Western disciplinary classification and displaying its failure to deal with cultural practices. Appreciation of the integrity and holism of Indigenous thought generates resistance to the act of isolating elements of cultural practices and to typifying them as mathematical.

4.4.3 Problematizing Schooling and Mathematics Education in Brazil

In Brazil, issues related to mathematics education for Indigenous peoples emerged linked to discussions about Indigenous schools. Although school was understood as a historical space of value imposition and assimilation for incorporation into the market economy, Indigenous communities began to claim schools as a place to build intercultural relationships based on political autonomy (Ministerio de Educação do Brasil, 2007). Therefore, Indigenous education, particularly mathematics education, cannot be divorced from Indigenous students' realities in seeking to meet the Indigenous communities' aspirations and respect for group cultural issues (Correa, 2002).

The discussions about mathematics education have been influenced by ethnomathematics research in contexts of Indigenous communities and also by processes of Indigenous teacher education in different projects in the country, such as those by Costa (2008) and by Mendes (2001). As stated by Sebastiani Ferreira (2004), the movement of teacher education with teachers as researchers in Indigenous cultures is crucial. Indigenous teachers assume a key role in research practices around language and cultural knowledge in their communities with the purpose of incorporating and articulating the community's knowledge in relation to school mathematics practices. Many Indigenous teachers have this perspective when questioned about why mathematics should be taught in Indigenous schools. One indigenous Xacriabá teacher asserted:

> We have to know the math also, since the activities are not just from textbooks, but in relation to our people. Mathematics in Indian schools is very important to our day-to-day. We are living in it. (Mendonça, 2006, p. 5)

The rights claimed by Indigenous communities call for an Indigenous School Education characterized by the "assertion of ethnic identities, the recovery of historical memory, the appreciation of languages, and knowledge of indigenous peoples and the revitalized association between school/society/identity, in accordance to societal projects defined autonomously by each indigenous people" (Ministerio de Educação do Brasil, 2007, p. 21). Indigenous schools are being proposed as intercultural spaces where situations of teaching and learning are related to the political and cultural identity of each Indigenous people. As said by a Guarani teacher:

> I see that this school should have all conditions of an Indigenous people, the Guarani-Kaiowá people, Terena, Xingu, Xavante […] whatever. But it has to be a school of that group, that nation, which teaches the language, dances, rituals, ceremonies, which is a school with proper autonomy of the local community. (Rodrigues, Ferreira, & Domite, 2009, p. 9)

In the same way, Indigenous peoples have linguistic rights to participate in school learning processes conducted in students' mother tongues. Those rights direct attention towards the community's sociolinguistic reality and language use in the community space as well as in school. Indigenous school education in an educational program that has no connection with the reality of Indigenous students can generate a weakening of a people's identity, making harder the struggle for survival. Therefore, mathematics education courses for Indigenous teachers have been anchored on issues related to Indigenous communities' aspirations and respect for cultural identity, to meet the needs of Indigenous peoples for building their own educational curriculum, according to their reality and consistent with the new demands that post-contact situations imposed (Correa, 2002).

Among studies with a focus on cultural practices of Indigenous groups, Indigenous languages have a key role in knowledge constitution. Ribeiro and Ferreira (2004) show that the Xerente people have an organization that is based on a dual logical thinking process. Numbers expressed in that Indigenous language have a dual construction because the number one is not a complete entity. A unity, for this community, occurs in the encounter between one half and the part that is lacking: "The whole number, reflecting the knowledge that comprises it, consists of the junction of two halves, which form an ethnomathematical dual system" Ribeiro and Ferreira (2004, p. 152). As this conception has no possible translation from the Xerente language into Portuguese, the authors underscore the need for vitalizing language to maintain the Xerente culture.

The case above exemplifies the fact that the encounter between Indigenous and non-Indigenous mathematics have generated tensions of a political and symbolic nature in relation to the uses of Indigenous languages and Portuguese. Such tensions have resulted in the development of new numerical terms, particularly in producing mathematical literacy materials in Indigenous languages, as a form of affirmation and maintenance of an ethnic identity as discussed by Mendes (2004, 2011). Following this trend, Indigenous schools are conceived of as intercultural spaces that can be conceptualized as "border schools"; that is, public spaces in which teaching and learning situations are related to the political and cultural identity of each Indigenous people. More generally, it can be argued that discussions of Indigenous school education in Brazil have questioned the relationship between society, culture, and school. Thus, intercultural and multilingual mathematics education draws on the social life of these groups in order to establish new meanings and functions for the school contexts.

4.5 Concluding Remarks

Since the 1980s, there has been a favorable advance of state policies towards the participation of Indigenous peoples regarding the recognition of their right to an appropriate education, according to their worldviews, cultures, and languages. In contrast to the colonial view of assimilation of Indigenous people through evangelization and monolingual/monocultural education, the advance towards the acknowledgement and respect of Indigenous cultures has been closely linked to the struggle of Indigenous organizations. In Brazil, Colombia, and Peru there are different routes for the implementation and development of educational models in the context of Indigenous peoples, which depend on the autonomy and dialogue that they establish with the state. In Peru, there is a diversified bilingual intercultural education provided by the government, with the participation of the Indigenous peoples. Education and professional development for bilingual and intercultural teachers are offered in pedagogic institutes and some universities. In Colombia, the communities are developing an entire educational system, following the principles of the *Educacion Propia* approach. In Brazil there is a differentiated, intercultural, and bilingual proposal for Indigenous schooling with emphasis on the initial education of Indigenous teachers in undergraduate programs at universities.

To attend to the needs of Indigenous populations, the three countries have diversified curricula, finding in different interpretations of ethnomathematics the theoretical and methodological elements necessary to support mathematics education in situations of bilingualism involving meetings of different cultures. Nevertheless, the implementation and development of mathematics education for Indigenous populations have generated tensions derived from the diverse worldviews, cultures and languages involved. Indeed, Indigenous peoples generally agree to establish a harmonious relationship with nature, as opposed to those who see nature as an object of exploitation. American Indigenous cultures seek the good life of all beings, including humans. Also, cultural practices associated with mathematics are present in the field of daily life and social relations and are interwoven with spirituality and cosmologies. Such ways of understanding themselves and the environment usually clash with dominant views of school mathematics belonging to the main culture, where school mathematics is seen as a knowledge that stands on its own, decontextualized of its origin and use. All this brings to the fore the epistemological debate about what counts as mathematical knowledge in the domain of schooling, as well as the criteria to select what deserves to be addressed in Indigenous education.

The production, dissemination, and use of bilingual educational materials in mathematics may lead to solving linguistic, social, and epistemological issues. In fact, creating neologisms and determining loanwords suitable for the intended objectives of mathematics teaching requires an in-depth study of the linguistic structures of Indigenous languages. It also demands the active participation of the community and institutional support to ensure the use of the terminologies proposed. A factor of complexity is the fact that these languages are predominantly oral, with all that this implies for the understanding of the symbolic language of traditional school mathematics, particularly when symbolic expressions of mathematical relations do not match the syntactic structure of the corresponding oral expressions in Indigenous languages.

Research on the mathematics education of Latin American Indigenous peoples is recent compared to the advances in other similar situations (e.g., Meaney, Trinick, & Fairhall, 2012). Research on the systematically low levels of achievement of Indigenous students in mathematics alerts us to the urgency of considering educational approaches that build on Indigenous traditional knowledge, ensuring basic conditions to enable a *pertinent* and *quality* mathematics education. It is also important to reconsider the relevance of applying standardized tests to Indigenous peoples or the weighting of their results to the policies to be implemented, since the exercise of citizenship is differentiated in these peoples. Not everyone wants to enter into the schemes of economic production that mainstream society builds, with its standardization in school mathematics. Multilingualism in Latin American schools is engaged in a broader political project for respect and protection of cultural diversity and Indigenous heritage.

References

Aroca, A. (2007). *Una propuesta de enseñanza de geometría desde una perspectiva cultural. Caso de estudio: Comunidad indígena Ika-Sierra Nevada de Santa Marta.* Unpublished master's dissertation, Universidad del Valle, Colombia.

Caicedo, N., Guegia, G., Parra, A., Guegia, A., Guegia, C., Calambas, L., Castro, H., Pacho, C., & Diaz, E. (2009). *Matemáticas en el mundo Nasa* (N. Caicedo & A. Parra, Eds.). Bogotá, Colombia: CIIIT.

Cauty, A. (1998). *Etnomatemáticas: el Laboratorio Kwibi Urraga de la Universidad de la Guajira.* Retrieved from http://etnomatematica.org/articulos/cauty4.pdf

Cauty, A., & Ramos, A. (1990). Vigilancia etnocultural: el caso de la numeración tradicional Nasayuwe. *Boletín de Lingüística Aborigen, 2*, 3–15.

Correa, R. A. (2002). A educação matemática nos cursos de licenciatura e a formação de professores indígenas. *Cadernos de Educação Escolar Indígena-3º grau indígena* (pp. 117). Barra do Bugres, Brazil: UNEMAT.

Costa, W. N. G. (2008). *A etnomatemática da alma A'uwe-xavante em suas relações com os mitos.* Unpublished doctoral dissertation, Universidade de São Paulo, Brazil.

Cueto, S., & Secada, W. (2003). Eficacia escolar en escuelas bilingües en Puno. *Revista Electrónica Iberoamericana sobre Calidad, Eficacia y Cambio en Educación, 1*(1), 19–20.

D'Ambrosio, U. (2011). *Educação para uma sociedade em transição* (2nd ed.). Natal, Brazil: EDUFRN.

Dirección General de Educación Intercultural Bilingüe y Rural (DIGEIBIR). (2012). *Hacia una educación intercultural bilingüe de calidad.* Lima, Peru: Ministerio de Educación de Perú.

Geertz, C. (1973). *The interpretation of cultures.* New York: Basic Books.

Instituto Brasileiro de Geografia e Estatística (IBGE). (2010). *Censo.* Brasilia, Brazil: Instituto Brasileiro de Geografia e Estatística. Retrieved from http://www.censo2010.ibge.gov.br/

Instituto Nacional de Estadística e Informática (INEI). (2009). *Censos Nacionales 2007: XI de Población y VI de Vivienda.Resumen Ejecutivo.* Lima, Peru: Dirección Nacional de Censos y Encuestas. Retrieved from http://www.inei.gob.pe/media/MenuRecursivo/publicaciones_digitales/Est/Lib0789/Libro.pdf

Knijnik, G. (1996). *Exclusão e resistência: Educação matemática e legitimidade cultural.* Porto Alegre, Brazil: Artes Médicas.

Landaburu, J. (2004). La situación de las lenguas indígenas de Colombia: Prolegómenos para una política lingüística viable. *Amérique Latine Histoire et Mémoire. Les Cahiers ALHIM, 10.* Retrieved from http://alhim.revues.org/125

Larson, M., Davis, P., & Ballena, M. (1979). *Educación bilingüe: Una experiencia en la amazonía peruana.* Lima, Peru: Ignacio Prado Pastor.

López, L. E. (1998). La eficacia y validez de lo obvio: Lecciones aprendidas desde la evaluación de procesos educativos bilingües. *Revista Iberoamericana de Educación, 17*, 51–89.

López, L. E., & Küper, W. (1999). La educación intercultural bilingüe en América Latina: Balance y perspectivas. *Revista Iberoamericana de Educación, 20*, 17–85.

Meaney, T., Trinick, T., & Fairhall, U. (2012). *Collaborating to meet language challenges in indigenous mathematics classrooms.* New York: Springer.

Mendes, J. R. (2001). *Ler, escrever e contar: práticas de numeramento-letramento dos Kaiabi no contexto de formação de professores Índios do Parque Indígena do Xingu.* Unpublished doctoral dissertation, Universidade Estadual de Campinas-UNICAMP, Brazil.

Mendes, J. R. (2004). Aspectos políticos e simbólicos na apropriação do discurso da etnomatemática: O caso dos professores Kaiabi do Parque Indígena do Xingu. In G. Knijnik, F. Wanderer, & C. J. de Oliveira (Eds.), *Etnomatemática, currículo e formação de professores* (pp. 364–376). Santa Cruz do Sul, Brazil: EDUNISC.

Mendes, J. R. (2007). Numeracy and literacy in a bilingual context: Indigenous teachers education in Brazil. *Educational Studies in Mathematics, 64*(2), 217–230.

Mendes, J. R. (2011). The meaning of ´number´ in Kaiabi: Identity and language in the context of Indigenous teacher education. In M. Setati, T. Nkambule & L. Goosen (Eds.), *Proceedings of the Study 21 Conference: Mathematics Education and Language Diversity* (pp. 218–223). São Paulo, Brazil.

Mendonça, A. N. (2006). *Investigando as práticas educativas dos professores Indígenas que ensinam Matemática na escola Xacriabá. Anais do EBRAPEM.* Belo Horizonte, Brazil: UFMG.

Ministerio de Educação do Brasil. (2007). *Educação Escolar Indígena: diversidade sociocultural indígena ressignificando a escola.* Brasilia, Brazil: Secretaria de Educação Continuada, Alfabetização e Diversidade. Retrieved from http://www.dominiopublico.gov.br/

Ministerio de Educación del Peru. (2013). *Documento nacional de lenguas originarias del Perú.* Lima, Peru: Ministerio de Educación.

Ochoa, R., & Peláez, J. (1995). *La matemática como elemento de reflexión comunitaria del pueblo Tule*. Medellín, Colombia: Lealon.

Parra, A. (2003). *Acercamiento a la etnomatemática*. Unpublished monograph, Universidad Nacional de Colombia Sede Bogotá, Colombia.

Parra, A. (2011). *Etnomatemática e educação própria*. Unpublished master's dissertation, São Paulo State University at Rio Claro, Brazil.

Parra, A. (2013). Linguagem escrita e matemática: Um viés etnomatemático. *Revista Latinoamericana de Etnomatemática, 6*(2), 24–34.

Queixalos, F. (1988). Numeración tradicional Sikuani. *Glotta, 3*, 28–31.

Ribeiro, J. P., & Ferreira, R. (2004). Educação escolar indígena e etnomatemática: um diálogo necessário. In M. C. S. Domite, J. P. Ribeiro, & R. Ferreira (Eds.), *Etnomatemática: Papel, valor e significado* (pp. 149–160). São Paulo, Brazil: Editora Zouk.

Rodrigues, M., Ferreira, R., & Domite, M. C. S. (2009). A formação de professores de professores e suas relações com cultura e sociedade: A educação escolar indígena no centro das atenções. *Bolema: Boletim de Educação Matemática, 22*(34), 263–282.

Roldán, R., & Gómez, J. (1994). *Fuero indígena colombiano*. Bogotá, Colombia: Gente Nueva Editorial.

Sebastiani Ferreira, E. (2004). Os Índios Waimiri-Atroari e a etnomatemática. In G. Knijnik, F. Wanderer, & C. J. de Oliveira (Eds.), *Etnomatemática, currículo e formação de professores* (pp. 70–88). Santa Cruz do Sul, Brazil: EDUNISC.

Tamayo, C. (2012). *(Re)significación del currículo escolar indígena, relativo al concimiento matemático, desde y para las prácticas sociales de la cestería y el cultivo del plátano: el caso de la comunidad Tule de Alto Caimán*. Unpublished master's dissertation, Universidad de Antioquia, Colombia.

UNICEF & FUNPROEIB. (2009). *Atlas sociolingüístico de pueblos indígenas en América Latina* (I. Sichra Ed.). Quito, Ecuador: FUNPROEIB Andes.

Villavicencio, M. (1983). *Numeración, algoritmos y aplicación de relaciones numéricas y geométricas en las comunidades rurales de Puno*. Puno, Perú: Ministerio de Educación. INIDE. DDE.

Villavicencio, M. (2013). *Matemáticas en educación intercultural bilingüe. Serie Matemáticas en EIB*. Lima, Peru: JERGIMPRESS E.I.R.L.

Viluche, J., & Yujo, S. (2006). *Nasawe'sx kiwaka fxi'zenxi êen*. Popayán, Colombia: Consejo Regional Indígena del Cauca, Programa de Educación Bilingüe; Asociación de Cabildos Ukawe Cxhab.

Woodward, K. (1997). *Identity and difference*. London: Sage.

Yáñez Cossío, C. (1990). *Representaciones y conceptos estructurantes*. Quito, Ecuador: Editorial Abya-Yala.

Chapter 5
Challenges and Opportunities for Second Language Learners in Undergraduate Mathematics

Viviane Durand-Guerrier, Mercy Kazima, Paul Libbrecht, Judith Njomgang Ngansop, Leila Salekhova, Nail Tuktamyshov, and Carl Winsløw

5.1 Introduction

Multilingual classrooms are nowadays common due to migration and student mobility. In many countries, university mathematics courses include students who are not native speakers of the language of instruction. We refer to such students as second language learners. The contexts are very different in each country, depending on history, policy,

V. Durand-Guerrier (✉)
University of Montpellier, Case Courrier 051, Montpellier Cedex 5 34095, France
e-mail: vdurand@math.univ-montp2.fr

M. Kazima
University of Malawi, Zomba, Malawi
e-mail: mkazima@cc.ac.mw

P. Libbrecht
Informatics, Weingarten University of Education Kirchplatz 2, Weingarten 88250, Germany
e-mail: paul@cermat.org

J.N. Ngansop
Université de Yaoundé 1, Yaoundé, Cameroun
e-mail: judithnjomg@yahoo.fr

L. Salekhova
Kazan Federal University, Kazan, Republic of Tatarstan, Russia
e-mail: salekhova2009@rambler.ru

N. Tuktamyshov
Kazan University of Architecture and Engineering, Kazan, Republic of Tatarstan, Russia
e-mail: nail1954@gmail.com

C. Winsløw
Institut for Naturfagenes Didaktik, University of Copenhagen, Øster Voldgade 3, Copenhagen 1350, Denmark
e-mail: winslow@ind.ku.dk

R. Barwell et al. (eds.), *Mathematics Education and Language Diversity*, New ICMI Study Series, https://doi.org/10.1007/978-3-319-14511-2_5

and so on. This diversity is evident in the five cases authors from different countries present in this chapter. Although it is a common assumption that language is not a crucial issue in the learning of advanced mathematics, there is research evidence that even in monolingual contexts, linguistic issues do appear with implications for equity (e.g., Zevenbergen, 2001). We also know that linguistic issues may be different, in nature and scope, for second language learners.

In this chapter, we first describe specific challenges related to mastery of the language of instruction that second language learners might face in learning undergraduate mathematics. We then describe five multilingual contexts, share insights from research, and present emergent proposals for supporting students to overcome difficulties.

5.2 Linguistic Difficulties in Advanced Mathematics for Multilingual Contexts

5.2.1 Lexical and Notational Issues

One important issue is the challenge in translating mathematical terms. It is easy to see that there are lexical differences across languages. These differences are sometimes augmented when we consider specialized uses of the languages, such as the vocabulary of academic mathematics. For example when considering "everyday language" the English word "field" could be translated into French as "domaine," "champs" or "terrain," but none of these translations are valid when considering the use of the term "field" in algebra, where the correct French translation is "corps." More delicate examples of these differences are pairs of expressions in two languages that look or sound similar but differ significantly in meaning. Such differences may have deeper consequences for understanding mathematical texts. For example, in French, a "nombre décimal" has a finite number of non-zero digits on the right hand side of the decimal separator, while in English, a "decimal number" can have infinitely many non-zero digits. As a consequence, even an Anglophone student with a good mastery of "everyday" French might translate a mathematically accurate and precise phrase in French into something mathematically confusing, such as "the set of decimal numbers is not a complete subset of the real number set."

For example, Kazima (2006) reports an experiment in which English words from the vocabulary of probability theory were presented with no context to a mathematics class of secondary level students whose first language was Chichewa. The findings support the claim that even when students used the same words as the teachers, the meaning was different. We know from research in mathematics education that this phenomenon is not specific to second language learners (Mathé, 2012), but as the study shows, it is complicated by the use of two languages. In this example, students moved from English to Chichewa (a language that does not have an established mathematics register), did their thinking in Chichewa, and then translated back into English (p. 187). This study raises another issue, the creation of mathematical terms in a language of instruction that does not have an established mathematics register, as is the case in Chichewa or Tatar. As a general phenomenon, this issue is not specific to mathematics, but as we shall see in this chapter, there are important implications for university-level mathematics.

The case of language-based differences in notation is specific to mathematics, and is also a widely underestimated issue. In his book *History of mathematical notations*, Cajori (1928) aimed "to give not only the first appearance of a symbol (whenever possible), but also the competition encountered and the spread of the symbols among writers in different countries" (p. 1), assuming that such a development could help with notational confusion in contemporary mathematics. Nevertheless, the belief that mathematical notation is "universal" and independent of language or location needs to be carefully considered.

Differences in notation appear even between countries sharing the same mathematical tradition. Libbrecht, Droujkova, and Melis (2011) gave two examples: the notation for half-open intervals in English, [a, b), French or German, [a, b[, and Dutch [a, b>; and the various notations for the greater common divisor, that correspond to an abbreviation of three words, and hence depend on language. Other issues arise when mathematical notations from one tradition are incorporated into a language with a different orthography or textual organization, such as Greek–Roman languages versus Arabic languages (i.e., Bebbouchi, 2011, pp. 529–530). In general, mathematical notations contain a lot of information, so that being able to decode or unpack such information is an important part of mathematical proficiency. For example, the notation $\lim_{x \to a} f(x) = L$ does not express simply an equality between elements of a space; the left part implies that f is a mapping that is defined in a neighborhood of a, and admitting in a, a unique limit.

Another issue concerns notations for which there is no general consensus, even within one country. For example, in most Western countries, if we are given a numerical function f, the expression f^n could refer to the product of function (multiplication), or to the functional power of function (composition). Such ambiguous notation is routinely used in calculus courses. If students are not aware of such subtle differences in the notation statements expressed in these courses, they may misunderstand exercises with this notation. Sometimes the mathematical context (e.g., trigonometry) can implicitly determine which of the two meanings are intended—but this can be confusing to students who (rightly) expect mathematical notation to be unambiguous and consistent across courses. Some of the specificity of mathematical notations arises because they are frequently abbreviations of words. For example, using either the beginning of the word (i.e., *sin* for "sinus of an angle" in many western languages, but *sen* for "seno de un ángulo" in Spanish), or the initials of the words in a group of words such as gcd (greater common divisor), pgcd (plus grand commun diviseur); or kgV (kleinstes gemeinsames Vielfache).

As the number of undergraduates in mathematics courses who use two or more languages increases, the need also increases for both student and instructor awareness of how crucial the context of production of a notation is for its adequate understanding and use. Such awareness would benefit instruction in both multilingual and monolingual contexts. An example of work that supports such awareness is a report (Libbrecht et al., 2011) on an initiative to collaboratively observe the diverging mathematical notations across cultures and languages, create a census of mathematical notations, and develop a collaborative database (Libbrecht et al., 2011, p. 166).

In addition, the undergraduate education phase of students coincides with a phase of independence where fetching knowledge from arbitrary sources is becoming a common practice and, often, a requirement. The most important sources of knowledge for undergraduate studies are the local library and the World Wide Web. The latter has increased largely in usage in recent years for its ease of access, its ubiquity, and the wealth of knowledge it allows access to.

However Wikipedia and most websites present to students a large diversity in ways of expressing mathematics. The issue with using these sources includes, but is not limited to, the dominance of a few languages in the most used (or simply available) online scientific resources for a given mathematical topic. It concerns also the conventional use of vocabulary and notation. In Wikipedia, often considered as a single coherent resource of knowledge as a normal Encyclopaedia would be, pages are often created or verified by multiple persons which leads to inconsistencies which are hard to remove and which a student needs to manage. This problem is similar to reconciling the language and notation used in multiple books. For example, in 2009, one could see the notation for the binomial coefficient written both as C_k^n and as C_n^k depending on the page.[1] Thus students using the web would need to reconstruct the context of the mathematical reasoning when reading this text; this was not necessary generally before university or with the use of paper books.

5.2.2 Logical Issues

Research has shown that tertiary students experience difficulties using logic, such as unpacking the logic of statements, interpreting formal statements, conducting argumentation or engaging in the proving process. For example, there is research evidence that undergraduate students experience difficulties with these aspects of logic that can impede their competencies in proof and proving (e.g., Durand-Guerrier, Boero, Douek, Epp, & Tanguay, 2012; Epp, 2011; Selden & Selden, 1995), and that these difficulties are aggravated for second language learners (e.g., Barton, Chan, King, Neville-Barton, & Sneddon, 2005 and this chapter).

Predicate Logic can be used to unpack the logic of a given statement by identifying the logical categories, connectors, quantifiers, and their respective scopes. The fundamental categories are *properties* modeled by one-place predicates such as "to be a primary number"; relationships modeled by two (or more)-place predicates

[1] The versions of these pages at these dates can be seen at http://en.wikipedia.org/w/index.php?title=Binomial_coefficient&oldid=309172219 and at http://en.wikipedia.org/w/index.php?title=Combination&oldid=278280354

such as "to be greater than" (binary relationship). In a given interpretation, arguments of various types can be assigned to properties and relationships: free variables (place-holders); singular elements; generic elements and according to the choices we get either singular propositions or open sentences. For example: "x is a prime number", where x is a free variable, is an open sentence; it is satisfied by some elements, and not satisfied by others: "29 is a primary number" is a singular sentence which is true, so 29 satisfies the open sentence, while "27 is a prime number" is false, so 27 does not satisfy the open sentence. In a relationship, it is possible to assign singular or generic elements to some places, and free variables to other places. It is also possible to close an open sentence by means of quantifiers, either universal or existential.

Moreover, while in propositional calculus the logical connectors are used to build complex propositions from elementary propositions and are defined by their truth tables, in Predicate calculus, the logical connectors can also be used to build complex open sentences, such as for example "if an integer is even, then its successor is prime," that could also be closed by quantifiers. As a matter of fact, these quantifiers often remain implicit, or could be hidden through linguistic means depending on the language, and it is common for ambiguities to arise due to the difficulty in identifying the respective scopes of quantifiers and connectors. A well-known difficulty is related to the meaning of the article "*a*" that can either refer to an individual, a generic element, or an implicit universal quantifier. Another difficulty concerns binary relations in universal statements: for example, in French, the statement "les faces du solide sont congrues deux à deux" (the faces of a given solid are pairwise congruent) can mean either "each time you take two faces, they are congruent," or "each time you take a face, there is another face which is congruent."

Another aspect concerns the distinction between open sentences and closed sentences, which plays a crucial role in proof and proving (Durand-Guerrier, 2003). In particular, in a given theory, a mathematical definition corresponds to an open sentence, satisfied exactly by the elements that it defines. A theorem is essentially a closed sentence that has been proved in the considered theory, either singular such as "π is transcendental," or general such as "For any positive integer k, there is a natural number N such that for all $n>N$, there are at least k primes between n and $2n$" (Erdös theorem). A conjecture, however, is a closed sentence that is thought to be true but has not been proved, such as Legendre's conjecture (which remains unproved): "For every $n>1$, there exists a prime p, such that $n^2<p<(n+1)^2$. In many cases, quantification remains implicit or is expressed by linguistic means that depend on each language, so that in some cases it may be difficult for students to identify the logical status of such statements.

5.3 Examples of Various Multilingual Contexts at University

In this section, we describe issues experienced by the authors in a number of countries, to show the variety of contexts.

5.3.1 A Multilingual Situation in Cameroonian Universities

Cameroon is a country in Central Africa; French and English are the two official languages. Teaching occurs in one of the two languages in primary and secondary school. To this end, the educational system is divided into two main subsystems: the *Francophone* subsystem and the *Anglophone* one. The two subsystems are so different that a student who started his education in one of those subsystems cannot easily continue studies in the other subsystem mainly because of the language barrier. This is because, even though the language of the other subsystem is taught in each subsystem, it is only with a very modest level of proficiency achievement. In higher education institutions, Cameroon has Anglo-Saxon style universities such as the universities of Bamenda (in the North-West region) and Buéa (in the South-West region). The other universities are bilingual; courses are given in French and English, depending on the teacher's dominant language.

In addition to the two official languages, there are about 230 ethnic groups, each with its own language (National Institute of Cartography, Cameroon). The current classification, based primarily on linguistic attributes, identifies six major groups. In the French-speaking area of Cameroon, the local languages are widely used in the family and they may strongly influence the way French is used. For instance, as part of her doctoral thesis, Tsoungui (1980) studied the interferences with French and the Ewondo language, spoken by the tribe of the same name. These interferences observed with the French language do not have much influence on English. On the other hand, the use of English by native English speakers is influenced by so called Pidgin-English, a composite language based on the English lexicon, but different than standard English. This is the main language in English-speaking areas of the country. It is also spoken in the contiguous zones of the Littoral and West Provinces, as well as in urban centers. We can say, in accordance with Tsoungui (1980), that "All Cameroons find themselves at least bilingual and often trilingual, even quadrilingual, French and English are never first languages" (p. 27). Consequently, we can find university students who must study in one or two languages—French or English, even though they are not fluent in either one.

5.3.2 The Case of Denmark

Over the past decades, internationalization has become an important priority for Danish higher education, and a priority that seems to have broad political consensus. As a member of the European Union, Denmark takes part in the student exchange program *Erasmus*, with about 2,416 outgoing and 6,186 incoming students in the academic year 2009/2010[2]—numbers which are continuously increasing. The total number of exchange students is about 16,000 (incoming) and 8,000

[2] Source: http://www.iu.dk/nyheder/kort-nyt/erasmus-fortsat-i-fremgang/

(outgoing) every year.[3] However, compared to the total number of students in higher education in Denmark (about 200,000), the numbers are not very high. The government aims to facilitate access to Danish universities for foreign students[4] as well as to encourage Danish students to go abroad.[5] We are mainly concerned with the first of these efforts here.

A main obstacle for foreign students can be the Danish language, which very few students have had the opportunity to learn abroad. However, within the last decades, more graduate courses are offered in English—particularly within scientific and technical fields. There is no common policy in this regard, even within a given university, and the question of the extent to which English should become a common language of instruction in advanced courses has been the subject of fierce debates, involving both practicalities and principles. On the one hand, it is hard to require short-term exchange students in a mathematics program to learn Danish well enough for academic instruction in this language. On the other hand, even in mathematics it is not straightforward for all students and teachers to "switch" languages. The current practices are mixed, with a tendency towards offering all texts in English and at least some of the teaching in English at the graduate level. From experience, this can work reasonably well, but with two major obstacles:

- Not all exchange students are fluent in English and in fact many are much less fluent in English than the average Danish student.
- In pure mathematics, we have a significant number of second-generation immigrant students (informal estimate: at least 10 % of the population), and this group seems to be particularly challenged with English—even when they are fluent in Danish.

In short, immigrant students have struggled to learn a second language (here, Danish) well, (and may have chosen mathematics or other technical subjects in part because they had difficulties with other subjects) and may now be asked to learn a third language, English. It would be a strange paradox if they were to lose out on Denmark's efforts to internationalize her educational systems.

5.3.3 Increasing Linguistic Diversity in France

In France, there is only one official language, French, which is used both in everyday life and for instruction in schools. For historical reasons, French is also an official language in some other countries in Europe, in numerous countries in Africa, in a few countries in the Americas or Oceania, alone or together with other languages (often English). There are also countries, mainly in Northern Africa, where French is not an official language but is still used as the language of instruction in

[3] Source: http://www.iu.dk/publikationer/2010-1/mobilitetsstatistik-for-de-videregaaende-uddannelser-2008-09/

[4] See, for example, the website http://studyindenmark.dk/

[5] See, for example, http://www.udiverden.dk/

mathematics from a certain grade on (e.g., secondary school in Tunisia). For this reason, France has historically attracted university students from a variety of countries. Recently, due to student mobility, the number of countries of origin of foreign students is increasing, as is the number of foreign students.

In a note of information,[6] the French Ministry of Tertiary Education and Research provides a synthesis in French of the results of a conjoint inquiry of the UNESCO, OECD, and Eurostat (OED). The results are given for 2007–2008. In France, there were 243,400 foreign tertiary students, representing 11.2 % of the total amount. The increase from 1999 to 2000 is about 75 %. Students from Africa are a majority (43.5 %), far more numerous than those from Europe (21.3 %) or from Asia (21.0 %). Students from China constituted in 2007–2008 the second most important contingent of foreign students in France. These foreign students are studying in France in a great variety of domains (e.g., Social Sciences, Trade and Law, Health and Social Sector, Engineering). If we consider the fact that mathematics is taught in a large spectrum of tertiary institutions, and that even some second language learners coming from countries where French is the official language may experience difficulties using French in academic contexts, it seems clear that many tertiary mathematics educators are likely to face linguistic diversity in their classrooms.

5.3.4 The Case of Malawi

Malawi is a southern African country with 17 local languages of which 4 are major languages in terms of number of speakers (Center for Language Studies, 2010). The most widely spoken language is the national language, Chichewa. English is the official language and the language of instruction from the fifth year of primary school through secondary school, at university, and in other further education. The Malawi school system has 8 years of primary school, 4 years of secondary school, and 4 or more years of university education. The first 4 years of primary school are taught in Chichewa or other local languages. Both English and Chichewa are also taught as subjects of study in schools. The situation in Malawi schools therefore is one where the classrooms are multilingual and students learn mathematics in English, a language that they are also learning and not yet fluent in.

The situation at university is not much different from elementary and secondary schools, in that all classrooms are multilingual and students often share one or more common local languages. University classrooms are more diverse in terms of the languages represented because students come from all parts of the country, unlike in elementary and secondary schools where students are local. Another difference is that at university there are cases where the lecturers or instructors are expatriates who may not share any of the students' local languages, while in elementary and secondary schools teachers are local and share at least one of the languages of the learners.

[6] Source MESR, 2011, Note d'information du 27 juillet 2011 http://media.enseignementsuprecherche.gouv.fr/file/2011/19/6/NIMESR1111_186196.pdf

Although by the time students come to university they would have had at least 8 years of English as the medium of instruction, from experience, it seems that problems with mathematical vocabulary, comprehension, and communication persist.

5.3.5 Bilingualism in Russia: The Case of Tatarstan

Tatar and Russian are the main languages used in the Tatarstan Republic. The Tatar language is used in family and daily life; it is taught as a subject in kindergartens, schools, and universities; newspapers and magazines are published in this language, and it is the language of instruction at all levels. At the same time, Russian is used in all spheres of society, including administration of the Tatar Republic, the courts, the highest authorities, etc., not to mention schools, family, and communication. Various legislative acts created the judicial base for creating a new language situation in Tatarstan and made it possible to design fundamental laws and values in the educational sphere.

There are currently 1,256 schools with one language of instruction (Tatar), and 410 bilingual secondary schools in Tatarstan. In 1990, 24 % of Tatars studied in their native language, mainly in rural areas. At present the rate is around 47 %. Beginning with Perestroika in the 1990s, Tatars had opportunities to acquire education in their native Tatar language not only in lower secondary schools, but also in high schools. At present a system of high-quality bilingual education using Russian and Tatar languages is developing. This process is based on the following principles: the integrity of the educational space of Russia; regional and ethnic needs connected with the social, economic, and political development of the society as a whole and the region in particular; the equal rights of the residents in acquiring higher education and freedom to choose the language of education. Bilingual educational programs exist at present in the Kazan State University of Architecture and Engineering, in the Kazan Federal University, in the Kazan State Technological University, the Kamskiy polytechnic institute, and in other universities and institutions.

One of the most pressing issues now in Tatar society is the creation of mathematical terminology in Tatar. The Mathematical Terminological Commission has developed more than 1,000 terms (Salimov & Tuktamyshov, 2000). There are many borrowed words from Arabic, Farsi, Russian, and West European languages used in Tatar in scientific contexts. Many scholars believe that if a borrowed scientific term is already widely used, it should be preserved.

5.4 Results of Research Studies

In this section, we present results of three selected studies in three linguistic contexts: the first one in New Zealand, where English is the language of instruction, the second in Cameroon where French is the language of instruction, and the third in Tatarstan where both Russian and Tatar are languages for instruction. The first two illustrate the difficulties second language learners experience and the third one presents a tentative approach to overcoming some of these difficulties.

5.4.1 The Case of Students in New Zealand Who Have English as an Additional Language

Like many similar research-intensive universities in the Western world, the University of Auckland in New Zealand receives an increasing number of students whose first language is not the language of instruction in the university. There are many immigrants from China and other non-Anglophone countries. All university instruction is in English, the first language of most New Zealanders. At this particular university, a series of studies were conducted by Barton and Neville-Barton (2003, 2004; Barton et al., 2005). The results, when coupled with the explanatory analysis, reveal phenomena of a more general nature. The studies aimed at comparing the performance of students with English as a first language (L1 students) to that of students with English as an Additional language (called "EAL students"), at various levels of the mathematics program (first and third year).

The first studies (Barton & Neville-Barton, 2003, 2004) considered students in their first year of undergraduate mathematics study. The main outcome was that EAL students at this level received better marks than L1 students, with an extra advantage for recently arrived students (more specifically, with less than 6 years of English learning experience). This was explained by the more solid technical knowledge of students who received primary and secondary mathematics instruction abroad (in countries such as China), and the advantage of recently arrived EAL students who did not have to struggle with the English language at the secondary level. The results are further explained by supplementary tests with mathematical questions phrased with more or less substantial use of natural language. As could be expected, L1 students understand, and perform better, on questions that are heavily based on the natural language (English), while the EAL students were inhibited in this context by the tendency to switch to symbolic modes of work. An interesting further result was the high level of self-reported understanding of textually rich questions by EAL learners, which sharply contrasts with their weaker performance on these questions. On the other hand, their data show that, in typical first-year courses (calculus and linear algebra with a focus on technical work) the EAL students appear to be less disadvantaged by their natural language capabilities, and in fact, any possible disadvantage is outweighed by their stronger capabilities in the symbolic mode.

However, the apparent success of EAL students in the first year does not continue into the second and third year. Barton et al. (2005, p. 728) note that in the second year, there was a significantly higher proportion of students with weak capacities in English than in the first and third year. The difference between the first and second year was explained by a mixture of factors, including a change of admission policies. The difference from second to third year seems to be mainly due to the students with weaker English fluency dropping out. This is confirmed by in-depth analyses of the language requirements for the more advanced courses in mathematics, and by a test of students' comprehension of linguistically and logically complex propositions, and their capacity to transform them (for example, given a mathematical phrase with the structure "If A then B", produce an equivalent phrase with structure "If not B then not A"). Here, a clear disadvantage for EAL learners is apparent, and it was demonstrated using a variety of methods that these difficulties constitute severe obstacles to the work proposed in more advanced courses. This also explains the difference with the results found with first-year students (Barton et al., 2005):

> ...the language used in lectures and texts at first-year level is repetitive, confirmatory, and predictable, and is mostly used to describe paradigmatic examples similar to those that students will be required to repeat. At third year level it changes to one-off explanations and logical trains of reasoning with examples that are drawn from a variety of unpredictable areas. In the third year examples are used to illustrate single aspects of a theorem, and are not designed to be copied or reproduced. Rather, students are required to reproduce the logical trains adapted to new situations. (p. 722)

Thus, the work required from third year students in advanced mathematics courses is significantly *more* dependent on discursive fluency in English that includes discernment of logical subtlety in phrases, rather than understanding statements that are mere frames around symbolic representations of algorithms and other computational techniques. So, while the first-year experience seems to confirm the folklore belief that success in mathematics is relatively less dependent on capacities related to the use of natural language, students who are not fluent in the language of instruction faced new difficulties in advanced mathematics in a more theoretical form.

Barton et al. (2005, p. 729) suggest that more effort needs to be put into supporting EAL students with "first language tutorials" or "specific mathematical English additional courses." To the extent that their results are generalizable, in particular if it is a universal trend that the specific need for such measures is more important at more advanced levels, then there is something paradoxical about the shift made to English at these levels in some contexts (such as the Danish one described above, where English is a second language to most learners). More research is needed to investigate the viability of alternatives such as earlier or more partial shifts of instructional language. As an example of such alternatives, it is common in Danish universities to base semi-advanced mathematics courses on textbooks in English, while the instruction is still in Danish. Such practices are usually motivated by the lack of specialized textbooks in a language like Danish, but it may also represent advantages for those students who will later on study or work with mathematics in English.

5.4.2 *Logical Issues: The Case of Negation of Quantified Statements*

At first glance, the concept of negation could appear as a very simple one, met and used early by children. This simplicity, however, holds only for singular statements, such as "π is a rational number", whose negation is "π is not a rational number." Indeed, as soon as quantifiers are involved, negation becomes more complex, as is well-known by logicians, from Aristotle, who insisted on the distinction between contradiction (logical negation in modern terms) and contradictory, to modern logicians such as Russell. Another serious difficulty concerns pairs of statements of type "Some *A* is *B*" (3) and "Some *A* is not *B*" (4). As already realized by Aristotle, while sentence (3) is affirmative and sentence (4) negative, there is no opposition between the two. In French, a standard way to build negation is the use of the locution "ne … pas". Applied to a quantified statement of the type "Tous les *A* sont *B*" (All *A* are *B*), it provides "Tous les *A* ne sont pas *B*" (All *A* are not *B*). However, as held by Fuchs (1996), such sentences are sometimes used to express "No *A* is *B*". It is important to notice that the standard interpretation in France of such sentences is not congruent with the interpretation suggested by the underlying logical structure which in a word-to-word formalization would lead to "for all *A*, not *B*", while the standard interpretation leads to "There exists *A* which is not *B*". The work of Ben Kilani in the Tunisian context (Ben Kilani, 2005; Durand-Guerrier & Ben Kilani, 2004) showed the difficulties students experienced in understanding *mathematical* negation (see also Edmonds-Wathen et al., this volume).

While at secondary school, dealing explicitly with the negation of quantified statements in mathematical activity is not so common, it is very common at university where students are likely to face negation of quantified statement in various mathematical practices, in particular in indirect proof such as proof by contradiction or proof by contraposition. In this respect, this topic is a specific challenge for second language learners in undergraduate mathematics.

In an ongoing research study in the francophone educational context in Cameroon, Judith Njomgang Ngansop identified the influence of a native language on the learning of mathematics in French, particularly on questions of logic (Njomgang Ngansop & Durand-Guerrier, 2012). The chosen language is Ewondo, one of the main languages in the area of Yaoundé. The work of Tsoungui (1980) who carried out a comparative study between French and Ewondo grammars served as a resource (see paragraph 2.1). As in French, the construction of negation in Ewondo uses a discontinuous morpheme "à … kig" in singular, and "be … kig" in plural; kig is used mainly to reinforce the negation and can be placed after the verb or at the end of the sentence. Other morphemes can be used according to the context (depending on the form of the statement and on the verb tense).

An interview with a fluent Ewondo speaker completed this description, showing significant differences with French: the form of the verb "to be" depends on the form affirmative/negative of the sentence (*bene/bèsé*); for statements of the type "all *A* are *B*", the morpheme "kig" is used after the predicate. For example: the word-to-word translation in French of its negation in Ewondo "be ndabe bese *bèsé* kig vié" is "les boules toutes ne sont pas rouges" that means "toutes les boules ne sont pas rouges" [all the balls are not red]. In standard Ewondo as in standard French, and unlike in standard Arabic, this means "some balls are red, not all." However, while there is ambiguity in French (such sentences are often used to express "No balls are red"), there is no ambiguity in Ewondo: the translation of "no ball is red" is "ndabe *zing* be se kig vié."

A preliminary experiment was carried out with three students who speak Ewondo fluently (but none of them were literate in Ewondo). These students had answered a questionnaire addressed to 80 first-year students comprising the 3 items involving negation and had follow-up sessions for volunteer students to evaluate responses to the test and to clarify some concepts of logic. After this, the students were interviewed by the researcher, who was assisted by a teacher of French language whose mother tongue is Ewondo and who speaks this language fluently. During the interview, students were given three statements including an everyday-life conditional statement; they were asked to translate each statement in Ewondo, then to give the negation in Ewondo, and translate in French (all questions and answers were oral).

The first results show that for the three students, the negative form (e.g., the syntax) is prominent. They translated "toutes les boules sont rouges" (1) correctly in Ewondo, then gave the correct negation in Ewondo, and finally provided "toutes les boules ne sont pas rouges" (2) (the french standard form) as the negation of (1). The three students negated "certains entiers sont pairs" [some integers are even] (3) by "certains nombres entiers ne sont pas pairs" (4) [some integers are not even], which, as already said, is not the negation.

In addition two university mathematics teachers were interviewed. The results with the university teachers support our hypothesis of an effect of Ewondo on French concerning logical statements involving negation. Following Ben Kilani (2005), we use the underlying word-to-word logical structure of the involved statements as an indicator of this complexity both a priori and a posteriori. These preliminary results show the complexity of negation in French for Ewondo speakers and illustrate the importance of logical issues in the dialectics between syntax and semantics (Durand-Guerrier, 2008).

This research study focused on the logical structure of negative sentences, but there are also lexical aspects to this issue. For example, in another study Kazima (2006) points out that some negatives in Chichewa are reverse of negatives in English: for example, in Chichewa, the word for *likely* is the negative of the word for *unlikely*, such that *likely* is literally interpreted as *not unlikely* (p. 172); the empirical results from a study in Malawi show that this affected students' understanding of probability terminology in English and their use in sentences, in particular when negation was involved (p. 187).

Theses results encourage research concerning the use of negation in mathematics throughout the curriculum in multilingual contexts, taking into account mother languages, languages of instruction, and mathematical and logical discourses, in order to identify opportunity for overcoming these difficulties from secondary level. At tertiary level, in addition, it would be worthwhile to study the impact of any difficulties with negation on indirect reasoning such as reasoning by contradiction or by contraposition on the one hand, and on mathematical conceptualization on the other hand.

5.4.3 An Ongoing Experiment for Teacher Training in Tatarstan

We now turn to discussing preliminary results from a teaching experiment in bilingual education using Russian and Tatar in a preservice mathematics teacher preparation program. The experiment was carried out at the Kazan Federal University in

Russia. Tatar is the students' mother tongue and Russian is used as the second language of instruction. Kazan is the capital of Tatarstan—one of the republics of Russia, 53 % of its population are Tatars.

Mathematical language in Tatar has its own peculiarities, especially at the lexical and grammatical level that is reflected in the special terminology, phraseology, and syntax, in terms of lexical and grammatical structure, and in the genre of mathematical texts. For example, there are general syntactic differences between word order in sentences in Russian and Tatar. The main difference lies in the fact that the verb in a Tatar sentence is always written at the end. Another example more specific to mathematics is the difference in the names of fractions in Russian and Tatar. In Russian the value of the numerator is said before the denominator, while in Tatar it is the opposite. For example, the fraction 1/2 in Russian is read as "odna vtoraia" (1 over 2), and in Tatar, "ikeden ber" (2 under 1).

Our research is based on the proposition that language and thought interact in the process of learning and teaching mathematics. Accordingly, it can be argued that cognitive mathematical activity is connected to speech activity, and that the learning of content occurs at the same time with mastering the means of expression in the second language. Based on the ideas of Vygotsky (1934) that thought is accomplished in the word, we propose to develop thinking in a second language with the help of speech-intellectual tasks, which can be divided into conceptual-lexical and mathematical tasks (Salekhova, 2007; Salekhova & Tuktamyshov, 2011). A conceptual-lexical task in mathematics is a task correlating a new mathematical term in the second language with a known mathematical concept and corresponding mathematical term in a student's native language. As a result, two equivalent mathematical terms (one in the native language and the second in the target language) are fixed in student's memory to denote one mathematical concept (Salekhova, 2007).

Conceptual-lexical tasks, which are given to students in the second language, provide an opportunity to introduce new concepts with the help of semantization, without translation. To achieve this goal, it is necessary to introduce the new mathematical concept through the logical development of familiar concepts, which are related with the unknown one, and through the context of the discussed topic. Mathematical tasks on a defined topic that are solved in the second language (Russian) make it possible to develop mental operations in the second language. It is rather difficult to control this process for the teacher. Nevertheless, the method of question-and-response in the second language during the solution to a problem makes it possible to retrace the steps in the reasoning of the student. Thus, solutions of these problems develop mental and verbal activity of students in a second language. This method of training directs students to the object of thought—elements of the problem and their interaction, rather than on the linguistic form of unfamiliar words in a foreign language.

The process of preparing future teachers of mathematics for bilingual schools is divided into three stages. At each stage a certain level of bilingualism is formed. Levels of bilingualism in the field of mathematical discourse are defined from considerations of speech as a tool for forming thought by means of native and second

Table 5.1 Model of bilingual education at the university

Stage of bilingual education	Level of bilingualism
I (1 year)	*The lowest level*
	Forming and wording of a thought with the help of the native language with its further translation into the second language
II (2 year)	*The intermediate level*
	Forming and wording of a thought by means of the native language and then with the help of the second language
III (3–4 years)	*The highest level*
	Forming and wording of a thought by means of the second language

languages. Solving various types of speech-intellectual problems is used as one of the bilingual teaching methods (Table 5.1).

We now analyze the solution of mathematical problems by students in the process of bilingual mathematics teaching. Ninety-one third-year students from the Faculty of Mathematics, of Kazan Federal University were involved in the pedagogical experiment. This group included students who studied mathematics in Tatar at school, and are studying higher mathematics in the university on a bilingual basis (by means of Tatar and Russian).

We used Bloom's Taxonomy (Bloom, Engelhart, Furst, Hill, & Krathwohl, 1956) of educational objectives in the cognitive domain to evaluate the results of the teaching experiment (knowledge, comprehension, application, analysis, synthesis, and evaluation). Based on observations of how students discussed and solved the ordinary first-order separable differential equations in Russian, we obtained the following results. The results show that 13 of the students did not master the necessary mathematical concepts and terminology in Russian, as they did not complete the first and second steps of solution. The other 35 students understood the problem, but they made some mistakes in the mathematical transformations. Forty-five students solved the ordinary differential equation correctly. Analysis of students' written works showed that about 50 % of the students mastered the 5 levels of Bloom's Taxonomy. Thus, this approach to bilingual instruction in mathematics, based on the solution of speech-intellectual tasks, seems to have had a positive effect. These results are encouraging and point to the need for future research on the cognitive costs and advantages of bilingual mathematics learning when the language of instruction and the language of application differ.

5.5 Conclusion

In this chapter, we have discussed several examples of the challenges that second language learners might face in advanced mathematics courses. In the first part of the chapter, we provided some a priori considerations to describe the specific forms of lexical challenges in university mathematics, described how notation and logical issues are crucial in advanced mathematics, and discussed access to web-based resources that are mainly in dominant languages. The cases we presented in the second part do not provide an exhaustive picture of the phenomena as they appear in practice. However, they do provide concrete examples of the variety of multilingual contexts at the university level: monolingual teaching with a significant number of second language learners, as in France or in Denmark; bilingual or multilingual systems of teaching (as in Malawi or Tatarstan schools, and university in Cameroon).

With the three case studies presented in the third section, we provided some evidence that second language learners face difficulties. The last example presented in Sect. 5.4 describes an attempt to overcome some of those difficulties. The second example, which focuses on negation, shows the necessity of studies across secondary/tertiary levels. Indeed, university level teaching should draw on what is known from research at the secondary level, and conversely, research at university level should enlighten what aspects to take into consideration at secondary school.

Among the open questions that should be considered in further research, a crucial issue concerns the effects of teaching mathematics in a "dominant language." Could such choices lead to unexpected exclusion phenomena? Which paths can be taken to avoid them? In which respect could multiple languages provide resources in the teaching and learning of mathematics for undergraduates? Due to the diversity of contexts that we described in Sect. 5.3, it is likely that possible answers will be strongly dependent on the linguistic context.

To sum up, contrary to popular belief, the study of advanced mathematics is indeed sensitive to language matters, and language diversity can impact the learning of a large number of mathematics students across the world. We hope that the mathematics education community involved in advanced mathematics, as well teachers and researchers, will become aware of this international issue in learning and teaching mathematics.

References

Barton, B., Chan, R., King, C., Neville-Barton, P., & Sneddon, J. (2005). EAL undergraduates learning mathematics. *International Journal of Mathematical Education in Science and Technology, 36*(7), 721–729.

Barton, B., & Nevillle-Barton, P. (2003). Language issues in undergraduate mathematics: A report of two studies. *New Zealand Journal of Mathematics, 32*(Suppl), 19–28.

Barton, B., & Nevillle-Barton, P. (2004). *Undergraduate mathematics learning in English by speakers of other languages*. Paper presented at the 10th International Congress on Mathematics Education, Copenhagen, Denmark.

Bebbouchi, R. (2011). Le passage des mathématiques en Arabe aux mathématiques en français en Algérie : difficultés et avantages. In M. Setati, T. Nkambule, & L. Goosen (Eds.), *Proceedings of the ICMI Study 21 Conference: Mathematics Education and Language Diversity* (pp. 528–533). São Paulo, Brazil: ICMI Study 21.

Ben Kilani, I. (2005). *Les effets didactiques des différences de fonctionnement de la négation dans la langue arabe, la langue française et le langage mathématique*. Thèse en co-tutelle de l'université de Tunis et de l'université Lyon 1, Tunisie et France.

Bloom, B. S., Engelhart, M. D., Furst, E. J., Hill, W. H., & Krathwohl, D. (1956). *Taxonomy of educational objectives: The cognitive domain*. New York: McKay.

Cajori, F. (1928). *A history of mathematical notations*. Chicago: Open Court.

Centre for language studies (2010). Language Mapping Survey for Malawi. University of Malawi. Report submitted to Open Society. Initiative for Southern Africa.

Durand-Guerrier, V. (2003). Which notion of implication is the right one? From logical considerations to a didactic perspective. *Educational Studies in Mathematics, 53*, 5–34.

Durand-Guerrier, V. (2008). Truth versus validity in mathematical proof. *ZDM Mathematics Education, 40*, 373–384.

Durand-Guerrier, V., & Ben Kilani, I. (2004). Négation grammaticale versus négation logique dans l'apprentissage des mathématiques. Exemple dans l'enseignement secondaire Tunisien. *Cahiers du Français Contemporain, 9*, 29–55.

Durand-Guerrier, V., Boero, P., Douek, N., Epp, S., & Tanguay, D. (2012). Examining the role of logic in teaching proof. In G. Hanna & M. De Villiers (Eds.), *Proof and proving in mathematics education* (pp. 369–389). New York: Springer.

Epp, S. (2011). Variables in mathematics education. In P. Blackburn, H. van Ditmarsch, M. Manzano, & F. Soler-Toscano (Eds.), *Tools for teaching logic* (pp. 54–61). New York: Springer.

Fuchs, E. (1996). *Les ambiguïtés du français*. Paris, France: Orphrys.

Kazima, M. (2006). Malawian students' meanings for probability vocabulary. *Educational Studies in Mathematics, 64*(2), 169–189.

Libbrecht, P., Droujkova, M., & Melis, E. (2011). Notations across cultures for teaching. In M. Setati, T. Nkambule, & L. Goosen (Eds.), *Proceedings of the ICMI Study 21 Conference: Mathematics Education and Language Diversity* (pp. 160–168). São Paulo, Brazil: ICMI Study 21.

Mathé, A. C. (2012). Jeux et enjeux de langage dans la construction de références partagées en classe de géométrie. *Recherches en Didactique des Mathématiques, 32*(2), 195–228.

Njomgang Ngansop, J., & Durand-Guerrier, V. (2012). Negation of mathematical statements in French in multilingual contexts: An example in Cameroon. In M. Setati, T. Nkambule, & L. Goosen (Eds.), *Proceedings of the ICMI Study 21 Conference: Mathematics Education and Language Diversity* (pp. 268–275). São Paulo, Brazil: ICMI Study 21.

Salekhova, L. (2007). *Didactic model of bilingual teaching mathematics in high school.* Unpublished doctoral dissertation, Kazan Federal University, Russia.

Salekhova, L., & Tuktamyshov, N. (2011). Bilingual mathematics teaching in conditions of higher educational establishment. In M. Setati, T. Nkambule, & L. Goosen (Eds.), *Proceedings of the ICMI Study 21 Conference: Mathematics Education and Language Diversity* (pp. 342–347). São Paulo, Brazil: ICMI Study 21.

Salimov, R. B., & Tuktamyshov, N. K. (2000). *Mathematics*. Kazan, Russia: Izdatelstvo "Magarif".

Selden, A., & Selden, J. (1995). Unpacking the logic of mathematical statements. *Educational Studies in Mathematics, 29*(2), 123–151.

Tsoungui, F. (1980) *Le français écrit en classe de 6ème à Yaoundé. Recherches des interférences de l'Ewondo dans le français et proposition pédagogiques*. Thèse de 3ème cycle, Université de la Sorbonne Nouvelle, France.

Vygotsky, L. S. (1934). *Myshlenie i rech. Psikhologicheskie issledovanija* [Thinking and speech. Psychological investigations]. Moscow-Leningrad, Russia: Gosudarstvennoe Sotsialno-Ekonomicheskoe Izdatel'stvo.

Zevenbergen, R. (2001). Changing contexts in tertiary mathematics: Implications for diversity and equity. In D. Holton (Ed.), *The teaching and learning of mathematics at university level: An ICMI study*. Dordrecht, The Netherlands: Kluwer.

Chapter 6
Language Diversity in Mathematics Teacher Education: Challenges Across Three Countries

Anthony A. Essien, Nancy Chitera, and Núria Planas

6.1 Different Countries, Similar Challenges

Do mathematics student teachers and mathematics teacher educators across countries face similar challenges when integrating language, didactical and pedagogical issues in the construction of professional knowledge? To unravel this question, we look at mathematics teacher educators' awareness of the context of their practice, the practices they use, and how they position themselves with regard to the linguistically diverse classrooms of the mathematics student teachers that they teach. Little work has been undertaken on teaching and learning in linguistically diverse mathematics teacher education settings, though there is evidence that the two dimensions are connected (Chitera, 2011; Civil, 2012). Most of the work accomplished has been carried out using classroom data and interviews with participants experiencing the same language policies and involved in programs from the same institution. As far as we know, no systematic research has examined data across countries with distinctive linguistically defined groups of mathematics student teachers and mathematics teacher educators in teacher education institutions. In this chapter we explore mathematics teacher education in linguistically diverse classrooms across three countries, South Africa, Malawi, and Catalonia-Spain, as a

A.A. Essien (✉)
University of the Witwatersrand, Johannesburg, South Africa
e-mail: Anthony.essien@wits.ac.za

N. Chitera
University of Malawi, Blantyre, Malawi
e-mail: nchitera@yahoo.com

N. Planas
Universitat Autònoma de Barcelona, Barcelona, Catalonia-Spain

University of South Africa, Pretoria, South Africa
e-mail: Nuria.Planas@uab.cat

R. Barwell et al. (eds.), *Mathematics Education and Language Diversity*, New ICMI Study Series, https://doi.org/10.1007/978-3-319-14511-2_6

way of opening a new line of important research. Despite the existence of similar challenges, the authors' experiences of working in teacher education classrooms in these countries show that they are different in many ways.

The following two questions that are the foci of this chapter:

- What do we know about mathematics teacher educators' awareness of the multilingual context of their practice?
- What practices are mathematics student teachers introduced to in the preservice classroom?

Unlike teaching mathematics to mathematics major students, teaching mathematics to mathematics student teachers is more multifarious because of the different facets involved in teacher education. In addition to being knowledgeable about the content they will teach, mathematics teacher educators also need to know how to teach it (content) in context, and have knowledge about instructional practices. A number of authors have argued for the integration of mathematics and language development in multilingual classrooms (Adler, 1995; Barwell, Barton, & Setati, 2007; Smit & van Eerde, 2011). These authors have argued against the avoidance of linguistic aspects in the teaching and learning of mathematics and for attention to be paid to the language needs of students in multilingual classrooms. This is why as part of the layers involved in teaching and learning mathematics in linguistically diverse teacher education classrooms, it is critical to examine how each of these classrooms pays attention to language issues. Our approach, therefore, allows provision for the analysis of evidence present in different classrooms (across the three countries) in support of the multiple and interacting layers in teacher education of: *becoming teachers of mathematics, becoming teachers of mathematics in multilingual classrooms, becoming learners of mathematics content, becoming learners of mathematical practices, and becoming proficient language users* for the purpose of teaching and learning mathematics (see Essien, 2014, for an elaboration of these multiple facets). This notion of multiple layers involved in teacher education provides the lens for examining practices into which mathematics student teachers are enculturated. The contexts of schooling regarding language and learning provide the background for the discussions on linguistically diverse teacher education classrooms across the three countries in focus. In particular, it sets the scene for examining the practices used in teacher education classrooms in order to delineate what challenges are similar or different and what lessons are to be learned across the three contexts.

6.2 The Three Contexts: South Africa, Malawi, and Catalonia-Spain

The choice of the contexts in this chapter is the result of, on the one hand, common interests of the authors on issues of mathematics teacher education and multilingualism and, on the other, the intention to show the presence of similar challenges in spite of different geographical and political conditions, particularly between Catalonia-Spain and the two African countries. The teacher education institutions in the three contexts are universities where both content and pedagogy are not separately taught by different departments/faculties, but within the schools of education where prospective teachers are educated. In what follows, we engage with the multilingual contexts around teacher education in the three countries.

South Africa presents a complex picture of multilingualism, due not only to its political history of Apartheid, but also to its distinct nature of multilingualism. Of the 11 official languages, 9 are indigenous African languages which can be grouped into 2 major families based on linguistic characteristics: the Sotho and the Nguni. The Sotho languages are comprised of South Sesotho, Sepedi, Setswana, while the Nguni languages are isiZulu, isiXhosa, siSwati, and isiNdebele. There is mutual intelligibility between the languages in these groups (this is not the case with the remaining two African languages, Tshivenda and Tsonga), so that it becomes relatively easy to learn other indigenous languages. Hence, it is not uncommon for students to discuss their mathematics task in their home languages during group work even though other members of the group may have different home languages. This picture is different from most African countries where the indigenous languages share little, if any vocabulary so that if the teacher used her/his language to teach, many students would not understand (Essien, 2010a).

To qualify as a teacher in South Africa, the *Minimum Requirements for Teacher Education Qualification* (DHET, 2011) stipulates that all teachers should

> be proficient in the use of *at least one* official language as a language of learning and teaching (LoLT), and partially proficient [that is, have conversational proficiency] in *at least one* other official language (including South African Sign Language). (pp. 15–16, italics in original)

But even though the South African Constitution and the Language-in-Education Policy (LiEP) make provision for learners to learn in any official language of their choice, research has shown that due to economic, political, and ideological factors, most learners prefer to learn mathematics in English, a language which for most is not their first or home language (Setati, 2008).

In Malawi, like in South Africa, English plays a significant role both in society and in the teacher education context. Malawi is divided into three regions (the Northern, Southern, and Central), each with its own main language. English is the official language and Chichewa, which is spoken by about half of the population (Baldauf & Kaplan, 2004), is the national language. There are 16 other indigenous languages. Hence, just like in South Africa, Malawi is a highly multilingual society where what is seen as an international language is more valued than the local languages in teacher education. In 1996, the Malawi LiEP required learners in the first 4 years of schooling to be taught in their home language (Secretary for Education's Letter 1996, No. IN/2/14). The LiEP for lower primary education supports multilingualism and allows primary teachers to be flexible and use home language(s) in their teaching. However, English remains the LoLT for upper primary classes and in secondary and tertiary education.

The present primary teacher education program, set up in 2005, runs for 2 years. The first year is residential training when mathematics student teachers stay in college and attend lectures and complete projects and assignments. The second year is the school-based education year which is carried out in the form of teaching practice in various primary schools. Mathematics student teachers are expected to become conversant with content related to pedagogical knowledge such as the technical skills of lesson planning, and teaching methods, as well as the specific learning areas that are offered in primary education: Numeracy and Mathematics, and Literacy and Languages (Chichewa and English), among others. Numeracy and Mathematics consists of subject content and methodology. It is organized in such a way that mathematics teacher educators first teach a particular topic followed by the pedagogy related to it.

The Catalonian context plays out differently from the South African and Malawian contexts in different ways, notably given the fact that a local language (Catalan) is the official LoLT. In Catalonia, an autonomous region in north-eastern Spain, the choice of Catalan as the LoLT was made after 1983 as a way to integrate the population that had arrived from other parts of Spain in successive immigration waves. At present, there was a sort of de facto bilingualism between the two major languages in the region (Catalan and Spanish), though in reality the society is now becoming more and more multilingual with the arrival of people from North Africa, Central Asia, and Latin America. So far, successive LiEPs have alternatively served the interests of the Spanish and the Catalan dominant groups, while other language groups, mostly represented by immigrant families, have not been considered to the same extent (Planas & Civil, 2009). This results in the invisibility of certain languages in the teacher education system, particularly in the preparation courses for mathematics student teachers to teach mathematics in primary and secondary schools.

In response to the above situation, at some institutions, the training for preservice teachers has recently included innovative courses on teaching and learning mathematics in multilingual classrooms. Despite the current LiEP and the Catalan Law of Universities ("Catalan is defined as the natural language of the universities in Catalonia", Official Diary of the Government of Catalonia, 2003, p. 3327), it has been possible to design and develop pilot courses in which mathematics student teachers' languages are planned to be flexibly used. A learning environment has been fostered to challenge mathematics student teachers to think about issues of language diversity and mathematics education, as well as to become designers of activities with an integrated focus on mathematics and language. While reflecting on the implementation of such courses (they are very few, the approach is not widespread, and the curriculum is experimental), some changes have been incorporated to facilitate greater insight into linguistically responsive practices within the groups of mathematics student teachers. To develop understandings of language as a crucial resource in the learning and teaching of mathematics, it is thought that mathematics student teachers need intentional and explicit learning environments.

6.3 What Do We Know About Mathematics Teacher Educators' Awareness of the Multilingual Context of their Practice?

We now address ways in which mathematics teacher educators think of language diversity in the teaching and learning of mathematics. Three teacher education scenarios, with slightly different pedagogical orientations in the construction of professional knowledge, are used to gain an understanding into mathematics teacher educators' awareness of the complexity of teaching mathematics to linguistically diverse mathematics student teachers who will, in turn, teach mathematics to linguistically diverse students. Our broad interpretation of awareness includes (1) the recognition of multilingualism as (potential) resource, rather than hindrance; (2) the attention to the linguistic structures in mathematical and everyday languages as compared to the structures in the home language(s) of the learners; (3) the attention to how learners use language as they engage with activities like making conjectures, examining constraints, making inferences, abstracting, inventing, explaining, justifying, and challenging; and (4) the use of appropriate mathematical language to respond to learners. This idea of awareness plays a key role in the practices that are privileged. In the discussion that follows, we highlight two common facts in the three contexts: first, the mathematics teacher educators are aware of the linguistically diverse contexts in which they teach; and second, the teacher education institutions rarely attend to the complexity of teaching mathematics to linguistically diverse mathematics student teachers in a structured way in their programs.

6.3.1 Mathematics Teacher Educators' Awareness in South Africa

In a study involving 4 teacher education institutions in South Africa and 12 teacher educators (Essien, 2010b), 3 categories emerged regarding mathematics teacher educators' awareness of the context of their practice. The first category consisted of mathematics teacher educators for whom awareness is about experience and gained through teaching second-language English learners and reflection on how mathematics student teachers use language during their teaching. The second category consisted of mathematics teacher educators for whom awareness came through as being about disciplinary knowledge and knowledge of language issues in the teaching and learning of mathematics. For the third category, awareness came through as something gained through research and experience of teaching in linguistically diverse classrooms. What is important to note is that the mathematics teacher educators were all aware of the challenges embedded in their context of practice even though they talk differently about awareness.

Even though the mathematics teacher educators were mostly aware of the general context of their practice, that they were teaching mathematics to preservice teachers who will, at the end of their training, teach mathematics to students, they admitted that at the institutional level, they did not receive enough support by way of courses aimed at introducing preservice teachers into the dynamics of teaching and learning in linguistically diverse classrooms. Rachel talked about how she prepared mathematics student teachers to teach mathematics in multilingual contexts:

Rachel: We don't prepare them in any structured way, we prepare them by taking up suggestions that may arise during practical teaching and talk about those during the methods class. And we take it up by making them more explicitly aware during their methods class to be aware that people don't necessarily understand their English, but nothing structured.

Rachel echoes the sentiment of other mathematics teacher educators who also indicated that courses were not aimed at creating awareness of what it entails to teach mathematics in multilingual contexts, despite the fact that most preservice teachers will teach mathematics to learners who are not English dominant. As argued in Essien (2010b), this concern is not new to preservice education in South Africa; neither is it new that in preservice education the importance of language in the teaching of mathematics is not well-attended to. In the early 1990s, a report by the National Education Policy Investigation (NEPI, 1993, p. 181) observed that the role of language in (mathematics) knowledge acquisition is not a focus area in primary and secondary teacher education. The conclusion was that, notwithstanding whether the language of instruction was the home language or the second language, these gaps in teacher education affect the professional ability to use the language of instruction in the best interest of the learners. Two decades after this finding, sadly this recommendation remains valid for teacher education institutions in South Africa.

6.3.2 Mathematics Teacher Educators' Awareness in Malawi

As with the contexts of South Africa described above and Catalonia (see next section), responses to interview questions and classroom observations in research conducted by Chitera (2009) involving four mathematics teacher educators who had different home languages (but could all speak Chichewa and English) indicated that mathematics teacher educators were aware of the multilingual context of their practice. Their responses indicated that even though they were aware of the multilingualism in their classrooms, they regarded multilingualism as a problem and the use of languages other than the LoLT in the preservice classroom as problematic. The mathematics teacher educators described their classrooms as multilingual, not because participants spoke different home languages, but because the mathematics student teachers were allowed to use their home languages when they were unable to express themselves adequately in English. For example, Otani explained that their classes were multilingual because there were mathematics student teachers who failed to communicate in English and were then allowed to use Chichewa. Hence their classes were deemed to be multilingual, not because student teachers spoke different home languages, but because they were allowed to speak Chichewa if they failed to speak English:

Otani: Yeah, that one is multilingual because there are some who cannot express themselves in English so we accept Chichewa.

Otani explained that a way of helping his students was to allow the use of home language(s). The overriding notion was that mathematics student teachers who were not able to express themselves in English could switch to their home languages and hence compensate for the deficiency. Skiba (2007) suggests that in circumstances where code-switching is used due to a presumed inability of the person to competently express, it serves for continuity in speech instead of presenting interference in language. In this respect, the use of home languages in a preservice classroom stands to be a supporting element in communication and interaction. Hence, the use of home languages occurs when mathematics student teachers are given the opportunity to use them by their mathematics teacher educators. It is not a spontaneous practice but rather a controlled practice. In Malawian classrooms, the use of local languages cannot be avoided, and therefore, student teachers need to graduate from the teacher education colleges better prepared to function productively when they begin to teach. Even though the mathematics teacher educators were aware of the benefits of using home languages, the exploration of how to use them in a mathematics classroom is not practised freely in the public domain, because every student teacher is supposed to speak English. Moreover, there are no structured courses aimed at inducting mathematics student teachers into the intricacies of teaching and learning in linguistically diverse classrooms.

6.3.3 Mathematics Teacher Educators' Awareness in Catalonia

In Catalonia, in 2010 and 2011 interviews were undertaken with five Catalan and Spanish bilingual mathematics teacher educators working in an institution in which a few lessons on multilingualism and mathematics education had been planned as part of a pilot and singular design research. The mathematics teacher educators were free to implement these lessons or not as they choose, and three of them declined. In conversation with those who declined, several reasons were provided like "working priorities" and "more interest in new technologies." All the mathematics teacher educators were also asked to talk about what a multilingual mathematics school classroom looks like, and what are some of its needs. In their responses, though they recognized multilingualism as a potential resource in the teaching and learning of mathematics, awareness appeared as rather weak. Most mathematics teacher educators referred to the need to create classroom conditions in which non-dominant Catalan mathematics student teachers did not experience their language "difference" as an obstacle, but none of them referred to the case of mathematics.

Unlike in the South African context, the mathematics teacher educators did not express the need to develop professional knowledge to use multiple languages in the teaching and learning of mathematics. For them, being bilingual was significant, but the experience of bilingualism was not problematized at a professional level. Their recitation of non-problematic bilingual stories framed their conceptualization of language diversity as pedagogically neutral. This is consistent with what was found in the work with a group of in-service mathematics teachers who did not initially see language issues as crucial for their teaching in multilingual classrooms. The professional development initiative, however, succeeded in creating optimal conditions for those teachers to work on awareness. After one academic year they became more attentive to how their students switch their languages depending on the complexity of the mathematics (Planas & Civil, 2009).

Teacher education programs in Catalonia have a field orientation in that mathematics student teachers spend long periods in schools to observe teaching practices. This organization opens a space in terms of visiting mathematics classrooms with immigrant learners who are in the early process of learning Catalan, and also with in-service teachers who are at different levels of awareness. When such classrooms were mentioned in the interviews, the impact of language issues was minimized by the mathematics teacher educators, even by the two mathematics teacher educators who had volunteered to teach the innovative lessons:

Judit: Language diversity is more present now than ever in the past. But it's not so much about math education. It's much more about creating the sensitivity to language and cultural differences in the class with the student teachers.

Magda: Many mathematics teachers have never seriously reflected on issues of multilingualism, nor have they ever had experience with language diverse backgrounds in their classrooms. But they still are fantastic teachers.

Judit and Magda shared awareness of the multilingual reality of many mathematics classrooms in the country and, like the mathematics teacher educators in Malawi, had the view that multilingualism was problematic when the students did not own the LoLT. Nevertheless, and differently to what happened with the mathematics teacher educators in South Africa and Malawi, they did not address the multilingualism and language practices that were present in the preservice classrooms, which were their more direct source of data. The mathematics student teachers coming from other parts of Spain tend to switch from Spanish (informally allowed) in small groups to Catalan (institutionally mandatory) in whole group. But practices of code-switching were not commented on by the mathematics teacher educators in the interviews, nor were they thought of as useful to model language practices for multilingual classrooms in schools.

6.4 What Practices Do Mathematics Student Teachers Get Exposed to in the Preservice Classroom?

To obtain an accurate picture of mathematics teacher education in our contexts, there is a need to examine the preservice classroom practices. We explore classroom practices that point to the (re)construction of meanings of language diversity and mathematics education with mathematics student teachers. More particularly, we report on practices that are oriented toward the multiple layers involved in teacher education. To reiterate, in a linguistically diverse teacher education classroom, mathematics student teachers (who will most likely teach in linguistically diverse classrooms) are at once becoming teachers of mathematics, teachers of mathematics in multilingual classrooms, learners of mathematics content, learners of mathematical practices, and proficient language users.

6.4.1 Teacher Education Practices in South Africa

Research involving observations of teacher education classrooms in two teacher education institutions (Essien, 2013) revealed that one of the institutions predominantly enculturated mathematics student teachers into becoming learners of mathematical content. Hence, the overriding facet of teacher education that was privileged was the acquisition of mathematical knowledge. The mathematics teacher educators perceived their responsibility to be that of fulfilling this role in the class and the mathematics student teachers saw themselves as recipients of this knowledge. This was clear in how the mathematics teacher educators used practices such as defining, explaining, and exemplifying to develop the mathematical knowledge of the mathematics student teachers through an authoritative communicative approach (Mortimer & Scott, 2003). On the other hand, at the second institution, not only was the acquisition of mathematical knowledge an important enterprise, but also the enculturation of the mathematics student teachers into becoming teachers of mathematics. Induction into this layer of teacher education was clear in how the mathematics teacher educators insisted that mathematics student teachers explained the thinking (as teachers would) behind solutions proffered by fellow mathematics student teachers and also required of them to explain terms as they would to learners who were encountering the terms for the first time. Through practices such as predicting, conjecturing, justifying, and critiquing conjectures, the mathematics teacher educators were conscious of the fact of not teaching mathematics solely for the purpose of content knowledge, but teaching would-be teachers, as is evident in the excerpt below:

Hendricks: Now before you do that and I'll repeat this, we're sitting here in a situation that is very different from what statisticians do. They go into the computer, they click trend line, it gives them the trend line they want ... If I ask you to not start with the trend line but background that knowledge of yours, I want you to do it, get the valuable skill of a teacher, background the knowledge that you have and pretend that you know only what the learners in your class know [...] How will you predict what the fuel consumption is for 2,000 kg and for 2,500 kg?

Throughout the lesson, Hendricks kept indicating to mathematics student teachers what statisticians do, and what they do not do, and more importantly, what they as mathematics student teachers need to become enculturated into. First, they must be able to think like learners who have never been introduced to the concept of a trend line and think of how they would be able to interpolate from a given set of data. Second, they must be able to draw the trend line accurately. A finding that comes to the fore from the analysis of the various practices in the preservice classrooms of the South African context is that the practices-in-use in their classrooms were mostly those that inducted mathematics student teachers into becoming learners of mathematics content. There were very limited practices aimed at inducting mathematics student teachers into becoming teachers of mathematics and even more limited ones that inducted them into becoming teachers of mathematics in multilingual contexts. Hence it appears that mathematics student teachers are not being prepared adequately to understand and deal with the challenges involved in teaching mathematics in linguistically diverse settings. This has huge implications for teacher education in South Africa where most classes are multilingual and where most learners, despite their low English language proficiency, choose to do mathematics in English (Setati, 2008).

6.4.2 Teacher Education Practices in Malawi

In the Malawian context, the classroom practices show the major interest is to prepare mathematics student teachers to become teachers of mathematics rather than teachers of mathematics in multilingual contexts. Most discussions are on how to teach procedures for arriving at correct answers. Otani was one of the mathematics teacher educators who generally tried to instill in mathematics student teachers how to become teachers of mathematics by rephrasing their presentations. In the extract below, Otani asks one of the mathematics student teachers in his class to explain how to teach the addition of 0.7 and 0.2 using a number line. We begin with what the mathematics student teacher explained and then show how Otani rephrased the student's presentation.

Mark: Then you write small lines, can I … (writing on the board) then first of all we are going to tell our learner to stand at this point where we have a zero and we are going to tell that one to at least move seven steps to this side of the line that means start from here [zero] move one, two, three, four, five, six, seven it means at this place going to start here and stop here, so this means that this is still a fraction because we haven't reached the whole number and thereafter we are also going to ask that learner to move two more steps from this one, so it means start from here to here and here and then we are going to ask him to how many has he jumped.

The student focused on how he was going to teach and the steps that he would follow if he was teaching in a primary classroom. The first thing to be noted is the way in which he used grammar and pronouns. His language illustrates his recognition of how to involve the learners in a mathematics classroom through the emphasis on "telling" and "asking" in relation to solving the problem. The learner in this instance is positioned as a passive recipient answering questions, whereas the teacher is positioned as a professional expert who has power over the learners. The student's language also portrays a controlling (or possessing) belief that it is expected teachers should have. This is the responsibility that teachers assume in a class: controlling the turns that students are given in to speak, and at what time. After the mathematics student teacher did most of the talking alone, Otani responded to the student's presentation as follows:

Otani: So instead of drawing a number line from here up to two here, you can just draw your number line unnhu then here you indicate that it is zero and here is what!

Chorus: One.

Otani: One, then between zero and one you mark how many points.

Chorus: Some say nine while some say ten.

Otani: Yah, nine points, so it's one, two, three, four, five, six, seven, eight, nine then this one simply means that each point is a, a fraction why because we are, we are in the range between zero and one which is a whole number so you let your pupil to stand on zero and ask him or her to move how many steps!

Chorus: Seven.

Otani: Seven steps, one, two, three up to …

Chorus: Seven.

Otani: Then ask him or her to add how many more steps?

Chorus: Two.

Otani: Then you ask him to say the number indicated on where he/she is standing. Seven, zero point, but first of all you should discuss this one (7 plus 2) eti, these are the things that they have done already, not so?

The steps followed by Otani were the same as those in the student's presentation. The mathematics teacher educator began by saying, "so instead of drawing a number line from here up to two here." This statement serves to correct the student's first step of drawing the number line from zero to two. He presented it as an alternative way to what the students did. Then he said to mark the nine equal points between the points zero and one. He pointed out that each mark represents a fraction because they are in the range between zero and one. Although not very explicit, Otani tried to give reasons for each of his teaching steps. Thus his language combined procedural and conceptual discourse, which was lacking in the student's language. Also just as the mathematics student teacher had done, Otani's language indicated how the students could involve their learners in solving this problem, and he explained how this could be done. This extract also reflects the students' involvement in the solving of the problem. Otani noted that "you let your pupil to stand on zero and ask him or her to move how many steps!" and then five lines further "then ask him or her to add how many more steps." He indicated to the mathematics student teachers that teaching mathematics extends beyond simply explaining some steps to the learners; he highlights what steps were necessary, and which were needed at each stage. Thus the emphasis is both on becoming learners of mathematics content and on becoming teachers of mathematics, while how to become teachers of mathematics in multilingual classrooms is left entirely to the mathematics student teachers to figure out (Chitera, 2012).

6.4.3 Teacher Education Practices in Catalonia

In the Catalonian context, it has been found that when mathematics teacher educators draw mathematics student teachers' attention to the fact that they are prepared to become teachers of mathematics in primary classrooms, most practices are oriented towards becoming learners of mathematics content. Discussions in preservice classrooms reflect some of the students' difficulties with the understanding of notions involving fractions, ratios, and proportions. When some of the Spanish-dominant mathematics student teachers try to use only Catalan in the discussion of these topics, they introduce questions about how to translate technical vocabulary from Spanish. In the extract below, Silvia, a mathematics student teacher, expresses a doubt about which is the appropriate Catalan translation and spelling for the Spanish words "extremos" and "medios" (the "extremes" and the "means" when equalling two ratios and getting a proportion, whose Catalan translation is "extrems" and "mitjos"). The intervention by Silvia appears as a reaction to an ongoing discussion on the distinction between ratio and proportion. The situation is that of having a second language being brought into the mathematical discussion:

Silvia: If you get *a* and *b*, you can only have a ratio, not a proportion, right? You need two ratios, and then the first ratio equals the second … I remember saying "extremos" and "medios" … But … how is it in Catalan? Extre … extrems? Do you spell it like that? And "medios" … Me … Medis … It sounds weird … What's the word?

Silvia is not sure about how to say the two words in Catalan. The mathematics teacher educator gives the literal translation and uses the appropriate Catalan words when referring to the inversion of the extremes and means in a proportion. In another lesson from the same preservice classroom, the interest on becoming proficient Catalan users in mathematical discussions also appears when a mathematics student teacher, Mónica, starts presenting her solution to a problem to the whole class. When she realizes that she is speaking Spanish, she switches to Catalan, repeats the same sentence that she has already said in Spanish (a language that all participants in the classroom are expected to know), and apologizes for not having used the LoLT:

Mónica: [Spanish] I have used a tree diagram to look for all possible combinations … [Catalan] Oh, sorry! [She repeats the same sentence now in Catalan, and keeps using Catalan] I have used a tree diagram to look for all possible combinations … and then I have made a multiplication.

Various extracts in this lesson, and others, point to the class talk as a place for revisiting the mathematics student teachers' mathematical knowledge and eventually dissuading the students from orienting the discussion toward language issues. In the extract below, Mónica and the mathematics teacher educator are talking about the mathematical problem that has been posed in the classroom. Although Mónica is Spanish-dominant and the mathematics teacher educator is Catalan and Spanish bilingual, the conversation takes place entirely in the LoLT (Catalan). When Mónica has completed her explanation (students need to find, in small groups, how many possible menus can be obtained by combining three appetizers, four main plates, and three desserts), she raises a question in relation to 'the' language:

Mónica: We have thought only a few ways [of resolution] because ... we've been doing at a low speed ... we've realized that we were writing everything down in Spanish and we have spent time translating it to Catalan.

When Mónica raises a language issue, the mathematics teacher educator does not take that issue up, but carries on with the conversation about the different ways to solve the problem. She does not seem to realize the importance of the fact that the mathematics student teachers in that group have spent time translating their explanations from Spanish into Catalan, and that this time, as reported by Mónica, has been detrimental to the development of alternative solutions to the problem. The mathematics teacher educator is trying to push Mónica to share the second solution with the other mathematics student teachers in the class, and this is actually an expected practice in a problem-solving environment. What is surprising is that the mathematics teacher educator does not comment on the implications of having put the focus on the language. This is one of the several examples that points to the emphasis on becoming learners of mathematics, teachers of mathematics, and becoming proficient Catalan language users in the preservice classroom practices.

6.5 Joint Discussion

The discussion of this chapter focuses on two areas: the awareness and its implication as well as the practices and their implications in the training of mathematics student teachers in multilingual contexts. Because mathematics is abstract science, it is carried in semiotic form and therefore accessed through some form of language. Thus, the awareness of and attention to language use is critical in any classroom as far as classroom discourse is concerned. This awareness becomes even more critical in linguistically diverse contexts where students (and sometimes teachers) learn/teach mathematics in a language other than their first or home language. Moreover, it must be noted that one cannot assume a causal relationship between awareness and practice. Nevertheless, a practice such as introducing the students' languages requires gaining awareness of them as useful, not merely in terms of achieving rights but for epistemological purposes in relation to the learning of mathematics (Planas & Setati, 2014).

6.5.1 Awareness and Implications

The research summarized in the previous section, shows that there is a certain awareness of multilingualism and at least there is something being done toward the move to multilingualism in mathematics teacher education. There is awareness by the mathematics teacher educators in all three countries that the students' (home) languages can be used as a resource for teaching and learning mathematics. However, even though the acknowledgement that languages other than the LoLT can serve as a resource, our studies show that South Africa, Malawi, and Catalonia do not have well-structured programs and courses for introducing mathematics student teachers into the complexity of teaching mathematics in linguistically diverse contexts. Ngu (2004) noted that mathematics teacher educators, especially in African countries, are being prepared to teach in languages that are not their own. Under such circumstances, teachers would be expected to find it challenging to cope with teaching in linguistically diverse classrooms, in which many learners are also coping with the use of a language that is not their own. In brief, what we have is that teacher education programs are inadequate in the training of mathematics student teachers for creating awareness regarding the needs of multilingual contexts and for developing their professional activity in them. Hence in-service teachers are not adequately prepared for facilitating the learning of mathematics of their students.

6.5.2 Practices and Implications

Research in the three countries under discussion reveals that mathematics teacher educators draw mathematics student teachers' attention to becoming mathematics teachers and to the acquisition of mathematical knowledge. The emphasis is mostly given to understanding of mathematical notions and constructing an identity as a mathematics teacher. This is crucial and critical for mathematics student teachers' development as mathematics teachers. However, we argue that preparation of mathematics student teachers needs to go beyond these focal points to include the challenges of teaching mathematics in linguistically diverse classrooms, the major challenge being to consider the students' languages as pedagogic resources. The studies in the three countries have shown that the mathematics teacher educators' practices do not focus on the practices that would induct mathematics student teachers into teaching mathematics in multilingual contexts. Further, these studies have shown that being aware of the multilingual context does not necessary imply the adoption of multilingual practices in the classrooms. For example, all the mathematics teacher educators are aware of the multilingual nature of their classrooms, but their practices do not include systematic responsive practices such as harnessing the diverse students' languages. Also, we feel that the absence of the well-structured programs and courses for introducing mathematics student teachers into the complexity of teaching in linguistically diverse classrooms has added to the lack of multilingual practices in the mathematics classrooms. Well-structured programs and courses might act as a bridge that would blend the awareness of multilingual context and the practices of the mathematics teacher educators.

6.6 Concluding Remarks

We have focused on the awareness and practices of mathematics teacher educators who are teaching in linguistically diverse classrooms across three countries. We have shown that the nature of multilingualism in these countries is distinct, yet the challenges faced by mathematics teacher educators are similar: those of preparing preservice teachers to deal with the complexity of teaching in linguistically diverse mathematics classrooms. In multilingual classrooms with learners whose home language is not the LoLT and who are not yet proficient in the language of instruction, teachers are faced with the triple challenge of striking a balance between attention to mathematics, to the LoLT and to mathematical language (Barwell, 2009; Planas, 2012). It is not a given that preservice teachers would acquire the knowledge required in dealing with this challenge by the mere experience of being in a multilingual environment, but rather through some form of teaching and enculturation. The reality from data across South Africa, Malawi and Catalonia-Spain is that, even though the mathematics teacher educators have an awareness of what it entails to teach multilingual mathematics student teachers who themselves will teach in multilingual contexts, this awareness is not reflected unequivocally in their practice. Their classroom practices are mostly those that induct mathematics student teachers into becoming learners of mathematics content and becoming teachers of mathematics. Hence we conclude that mathematics teacher educators need to have knowledge pertinent to the teaching of mathematics in multilingual contexts, as well as to the role of the students' languages in the mathematics learning. One way of equipping mathematics student teachers with this knowledge is creating awareness of the intertwinement between language and mathematics. These should never be separate issues taught in isolation. Creating awareness of the multilingual context of teaching and what it entails should be a thread that runs through the entire (mathematics) teacher education curriculum. But beyond creating such awareness, the teacher educator needs to actively draw on mathematics student teachers' multilingualism by tapping into and exploiting the different languages available in the multilingual classroom.

This chapter has been diagnostic and explorative in delineating the issues that are present in mathematics teacher education classrooms across three different countries in a bid to explain what *is* in mathematics teacher education. It is, first and foremost, a step towards delineating what *could/should* be in our preservice mathematics teacher education programs. An immediate future challenge of the three authors is precisely to develop research that informs linguistically responsive mathematics teacher education practices for the ultimate benefit of the learning of mathematics. At this stage, however, it is difficult to understand what *could/should* be without an international and practical understanding of what is.

References

Adler, J. (1995). Dilemmas and a paradox: Secondary mathematics teachers' knowledge of their teaching in multilingual classrooms. *Teaching and Teacher Education, 11*(3), 263–274.

Baldauf, R. B., & Kaplan, R. B. (2004). Language-in-education policy and planning. In E. Hinkel (Ed.), *Handbook of research in second language teaching and learning* (pp. 1013–1034). Mahwah, NJ: Erlbaum.

Barwell, R. (2009). Multilingualism in mathematics classrooms: An introductory discussion. In R. Barwell (Ed.), *Multilingualism in mathematics classrooms: Global perspectives* (pp. 1–13). Bristol, England: Multilingual Matters.

Barwell, R., Barton, B., & Setati, M. (2007). Multilingual issues in mathematics education: Introduction. *Educational Studies in Mathematics, 64*, 113–119.

Chitera, N. (2009). Code-switching in a college mathematics classroom. *International Journal of Multilingualism, 6*(4), 426–442.

Chitera, N. (2011). Language of learning and teaching in schools: An issue for research in mathematics teacher education? *Journal of Mathematics Teacher Education, 14*(3), 231–246.

Chitera, N. (2012). Language-in-education policies in conflict: Lessons from Malawian mathematics teacher training classrooms. *African Journal for Research in Mathematics, Science and Technology Education, 16*(1), 60–70.

Civil, M. (2012). Mathematics teaching and learning of immigrant students: An overview of the research field across multiple settings. In B. Greer & O. Skovsmose (Eds.), *Opening the cage: Critique and politics of mathematics education* (pp. 127–142). Rotterdam, The Netherlands: Sense.

DHET (Department of Higher Education and Training). (2011). *Minimum requirements for teacher education qualification* (Government Gazette No. 583). Pretoria, South Africa: DHET.

Essien, A. (2010a). What teacher educators consider as best practices in preparing pre-service teachers for teaching mathematics in multilingual classrooms. *Perspectives in Education, 28*(4), 32–42.

Essien, A. (2010b). Mathematics teacher educators' account of preparing pre-service teachers for teaching mathematics in multilingual classroom: The case of South Africa. *The International Journal of Interdisciplinary Social Sciences, 5*(2), 33–44.

Essien, A. (2013). *Preparing pre-service mathematics teachers for teaching in multilingual classrooms: A community of practice perspective.* Unpublished PhD thesis, University of the Witwatersrand, South Africa.

Essien, A. (2014). Examining opportunities for the development of interacting identities within pre-service teacher education classrooms. *Perspectives in Education, 32*(3), 62–77.

Mortimer, E. F., & Scott, P. H. (2003). *Meaning making in secondary science classrooms.* Philadelphia, PA: Open University Press.

NEPI (National Education Policy Investigation). (1993). *The framework report and final report summaries.* Cape Town, South Africa: Oxford University Press.

Ngu, J. (2004). *Initiative teacher education in Sub-Saharan Africa.* Paris, France: UNESCO.

Official Diary of the Government of Catalonia. (2003). *Law 1/2003 of Universities* (Reference No. 3826). Barcelona, Catalonia-Spain: Department of Universities, Research and Information Society.

Planas, N. (2012). Heteroglossia and "orchestration" in multilingual mathematics classrooms. In H. Forgasz & F. Rivera (Eds.), *Advances in mathematics education. Toward equity: Gender, culture, and diversity* (pp. 333–338). New York: Springer.

Planas, N., & Civil, M. (2009). Working with mathematics teachers and immigrant students: An empowerment perspective. *Journal of Mathematics Teacher Education, 12*(6), 391–409.

Planas, N., & Setati-Phakeng, M. (2014). On the process of gaining language as a resource in mathematics education. *ZDM Mathematics Education, 46*(6), 883–893.

Secretary for Education's Letter. (1996). *Circular about teaching standards 1 to 4 in local languages* (Reference No. IN/2/14). Lilongwe, Malawi: Ministry of Education.

Setati, M. (2008). Access to mathematics versus access to the language of power: The struggle in multilingual mathematics classrooms. *South African Journal of Education, 28*, 103–116.

Skiba, R. (2007). Code-switching as a countenance of language interference. *The Internet TESL Journal (for Teachers of English as a Second Language), XIII*(3).

Smit, J., & van Eerde, H. (2011). A teacher's learning process in dual design research: Learning to scaffold language in a multilingual mathematics classroom. *ZDM Mathematics Education, 43*(6), 889–900.

Chapter 7
Addressing Multi-language Diversity in Mathematics Teacher Education Programs

Denisse R. Thompson, Gladis Kersaint, Hannatjie Vorster, Lyn Webb, and Marthie S. Van der Walt

7.1 Introduction

Try to answer the following questions from a numeracy test:

1.1 Bhala esi sivakalisi sibe linani. Amawaka angamashumi amabini anamakhulu amabini anesithandathu.
1.2 102 − 36 =
1.3 1 048 + 21 376 =
1.4 23 × 145 =
1.5 168 ÷ 12 =
1.6 Dibanisa olu luhlu lwamanani lulandelayo.
213, 4 017, 1 273, 2 198, 21
(Webb, 2012)

Reflect on how you felt as you attempted to answer these questions in a language (isiXhosa) that is likely not the language of most readers of this chapter. Did you wonder if 1.1 was a set of directions for 1.2–1.5 or a separate problem? Did you wonder whether you were supposed to add the numbers in 1.6 or put them in sequence? What supports, if any, did the questions provide that enabled you to attempt the problems?

D.R. Thompson (✉) • G. Kersaint
University of South Florida, Tampa, FL, USA
e-mail: denisse@usf.edu; Kersaint@usf.edu

H. Vorster • M.S. Van der Walt
North-West University, Potchefstroom, South Africa
e-mail: hannatjie.vorster@nwu.ac.za; marthie.vanderwalt@nwu.ac.za

L. Webb
Nelson Mandela Metropolitan University, Port Elizabeth, South Africa
e-mail: lyn.webb@nmmu.ac.za

R. Barwell et al. (eds.), *Mathematics Education and Language Diversity*, New ICMI Study Series, https://doi.org/10.1007/978-3-319-14511-2_7

The frustrations you likely felt are not unlike those experienced by primary and secondary students in mathematics classrooms in many countries. Increasingly, English is used as the language of mathematics instruction in many countries, regardless of the social or home language spoken by teachers and their students (ICMI Study 21 discussion document, this volume, pp. 297–308). What preparation do mathematics teachers need in order to address the language diversity of their students? What is the role of mathematics teacher educators (MTEs) in this preparation?

In this chapter, MTEs from two different environments join together to share insights on the role that MTEs might play in this preparation. Throughout, we assume that English is the language of learning and teaching (LoLT) in mathematics classes, even though many of the students may speak another language or languages, at home and in their community. The authors from South Africa work in an environment with 11 official languages, but in which English is the language of instruction because it is the academic language and is used as common language in multilingual contexts. In this setting, both teachers and students may have difficulty transitioning from informal use of mathematical language, often in the students' home language, to formal mathematical language in English (Webb, 2012). In contrast, the authors from the United States work in an environment where English is the primary academic and social language. However, US school systems face an increasingly diverse student population, with approximately 11 % being designated as English language learners (ELLs) (NCELA, 2011) who maintain the use of their mother tongue at home or in social/cultural settings. As a result, American classrooms may include students with varying levels of English proficiency; in fact, in some school districts, more than 100 different languages may be spoken by students.

We believe there are more similarities than differences in the issues and challenges we face as MTEs who prepare and support teachers.[1] Hence, we begin by raising awareness of some issues involved in helping students learn to read, write, speak, and listen to mathematics—a foreign language for most students, regardless of their English language proficiency. We then discuss issues related to orchestrating classroom discourse in such settings. We end by sharing strategies MTEs might use in teacher preparation programs to prepare teachers to teach students from linguistically diverse backgrounds.

7.2 Raising Awareness of Issues Related to Teaching the English Mathematics Register

MTEs must address a variety of issues when working with teachers, including mathematics content knowledge, content-specific pedagogical knowledge, and general aspects of pedagogy advocated as part of educational reform (e.g., inquiry learning, high expectations, and tasks with high levels of cognitive demand). Given these multiple areas of responsibility, it might be difficult to consider who might

[1] In this chapter, we use the word *teachers* to refer to both prospective teachers and practicing teachers enrolled in either undergraduate or graduate teacher education programs, respectively.

address additional needs, such as teaching students who do not speak the language of instruction. We take the position that mathematics is a sign system that includes language aspects unique to the mathematics register (e.g., words, symbols, sentences, graphs) (Halliday, 1978). Hence, we believe that MTEs must help teachers understand the features of this sign system that may influence student learning, including students who are learning the language of mathematics concurrently with English. Supporting such language development involves specific expertise of those who have knowledge of mathematics, as well as knowledge of language difficulties that students often face when studying mathematics.

Students engage in mathematical discourse through the language of instruction, in this case English. When the information to be conveyed is mathematical in nature, the context is complex because there is always an interplay of at least two languages—mathematics, thought of as a unique language (Usiskin, 1996), and the language of the classroom. Because the development of mathematics language occurs primarily within the confines of the classroom, all students, regardless of their home language, are mathematics language learners[2] (Thompson, Kersaint, Richards, Hunsader, & Rubenstein, 2008). This notion is supported by curriculum recommendations in many countries (e.g., Department of Basic Education, 2011; NCTM, 2000) that emphasize the importance of communication because learning and teaching mathematics is conducted largely through interaction, including talk, as well as the use of written symbols, diagrams, charts, and other texts. So, if individuals are to develop the ability to communicate mathematically, they need opportunities to communicate as a regular and ongoing part of their mathematics classes. This suggests that mathematics teachers must be prepared to help students learn and master mathematical language. To this end, MTEs need awareness of the issues that teachers face and about which teachers might need to be sensitized.

7.2.1 Understanding the Language Context in Which Teachers Work

Two different aspects of language learning should be considered by MTEs and teachers. Individuals first develop Basic Interpersonal Communication Skills, which are skills for personal and social communication. However, for mathematics they also need Cognitive/Academic Language Proficiency, that is, the academic language needed to communicate mathematically (Cummins, 1981). Individuals may be fluent in terms of Basic Interpersonal Communication Skills in either English or their home language, and yet lack the proficiency in the academic register needed to communicate mathematically.

Classroom demographics vary, both for the classrooms of MTEs and for the classrooms of their teachers. In some contexts, individuals are still learning English

[2] For simplicity in this argument, we ignore the fact that students' home language may also lead to a mathematics that may well be quite different from the school mathematics with which they engage.

(ELL students), while their peers are English mother tongue speakers. In such cases, teachers should help to develop the English of these students so they can effectively function on the same level as their native English-speaking peers. In this English-only context, teachers will introduce the academic language of mathematics in English to all students with appropriate support and scaffolds for those still learning English. Teachers consequently have to be sensitive to whether an ELL student is a "new arrival" with low expertise in English, has been born in the country but is more proficient in a community language other than English, or is a "high achieving" multilingual with high proficiency in English (Harris & Leung, 2007).

In other contexts, for example South Africa, most of the students in multilingual classrooms are English as Additional Language learners (EAL learners), adding English to a repertoire of various other languages. In this context, students need their home language as well as English to facilitate understanding while proficiency in English is developing as an ongoing process. Adler (2001) identified three different environments in South African multilingual classrooms that may be applicable elsewhere as well. First, the urban-suburban environment is basically the same as the English-only context described previously. Second, in the Urban/Township contexts there is a strong regional language that coexists with different other home languages and many students do not have high English proficiency. Third there is the rural context, where students hear English mainly at school and most of the students have the same home language. In the last two contexts, teachers often code-switch—in other words move back and forth between English and the students' home language (Vorster, 2009). MTEs should help teachers recognize the differences that can exist within these contexts so that teachers are able to choose language strategies and mathematical language teaching practices that are most conducive to students' success in each context.

7.2.2 Understanding Potential Difficulties with Mathematics Language

Despite the differences in English fluency that may exist, all students who learn mathematics in English must learn the mathematics register to communicate effectively in mathematics. The mathematics register includes "unique vocabulary, syntax (sentence structure), semantic properties (truth conditions), and discourse (oral and written text) features" (Kersaint, Thompson, & Petkova, 2013, p. 43). Because of this, teachers must become aware of features of the mathematics register that should be addressed explicitly as part of instruction, such as the following:

- Words are used differently in mathematics than in social English (e.g., *difference* between products vs. *difference* in mathematics).
- Words may have different meanings in various disciplines (e.g., *radical* in mathematics vs. in science vs. in social studies vs. in English).

- Words may have different meanings within mathematics (e.g., *base* of a triangle vs. *base* of a power, or *median* of a data set vs. *median* of a triangle).
- Phrases have meanings separate from the meanings of the individual words (e.g., *if-then*, *polygon* vs. *regular polygon*, *root* vs. *cube root*).
- Syntax can create issues, particularly when more complex sentence structures are used, such as passive voice or *if-then* constructions.
- Semantics is essential to draw meaning from language (e.g., 3 times a number is 5 more than the number meaning $3x=x+5$ so that "a number" and "the number" are represented by the same variable).
- Cultural references are often embedded within word problems that may influence students' ability to comprehend (e.g., "in the red" to mean a deficit).
- Specific language groups have specific problems with English, which especially impact on mathematical language where every word has to be understood correctly (e.g., some South African indigenous language speakers specifically have a problem with connectives such as "or").

Symbols have their own issues but are crucial for understanding:

- Multiple words may be needed to verbalize a symbol (e.g., $\sqrt{\ }$ as *square root of*).
- Multiple verbalizations are possible for a single symbol (e.g., + as *plus*, *increased by*).
- In contrast to vocabulary words for which phonetic clues can be used to help verbalize the word, no clues are embodied within the symbol to help a reader verbalize it. Verbalizations, even for simple symbols, must be explicitly taught.

The goal is to prepare mathematics teachers to attend to language issues they themselves may not be aware of, but that can have a significant impact on how students make meaning in the classroom (Pimm, 1987). Once teachers have this basic knowledge, MTEs need to help them learn *how* to integrate mathematical literacy into their instruction to facilitate both mathematical language and mathematical understanding.

7.3 Facilitating Discourse in English as the LoLT in Mathematics Classrooms

In contrast to ordinary language that is used in many settings, individuals tend to use mathematics language primarily within the narrow setting of the mathematics classroom. Thus, teachers need to provide many opportunities for students to engage in the entire mathematics literacy spectrum (e.g., read, write, speak, listen to, interpret) if they are to become fluent. This section highlights a number of issues in preparing teachers to orchestrate discourse in classrooms with EAL students: using dialogic practices, means of questioning, and engaging in code-switching.

7.3.1 Dialogic Practices

Vygotsky (1978) maintained that learning is constituted through dialogic practices. An interpersonal dialogue is defined as:

> a discursive relationship between two or more participants characterized by thought-provoking activities such as questioning, interpreting, explaining and rethinking [...] in an interaction, either teacher-student or student-student. (Gorsky, Caspi, & Trumper, 2006, p. 74)

Gorsky et al. (2006) maintained that learning is facilitated by interpersonal dialogue. They divide dialogue into two models: intrapersonal and interpersonal dialogue. *Intrapersonal dialogue* mediates learning and refers to the interaction between the student, individually, and the subject matter that the student is attempting to learn, in this case mathematics. The structural resources that enable intrapersonal dialogue are the materials from which the student is learning, either textbooks or previous examination papers written in English. *Interpersonal dialogue* facilitates learning and is enabled by the human resources of the teacher and fellow students in the mathematics class. The teacher often teaches mathematics in English or code-switches between English and the home language. Gorsky et al. noted that if students were faced with an insoluble problem, they first turned to intrapersonal dialogue, i.e. they relied on self-study mediated by texts, but if that failed they turned to student–student interpersonal dialogue, and seldom asked the teacher for help. In contexts where the structural resources are inaccessible because of language, students should be able to access interpersonal dialogue with fellow students in the class, in the form of exploratory talk in their home language.

The teacher plays a vital role in creating and maintaining this dialogue. In language diverse classrooms, the interpersonal dialogue between student and student should, as far as possible, be conducted in a language in which the students are proficient. However, as teachers aim to facilitate dialogue in classrooms with students whose English fluency may be at varying levels, they must confront difficulties caused by tensions between informal and formal language such as:

> how to encourage movement in their learners from the predominantly informal spoken language in which they are fluent [the home language], to the formal written language [mathematics in English] that is frequently perceived to be the landmark of mathematical activity. (Pimm, 1991, p. 21)

Pimm suggests three routes:

1. A direct route from informal spoken language to formal written language.
2. An indirect route from informal spoken language through more formal spoken language to formal written language.
3. An indirect route from informal spoken language through informal written language to formal written language.

Because of the added complexity of an additional language (in the South African case, English), Setati (2005, p. 84) adds steps along the way from informal spoken

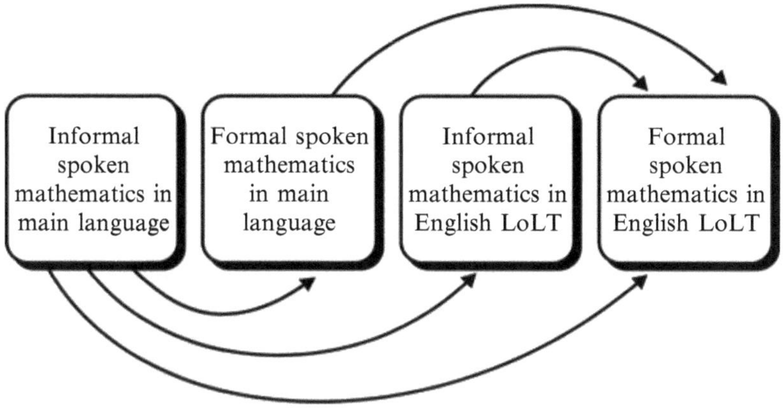

Fig. 7.1 Routes to formal spoken mathematics in English LoLT (adapted from Setati & Adler, 2000, p. 250)

mathematics in the students' home language to formal spoken mathematics in English. The route could be expanded to include: Informal spoken language in home language—formal spoken mathematics language in home language—informal spoken mathematics in English LoLT—formal spoken mathematics in English LoLT. These routes can be visualized as in Fig. 7.1.

The routes are varied and convoluted. Barwell and Kaiser (2005) argue that if students can be encouraged to talk informally about their mathematical reasoning in their home language, there is more chance that they will be able to develop formal mathematical discourse. In order to talk either formally or informally about mathematics, students have to acquire the mathematical words in the LoLT to use in sentences to develop a meaningful dialogue.

However, the mere presence of dialogue does not constitute meaningful talk and does not necessarily lead to understanding; rather, the quality and type of discourse are crucial in leading to conceptual understanding of mathematics. Mercer and Littleton (2007) analyzed talk and classified it into three types: *disputational talk* where participants agree to disagree, but where no reasons for decisions are given; *cumulative talk* when participants simply agree with each other's opinions without engaging with the issue; *exploratory talk* which is the most educationally sound method of communication. Mercer and Littleton structured dialogue as exploratory talk for primary school classes and provided teachers with specific guidelines for its implementation so teachers could negotiate with the class for the development of dialogue in groups. For example, students should share relevant ideas and help each other to understand the problems; they should listen to each other's contributions and respect their ideas, even if they disagree; they can challenge and counter-challenge arguments, but they should give reasons and substantiate their challenges with sentences such as, "I think … because …." If possible the groups should work towards an equitable consensus.

If the journey towards mathematical understanding can be smoothed by facilitating dialogue, in the form of exploratory talk, among and between students in the language in which they are most proficient, can the promotion of structured dialogue also facilitate the development of reasoning skills and language? If so, then what principles could teachers use to encourage their multilingual students to engage in dialogic practices?

Rojas-Drummond and Mercer (2004) studied interactions in Mexican classrooms and found that teachers whose pupils achieved the highest results either treated learning as a social communicative process or used judicious questioning. The teachers were observed organizing interchanges of ideas and mutual support amongst students and generally encouraging students to take a more active, vocal role in classroom events. They used question-and-answer sequences not just to test knowledge but also to guide the development of understanding. These teachers often used questions to discover the initial levels of students' understanding and adjusted their teaching accordingly, and used "why" questions to get students to reason and reflect about what they were doing (Mercer & Littleton, 2007). Thus, teachers play an active role in guiding their students in dialogic interactions.

7.3.2 *The Role of Questioning*

Although teachers use questions as a matter of course to monitor progress, the skill of asking higher-order questions that focus on communication and conceptual understanding is not a trivial one. MTEs need to help teachers develop skill at questioning techniques. Questions can serve many communicative roles: to test students' knowledge; to manage classroom activities; to assess students' understanding; or some combination of these roles (Mercer & Littleton, 2007). Teacher questioning can be used in the development of students' learning and their own use of language as a tool for reasoning. Teachers can encourage students to make explicit their thoughts, reasons and knowledge and share them with the class; teachers can model useful ways of using language that students can appropriate for themselves in peer group discussions; and teachers can provide opportunities for students to make longer contributions in which they express their current state of understanding, articulate ideas, and reveal problems they are encountering (Mercer & Littleton, 2007, p. 36). In many language diverse classroom settings, the discussion around the problem solving can be done in the students' home language; in the wrap-up phase, the teacher can rephrase and revoice the mathematical ideas in English, consolidating the learning process by writing the solution and the English terms on the board so that the spoken word in the home language is both heard and read in English.

During Socratic dialogue, Socrates took the part of a critical friend who questioned his students to develop their reasoning and argumentation skills. He continuously posed questions but did not provide answers or solutions. Although he did not openly disagree with his students, his questions were designed to help students arrive at their own conclusions (Frick, Albertyn, & Rutgers, 2010). A question is answered with a question in order to tease out the reasoning behind it. In multilingual mathematics classrooms, the teacher can force the student to defend his/her position by offering arguments against it. Very often there is no correct answer but the reasoning behind the stance is probed and critical thinking is engendered. It is not only the teacher who is responsible for judicious questioning in the classroom, but the role can be played by students among themselves in group interactions. To engender confidence in students, the answers (and the questions) can be posed and answered in students' home language. It is incumbent on the teacher to intersperse terms and phrases in English so that students are guided along the journey towards formal written mathematics in English. The practice of code-switching is widely used to facilitate this process, for example, Muke (2012) shows how the use of code-borrowing[3] within an explanatory indigenous sentence could empower learners to understand and use the English terminology.

7.3.3 *The Practice of Using Code-Switching to Engage in Mathematical Discourse*

Code-switching in sociolinguistics refers to the practice of using two or more linguistic varieties in a single communicative sequence (Moschkovich, 2007). Moschkovich (2007) views code-switching as a complex language practice which, while using the official LoLT, allows for more extensive use of the main language. She disagrees with the view that it stems from a deficit model where the speakers use code-switching when they are unable to recall suitable phrases in the language being spoken and sees it as the mark of fluency in two languages. In fact, Clarkson (2007) suggests that switching between languages is a distinct advantage as it gives students access to alternate meanings and relationships.

Code-switching in mathematics classrooms can be described as the intuitive use of both English and the students' home language to facilitate mathematical understanding. Although teachers may sometimes be unaware of students' use of code-switching, either overtly by talking with peers, or privately in their own thinking (Clarkson, 1996), teachers in different parts of the world actively use this language practice to try and ensure better communication with students during mathematics sessions (e.g., in Papua New Guinea, Muke & Clarkson, 2011; in Iran, Parvanehnezhad & Clarkson, 2008; in South Africa, Setati & Adler, 2000).

When teachers do actively encourage code-switching, this normally ensures that the percentage of main language usage increases and that an additive model is employed, with the resultant transfer of mathematical concepts from one language to the other. Students are usually allowed to communicate about mathematics in the language of their choice. Code-switching is therefore front staged to facilitate mathematics and not back staged only to give instructions and for disciplinary purposes

---

[3] In this chapter, code-borrowing refers to the use of English terminology in an indigenous sentence, and has to be distinguished from transliteration, discussed under Section 7.3.3.

(Heller & Martin-Jones, 2001). This presupposes that code-switching is a technique that comes naturally to multilingual teachers. However, there are two inhibiting factors. First, many educators feel guilty if they code-switch as they feel they are depriving their students of an opportunity to acquire English (Setati, 2005). Second, most teachers have been educated in English and they find the indigenous terminology difficult. This results in teachers tending to use either the English terminology or transliterated words while communicating in an indigenous language. Transliterated borrowed words may not facilitate understanding of a concept. In transliteration, the English sounds are directly transferred into the indigenous language, inflected to suit the structure of the language, but without relation to the meaning of the concept, e.g., "Square" becomes "sekwere"*: (**sq** - sêk, **ua** - wê, **re** – rê)* in Setswana. In contrast the original Setswana word "khutlonnetsepa" can be linked to the definition of a square *(khutlo-angle, nne-four, tsepa-straight up).* A new transliterated borrowed word therefore still has to be fully explained in the indigenous language as was the practice noted by Muke (2012) about English borrowed words.

Code-switching is only usable in contexts where the class's language profile allows a strong regional language to facilitate better understanding of mathematics, with the prerequisite that the teacher is also fluent in that language. In some schools, this may be possible in one mathematics class and not in the next. In rural contexts in South Africa, code-switching is often feasible and necessary because of students' low English proficiency. MTEs should sensitize teachers to the importance of determining the language profile of each mathematics class in order to consciously decide on the best language practice for a specific group.

Code-switching as a practice developed informally, with teachers practicing it in different ways according to their perceptions of when students need their home language for better understanding of the mathematics. Although limited research has been done on best practices in code-switching (Muke, 2012), there are some directives that MTEs can discuss with teachers.

Language is important as a facilitating medium of understanding. It is crucial that teachers facilitate opportunities for students to improve fluency in both English and their home language, and more specifically also in the mathematics register of their home language in so far as it is developed. Teachers have to take cognizance of the threshold theory of bilingualism that proposes in general that "there may be a threshold level of linguistic competence" that bilingual students have to attain in *both* languages "*to* influence cognitive functioning" positively (Cummins & Swain, 1986). Clarkson and Galbraith (1992) in Papua New Guinea and Clarkson (1996) in Australia found evidence in a mathematical environment that supports Cummins' threshold theory. This implies that if EAL learners' main language is allowed to lapse, it will influence their cognition negatively. Gaoshubelwe (2011) noted in his analysis of mathematics lessons that some teachers mixed languages in a way that does not model the correct sentence construction/grammar of either the English or the home language's mathematical register. Teachers have to facilitate grammatically and mathematically correct language both in English and the home language.

Although teaching mathematical language is essential, it is important to balance visibility and invisibility of mathematics language teaching (Adler, 1999). The visibility of mathematics language teaching can be illustrated by the use of the morphology of the indigenous term to explain a concept, for example "adjacent angles": dikhutlomabapi, di- many, kuthlo - angle, mabapi—sit beside each other (Setswana). The mathematical explanation is interrupted to teach mathematical language or explain terminology. This should not be so extensive that it interrupts the argument of the mathematical reasoning. Invisible language teaching occurs where language teaching techniques are used that do not interfere with the flow of the mathematical reasoning. For example, a teacher may be modeling correct mathematical language through re-voicing when reformulating a student's sentence in correct mathematical language (Herbel-Eisenmann, Drake, & Cirillo, 2009; Setati & Adler, 2000), recasting when using a word in different sentences and contexts (Khisty, 1995), or through the use of synonyms for the same word. In multilingual settings, it is important for students to hear different English synonyms so they can recognize concepts as similar, because their "word sense" in English (Vygotsky, 1962) is not well developed and they do not automatically link synonyms to each other. Using a term first in the students' main language and then saying the correct term in English can be considered an extension of recasting.

Bilingual written text in explanations, assignments, and class tests can enhance understanding because students can oscillate between the languages to negotiate meaning and they are able to revisit the texts again (Vorster, 2008; Vorster & Zerwick, 2011). Available bilingual mathematics dictionaries can help to provide definitions in indigenous languages. Such dictionaries or modified bilingual terminology lists can be made available during tests (analogous to adding formula sheets).

Teachers should be cognizant of debates on terminology: there is a difference of opinion on whether teachers should use English terminology when they code-switch to the home language, use transliterated words, or use the correct indigenous terminology. The question is whether the bilingual use of mathematical terminology would add to better understanding or add to the workload of students. Countries have chosen different paths in standardizing terminology. While Tanzania purposefully chose terminology that conveys meaning, Malawi chose to use transliterated terminology (Kazima, 2008), and teachers in South Africa have to make their own choice. Another debate is whether new terminology should be coined for terms in cases where the indigenous terminology does not exist (Schäfer, 2010; see also Meaney, Trinick, & Fairhall, 2011 for a discussion on this topic and the successful development of *te reo Māori* mathematical terminology).

Examples of negative practices include "ritualization" where students chant answers as a group (Heller & Martin-Jones, 2001, p. 13), providing "safe time" for students who cannot express themselves, and circumvention of language. Teachers sometimes use only one-word instructions, for example *solve*, *factorize*, etc., or ask mainly procedural or algorithmic problems to avoid language issues. These practices do not help students build mathematical literacy, which has become important in the current constructivist teaching and learning environment.

7.4 Incorporating Mathematics Language and Literacy into the Teacher Preparation Program

With the goal for cooperative learning and more discursive practices within mathematics classrooms, the need to communicate mathematically and to comprehend mathematics language (both verbal and written words and symbols) becomes essential. Thus, within teacher education programs, MTEs need to help teachers understand the influence of language in supporting students' ability to interpret information conveyed and communicated in the mathematics class. Once teachers have been sensitized to the issues and language practices identified in the preceding sections, many teachers might question how they can engage students in these literacy practices while still teaching "all the content they need to cover [as mandated by state or national curriculums]." Thus, our task as MTEs is to help them understand that "language is a tool, whereas discourse is an activity in which the tool is used or mediates" and that they need to "embrace the complex linguistic nature of mathematical activity" (Gutiérrez, Sengupta-Irving, & Dieckmann, 2010, p. 34). Put simply this is a way of teaching, not an extra topic that is to be added to the amount of content that is to be taught.

A challenge for MTEs is how to foster the knowledge and skills of prospective teachers regarding the effective teaching–learning of mathematics in multilingual classrooms (Graham & Phelps, 2003). Teachers need multiple opportunities to consider how to incorporate the development of mathematics language and literacy skills as part of their regular curriculum. It is one thing to provide teachers with information about language features that need to be considered and a range of practices to address them and to give teachers opportunities to experience these practices in their own learning. It is another thing to have teachers plan to implement these practices into their classroom in a way that becomes an integral part of their teaching and not considered a supplementary activity that can be ignored. In this section, we share strategies we have used in our teacher education programs to help teachers begin to consider implementing literacy into their own classrooms. MTEs can highlight and engage teachers in discussions about different instructional practices that can be used to emphasize language and concept knowledge development. Teachers can then be given opportunities to integrate these practices in lessons and practice implementing them with each other or with groups of linguistically diverse students in small group or whole class settings as part of practical teaching experiences in schools.

7.4.1 Developing Language Modules to Integrate into Methods Courses

MTEs might create modules dedicated to mathematical language and language practices in multilingual classrooms. Such modules should include experiential learning where dialogic practices, including exploratory talk and different mathematical language teaching aids and techniques, can be applied and practiced.

Alternatively, MTEs might choose one mathematical topic for discussion in the course and model how explicitly teaching the language of mathematics could be incorporated in the planning and teaching of that topic. Issues around teaching mathematical language would then be addressed during this time.

For instance, one module might focus on helping teachers learn to engage students in mathematics discourse as a means to address the entire mathematics literacy spectrum. By making direct connections to students' lived experiences, it might be possible to connect academic language to social language (e.g., an intersection of two roads can help provide meaning for the intersection of two lines). Through such connections, students can make meaning using insights from their social or home experiences. Another approach is to build common experiences as part of the mathematics class by engaging in brief conversations about unfamiliar contexts found in mathematics passages or word problems. When students engage in such discussion, teachers are ensuring that all students, regardless of class or social experiences, are interpreting the information in the same way.

A second module might focus on incorporating the use of visual representations and graphic organizers into mathematics instruction on a regular basis. The module could help teachers understand how the use of visual representations and graphic organizers can scaffold EAL students' learning of English, helping them make connections between and among concepts being studied. The use of visual representations to convey mathematics and English ideas allows students to examine similarities and differences between how mathematics language (words, symbols, and diagrams) is used to represent concepts and to explore different ways to convey mathematics ideas. For example, students can be encouraged to draw comparisons and contrasts between concepts (e.g., prisms and pyramids, rhombus and square) so that they see similarities and differences in order to develop a thorough understanding. Such discussions can be supported by the use of graphic organizers, such as Venn diagrams or concept maps, so students can visually see the connections and attend to ways to communicate these similarities or differences using the mathematics register. When such visual representations are used, teachers can include information to help students express mathematics ideas. For example, in addition to writing the symbol >, a teacher might say it, and then write the spoken language, such as "is greater than".

A third module could focus on helping teachers learn to adapt the use of regular English reading and language strategies to mathematics. For instance, many mathematics textbooks have headings within a lesson; students can learn to read the heading, convert it to a question, and then attempt to answer the question as they read the lesson (e.g., heading: Solving with a Table and a Graph; question: How do you solve an equation with a table? How is solving an equation with a table like solving with a graph?). Thus students learn how to use the textbook to support their own learning.

In addition, teachers can encourage students to develop personal dictionaries of mathematical terms, with definitions in students' own words, even in their home language, with diagrams and/or examples as appropriate. Tied to dictionaries can be the use of etymology and morphology. Etymology focuses on the origin of a word or symbol, e.g., the Greek symbol Σ for the capital letter S (used as the symbol for *sum* in series). Morphology is how a word is put together, e.g., trilateral = three sides. Morphology can be used with prefixes and suffixes to help students make sense of new words; if students know *tri* means three, they have a start on understanding *triangular*.

Another possible avenue for teachers is to set language aims for each lesson where applicable. This includes identifying any of the potential difficulties mentioned in Sect. 7.2.2; using different techniques to explain new terminology or linking it to the home language of the students; or practising correct grammar and sentence construction, for example, to formulate conjectures where concepts, relationships, and conditions have to be expressed.

7.4.2 Simultaneous Interpreting Between English and an Indigenous Language as a Tool in Teacher Education

In cases where teachers have been educated in English but will have to teach or code-switch to an indigenous language when teaching, as is the case in Malawi (Chitera, 2011) and South Africa, it can be advantageous if MTEs make use of simultaneous interpreting instead of teaching only through the medium of English. If the technique is used where the teachers listen to the interpreter in the indigenous language, using the headphone in only one ear while also listening to the lecturer, the teachers hear the correct mathematical terminology, as well as formulation of expressions in both English and the indigenous language. Simultaneous interpreting will benefit them when they themselves have to alternate between languages when code-switching, because they become better acquainted with the mathematics register in both English and the indigenous language. Furthermore, teachers become more aware of the necessity to teach mathematical language, both in the indigenous language and in English. They gain understanding of their students' problems to cope with the English mathematical register and to understand concepts when English is the LoLT. Furthermore, if teachers' study guides or workbooks are also bilingual, their expertise in writing mathematical language in both English and the indigenous language text is enhanced, and they are empowered to use written text in the indigenous language alongside English notes when teaching (Vorster & Zerwick, 2011).

7.4.3 Using Mathematics Educator Reflective Communities to Collaboratively Plan to Integrate Language in Mathematics Instruction

When teachers are empowered to determine for themselves those language practices they are able to integrate into their mathematics classroom, there is a greater likelihood such practices will be translated from planning into actual implementation. Thus, groups of teachers might work together to determine how they would incorporate mathematical literacy or other dialogic practices into the curriculum for a mathematics course of their choice. The goal is for teachers to consider how they will address the mathematics language issues (i.e., vocabulary, symbols) for a specific instructional segment, engage students in all aspects of the literacy spectrum (reading, writing, interpreting, speaking, and listening), determine the types of questions to use, and assess students so that insights about their mathematics language development can be ascertained. Engaging in such a project has the benefit of allowing teachers to consider instructional approaches that support mathematics and English language development without sacrificing a focus on rigorous content.

Rather than plan lessons to facilitate the knowledge and skills to teach in multilingual mathematics classrooms for an entire curriculum, an alternative model is *adapted lesson study* (see Fig. 7.2). Lesson study is a cyclical process used in Japan to professionally develop and focus the effectiveness of practicing teachers' teaching–learning experiences around students' learning (Lewis, Perry, & Murata, 2006). Internationally, teacher educators also use an adapted form of lesson study (Mathematics Educator Reflective Communities) for fostering/developing different aspects of mathematics education in their preservice mathematics teachers' classrooms (Fernandez, 2010; Murata & Pothen, 2011; Van der Walt, 2012). Lesson study has the potential to facilitate the knowledge, skills, and awareness of what multilingual classrooms require from teachers and to implement the various practices suggested in this chapter.

During the planning phase of the lesson (or unit of lessons), a group of teachers work collaboratively and cooperatively, taking into account the aims, including mathematical language aims, the school has set for multilingual students, focusing on multilingual students' learning and conceptual understanding. The lesson study group anticipates multilingual students' responses and reactions to the planned activities, problems, and exercises, including the activities, problems and exercises planned to accommodate and support students' language needs.

During the teaching of a lesson by one member of the group in one classroom of the school, the rest of the study group observes the lesson and collects data regarding students' thinking, understanding, and learning, with the aim to revise and refine

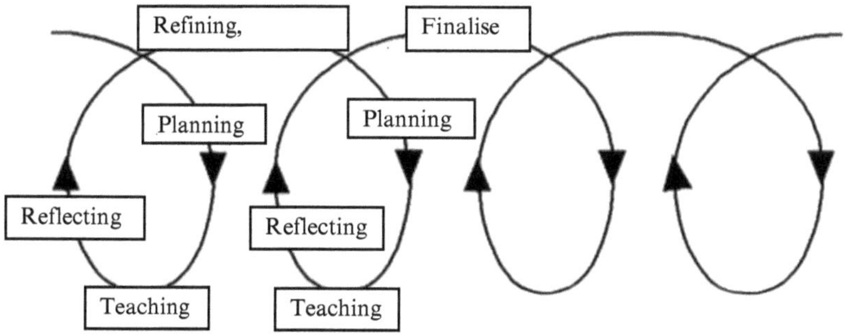

Fig. 7.2 Adapted lesson study (adapted from Van den Akker, Gravemeijer, McKenney, & Nieveen, 2006)

the lesson. The teachers facilitate discussions, also in multilingual classes to ensure students' engagement (Berliner, 2001), while the teacher models his/her own thinking to improve understanding and poses questions connecting students' developing mathematical ideas with mathematical language and symbols (Goos, 2004).

Lastly, the lesson study group comes together to reflect on and discuss the effectiveness of the lesson for students, using the data they collected and the experiences of the teacher who presented the lesson. Adaptations can be made to the lesson (or other lessons in the unit), and if necessary, the lesson can be taught again by another group member and observed again by the rest of the group. The lesson study cycle continues if necessary. To empower teachers to use Mathematics Educator Reflective Communities, MTEs can use this method in their course, for example with a group of teachers planning for a session of practical teaching.

7.5 Conclusion

Throughout this chapter, we have highlighted features of mathematical language that MTEs need to ensure their teachers know and we have shared approaches we have used in our teacher education programs to prepare mathematics teachers for addressing multilingual classrooms. However, we have little empirical data related to the effectiveness of these strategies, either from the perspective of the teacher and his/her willingness to implement the strategies in classrooms or from the perspective of the extent to which they help elementary and secondary students be successful with mathematics. Thus, there is a need to engage in studies that follow teachers from preparation programs in which practices for language diversity have been a focus into the field, in order to understand what practices are easily implementable and what effect those practices have. If different teacher preparation programs engage in different practices, we might begin to develop a body of research that suggests which practices work best with which teachers for which students in which contexts. The work described in this paper is appropriate for delivery by MTEs, not generalists, so that mathematics teachers have explicit instruction in applying these practices to support mathematics instruction. Thus, we advocate the need for MTEs to become more engaged with language issues as they prepare to support their teachers.

References

Adler, J. (1999). The dilemma of transparency: Seeing and seeing through talk in the mathematics classroom. *Journal for Research in Mathematics Education, 30*(1), 47–65.

Adler, J. (2001). *Teaching mathematics in multilingual classrooms.* Dordrecht, The Netherlands: Kluwer.

Barwell, R., & Kaiser, G. (2005). Introduction: Mathematics education in culturally diverse classrooms. *Zentralblatt für Didaktik der Mathematik, 37*(2), 61–63.

Berliner, D. C. (2001). Learning about and learning from expert teachers. *International Journal of Educational Research, 35*, 463–548.

Chitera, N. (2011). Helping student teachers to teach mathematics in local languages: Challenges for mathematics teacher educators. In M. Setati, T. Nkambule, & L. Goosen (Eds.), *Proceedings of the ICMI Study 21 Conference: Mathematics Education and Language Diversity* (pp. 20–27). São Paulo, Brazil: ICMI Study 21.

Clarkson, P. C. (1996). NESB migrant students studying mathematics: Vietnamese and Italian students in Melbourne. In L. Puig & A. Gutierrez (Eds.), *Proceedings of the 20th Conference of the International Group for the Psychology of Mathematics Education* (Vol. 2, pp. 225–232). Valencia, Spain: PME.

Clarkson, P. C. (2007). Australian Vietnamese students learning mathematics: High ability bilinguals and their use of their languages. *Educational Studies in Mathematics, 64*, 191–215.

Clarkson, P. C., & Galbraith, P. (1992). Bilingualism and mathematics learning: Another perspective. *Journal for Research in Mathematics Education, 23*(1), 34–44.

Cummins, J. (1981). *Bilingualism and minority language children.* Ontario, Canada: Ontario Institute for Studies in Education.

Cummins, J., & Swain, M. (1986). *Bilingualism in education: Aspects of theory, research and practice*. London: Longman.

Department of Basic Education (South Africa). (2011). *Curriculum assessment statements. Grades 10-12 mathematics (CAPS)*. Pretoria, Gauteng, South Africa: Department of Basic Education.

Fernandez, M. L. (2010). Investigating how and what prospective teachers learn through microteaching lesson study. *Teaching and Teacher Education, 26*, 351–362.

Frick, L., Albertyn, R., & Rutgers, L. (2010). The Socratic method: Adult education theories. *Acta Academica Suplementum, 1*, 75–102.

Gaoshubelwe, S. (2011). *Language practices in the teaching and learning of mathematics: A case study of three mathematics teachers in multilingual schools.* Unpublished MEd thesis, North-West University, South Africa.

Goos, M. (2004). Learning mathematics in a classroom community of inquiry. *Journal for Research in Mathematics Education, 35*, 258–291.

Gorsky, P., Caspi, A., & Trumper, R. (2006). Campus-based university students' use of dialogue. *Studies in Higher Education, 31*(1), 71–87.

Graham, A., & Phelps, R. (2003). Being a teacher: Developing teacher identity and enhancing practice through metacognitive and reflective processes. *Australian Journal of Teacher Education, 27*(2), 11–24.

Gutiérrez, K. D., Sengupta-Irving, T., & Dieckmann, J. (2010). Developing a mathematical vision: Mathematics as a discursive and embodied practice. In J. N. Moschkovich (Ed.), *Language and mathematics education: Multiple perspectives and directions for research* (pp. 29–71). Charlotte, NC: Information Age.

Halliday, M. A. K. (1978). *Language as a social semiotic: The social interpretation of language and meaning*. London: Edward Arnold.

Harris, R., & Leung, C. (2007). English as an additional language: Challenges of language and identity in the multilingual and multi-ethnic classroom. In J. Dillon & M. Maguire (Eds.), *Becoming a teacher: Issues in secondary teaching* (pp. 237–252). Maidenhead, England: Open University Press.

Heller, M., & Martin-Jones, M. (Eds.). (2001). *Voices of authority: Education and linguistic difference*. Westport, CT: Alex.

Herbel-Eisenmann, B., Drake, C., & Cirillo, M. (2009). "Muddying the clear waters": Teachers' take-up of the linguistic idea of revoicing. *Teaching and Teacher Education, 25*(2), 268–277.

Kazima, M. (2008). Mother tongue policies and the mathematical terminology in the teaching of mathematics. *Pythagoras, 67*, 56–63.

Kersaint, G., Thompson, D. R., & Petkova, M. (2013). *Teaching mathematics to English language learners* (2nd ed.). New York: Routledge.

Khisty, L. L. (1995). Making inequality: Issues of language and meanings in mathematics teaching with Hispanic students. In G. Secada, E. Fennema, & L. Adajian (Eds.), *New directions for equity in mathematics education* (pp. 279–285). New York: Cambridge University Press.

Lewis, C., Perry, R., & Murata, A. (2006). How should research contribute to instructional improvement? The case of lesson study. *Educational Researcher, 35*(3), 3–4.

Meaney, T., Trinick, T., & Fairhall, U. (2011). *Collaborating to meet language challenges in indigenous mathematics classrooms*. Dordrecht, The Netherlands: Springer.

Mercer, N., & Littleton, K. (2007). *Dialogue and the development of children's thinking: A sociocultural approach*. London: Routledge.

Moschkovich, J. (2007). Using two languages when learning mathematics. *Educational Studies in Mathematics, 64*(1), 121–144.

Muke, C. (2012). *The role of local language in teaching mathematics bridging class (grade 3) within South Wahgi area of Jiwaka Province, Papua New Guinea.* Unpublished doctoral dissertation, Australian Catholic University, Australia

Muke, C., & Clarkson, P. (2011). Teaching mathematics in the Papua New Guinea Highlands: A complex multilingual context. In J. Clark, B. Kissane, J. Mousley, T. Spencer, & S. Thornton (Eds.), *Proceedings of Mathematics Education Research Group of Australasia, Australian Association of Mathematics Teachers Joint Conference* (pp. 540–547). Adelaide, Australia: AAMT/MERGA.

Murata, A., & Pothen, B. E. (2011). Lesson study in pre-service elementary mathematics methods courses: Connecting emerging practice and understanding. In L. C. Hart, A. S. Alston, & A. Murata (Eds.), *Lesson study research and practice in mathematics education (part 2)* (pp. 103–116). New York: Springer.

National Clearing House for English Language Acquisition (NCELA). (2011). The growing numbers of English learner students. Retrieved from http://www.ncela.gwu.edu/files/uploads/9/growingLEP_0708.pdf

National Council of Teachers of Mathematics (NCTM). (2000). *Principles and standards for school mathematics*. Reston, VA: NCTM.

Parvanehnezhad, Z., & Clarkson, P. (2008). Iranian bilingual students reported use of language switching when doing mathematics. *Mathematics Education Research Journal, 20*(1), 52–81.

Pimm, D. (1987). *Speaking mathematically: Communication in mathematics classrooms*. London: Routledge.

Pimm, D. (1991). Communicating mathematically. In K. Durkin & B. Shire (Eds.), *Language in mathematical education: Research and practice* (pp. 17–23). Milton Keynes, England: Open University Press.

Rojas-Drummond, S., & Mercer, N. (2004). Scaffolding the development of effective collaboration and learning. *International Journal of Educational Research, 39*, 99–111.

Schäfer, M. (2010). Mathematics registers in indigenous languages: Experiences from South Africa. In L. Sparrow, B. Kissane, & C. Hurst (Eds.), *Proceedings of the 33rd Annual Conference of the Mathematics Education Research Group of Australasia* (pp. 509–514). Fremantle, Australia: MERGA.

Setati, M. (2005). Teaching mathematics in a primary multilingual classroom. *Journal for Research in Mathematics Education, 36*(5), 447–466.

Setati, M., & Adler, J. (2000). Between languages and discourses: Language practices in primary multilingual mathematics classrooms in South Africa. *Educational Studies in Mathematics, 43*, 243–269.

Thompson, D. R., Kersaint, G., Richards, J. C., Hunsader, P. D., & Rubenstein, R. N. (2008). *Mathematical literacy: Helping students make meaning in the middle grades*. Portsmouth, NH: Heinemann.

Usiskin, Z. (1996). Mathematics as a language. In P. C. Elliott & M. J. Kenney (Eds.), *Communication in mathematics, K–12 and beyond* (pp. 231–243). Reston, VA: NCTM.

Van den Akker, J., Gravemeijer, K., McKenney, S., & Nieveen, N. (2006). *Educational design research*. New York: Routledge.

Van der Walt, M. S. (2012). Voornemende wiskunde-onderwysers se metakognitiewe vaardighede tydens lesstudie in mikro-onderrig (MLS) [Prospective mathematics teachers' metacognitive skills during micro-teaching lesson study]. *Suid-Afrikaanse Tydskrif vir Natuurwetenskap en Tegnologie* [*South African Journal for Science and Technology*], *30*(1), 1–8.

Vorster, H. (2008). Investigating a scaffold to code-switching as strategy in multilingual classrooms. *Pythagoras, 67*, 33–41.

Vorster, H. (2009). A comparative study of different strategies to overcome language barriers in the teaching and learning of mathematics in countries where English is the LoLT. In A. Bilsel & M. U. Garip (Eds.), *Frontiers in science education research* (pp. 605–614). Famagusta, North Cyprus: Eastern Mediterranean University Press.

Vorster, H., & Zerwick, J. (2011). Exploring the use of simultaneous interpreting in the training of mathematics teachers. In M. Setati, T. Nkambule & L. Goosen (Eds.), *Proceedings of the ICMI Study 21 Conference: Mathematics Education and Language Diversity* (pp. 429–437). São Paulo, Brazil: ICMI Study 21.

Vygotsky, L. S. (1962). *Thought and language*. Cambridge, MA: MIT Press.

Vygotsky, L. (1978). *Mind in society: The development of higher psychological processes*. Cambridge, MA: Harvard University Press.

Webb, L. (2012, July). *Experiential learning: The introduction of exploratory talk to multilingual mathematics teachers*. Paper presented at the 12th International Congress on Mathematical Education, Seoul, South Korea.

Chapter 8
Mathematics in the Hands of Deaf Learners and Blind Learners: Visual–Gestural–Somatic Means of Doing and Expressing Mathematics

Lulu Healy, Elizabeth Becerra Ramos, Solange Hassan Ahmad Ali Fernandes, and Jurema Lindote Botelho Peixoto

8.1 Introduction

In this chapter, we focus on the linguistic resources used in the mathematics practices of students who lack, or who have very limited, access to one or other sensory field. Attention to students with disabilities is still relatively rare within the field of mathematics education as a whole. In the area of language diversity and mathematics learning, research has tended to consider diversity in relation to spoken languages rather than languages and linguistic resources expressed through other modalities, such as signed languages, or Sign,[1] and gestures. This scenario is beginning to change, at least with respect to those visual–gestural–somatic expressions described as gestures, with recent years bringing an increased attention to their communicative and cognitive functions in mathematical activities (see, for example, Edwards, Ferrera, & Russo-Moore, 2014; Nemirovsky, Kelton, & Rhodehamel, 2013; Radford, Edwards, & Arzarello, 2009) and an accompanying recognition of the multimodal nature of mathematical understandings (Radford, 2009; Roth, 2010; Roth & Thom, 2009). From these perspectives, the ways that

[1]Following Sacks (2000), we use the term Sign to denote all indigenous signed languages, though in this paper we limit our attention to MSL (Mexican Sign Language) and Libras (Brazilian Sign Language).

L. Healy (✉) • S.H.A.A. Fernandes
Anhanguera University of São Paulo, São Paulo, SP, Brazil
e-mail: lulu@baquara.com; solangehf@gmail.com

E.B. Ramos
Cinvestav-IPN, Mexico City, Mexico
e-mail: ebecerra@cinvestav.mx

J.L.B. Peixoto
State University of Santa Cruz, Illheus, Bahia, Brazil
e-mail: peixotojurema@gmail.com

R. Barwell et al. (eds.), *Mathematics Education and Language Diversity*, New ICMI Study Series, https://doi.org/10.1007/978-3-319-14511-2_8

linguistic resources are appropriated and used have their basis in the human body, its sensory and motor capacities and its location in space and time. For us, this suggests that a focus on the linguistic practices of those who do not hear with their ears or see with their eyes may open particular windows on mathematical cognition.

In case it may seem that we are treating mathematical cognition as an individual affair, bounded only by the body, we should stress that we see learners as essentially cultural beings and their participation in linguistic practices makes this particularly evident. Although bodily grounded, these practices are a central aspect in defining cultures, as every language, be it verbal, gestural, pictorial or of any another nature, is representative of a particular social group (Gee, 2014). The complexities associated with attending to different linguistic practices can result in tensions within the school context, especially if those practices that are emphasized do not take into account the specificities of the learners in question.

This has certainly been the case for deaf learners, whose history of participation in educational activities has been marked by a veritable battlefield related to linguistic modality. In the case of blind mathematics learners, on the other hand, considering linguistic resources of a visual-spatial nature, such as gestures, might seem misguided, since such resources will not be seen. However, there is evidence to suggest that even those who are born blind make use of gestures when speaking (Iverson & Goldin-Meadow, 1998) and when engaging in mathematics (Healy & Fernandes, 2011, 2014).

Whether we are speaking of Sign or gesture, Rotman (2009) has pointed to a general tendency to devalue communication systems which make use of the visual modality as compared to orally based ones and to assume that language should be identified with speaking, while communications using body movements are judged more primitive and nonintellectual. Perhaps the clearest example of this devaluation can be found in the history of Sign in the education of the deaf. In the following section, to consider how the attention given to visual rather than verbal communication forms has changed over time, we begin by recapitulating moments from this history.

8.2 The Rise, Fall and Rise of Sign and Gesture

For those who are born deaf into a hearing world, it is not so much the absence of sound but the consequences which derive from this which have dictated the ways in which they have been positioned and defined (Sacks, 2000). Before the mid-1700s, the deaf were treated as uneducable, their "inability" to speak indicative of an intellectual disability. It was only when attention began to be paid to signed languages that this view was challenged, initiating attempts to include deaf learners in the education system and also giving rise to a still ongoing debate about the type of language to be used in educational practice. The debate is frequently polarized in terms of use of signed languages verses oral methods.

On one side was the Frenchman Abbé de L'Epée, who, based on the signs used by the deaf people of Paris, elaborated the sign system, Methodical Sign. In 1760, he founded the first public school for the deaf in France, the "Institution Nationale des sourds-muets" in which this system was used. His interest in manual communication forms occurred as French philosophers were debating the origins of human languages and the sensualist Condillac was arguing that the language of action constituted the original, natural language. Condillac's view was that the transformation of embodied experiences into shared material signs (the "transformed sensations" that represent this language of action) holds the key to human knowledge.[2] Such a position implies no intellectual hierarchy between manual or verbal forms of communication.

At the same time, in Germany, the educator Samuel Heinicke was developing an oral/aural method to teach deaf people to speak. He was strongly opposed to the methods used by L'Epée and the ideological split was born (Moura, 2000). Almost a century later, in 1880, at the Congress of Milan participants "voted to proclaim that the German oral method should be the official method used in the schools of many nations" (Lang, 2003, p. 15). Deaf people were excluded from this vote. In announcing the congress recommendations, one of the organizers put it thus:

> Gesture is not the true language of man which suits the dignity of his nature. Gesture, instead of addressing the mind, addresses the imagination and the senses. Thus for us, it is an absolute necessity to prohibit that language and replace it with living speech...The fantastic language of signs exalts the senses and foments the passions, whereas speech elevates the mind much more naturally, with calm and truth. (Guilio Tarra quoted in Lane, 1984, pp. 391–394)

Following the Congress of Milan, Oralism, with its supposed intellectual and even moral superiority, dominated and, for many years, deaf learners were discouraged, and frequently physically forbidden, from using sign language during schooling.

It was only in the 1960s and 1970s that this dominance began to be challenged, particularly after the scientific recognition of American Sign Language (and consequently the sign languages of other countries throughout the world) as a true and natural language (Stokoe, 1960/2005). When the studies into the structure of sign languages began to emerge, formalist models of language drawn from structural linguistics were at their height. In this context, signs had to be shown to be equivalent to lexical items in spoken languages and phonological structures in sign languages corresponding to those of spoken languages needed to be identified (Armstrong & Wilcox, 2003). Clearly, in relation to deaf education and culture, recognition that the visual–gestural systems of communication used by the deaf are proper languages was (and is) fundamental. Yet, there were two perhaps unintended consequences of these attempts to demonstrate that the structures underlying signed languages were the same as spoken ones.

First, the issue of how the modality of Sign might impact on its nature has tended to be de-emphasized and, in particular, the iconicity of many signs downplayed, as

[2] According to Kendon (2008), at this point in time, distinctions were not drawn between gestures and Sign.

signifiers have had to be shown to be arbitrary in relation to what they signify. Second, though gestures might be produced in the same medium as signs, in the formalist view, they differ in that they are spontaneous and idiosyncratic and not constructed according to any standard forms, which meant, it could be argued, that they are not part of language. In order for Sign to be attributed the same status as spoken language, Sign and gesture were hence organized as distinct categories: Sign regarded as something linguistic and gesture seen as external to language, paralinguistic (Kendon, 2008; Wilcox & Morford, 2007).

It was only in the mid-1980s that gestures began to be seen as an integral part of language, with McNeill (1985) showing how they are used during speaking to constitute the conceptual content of the utterance. For Kendon (2008), though, while McNeill's work led to an increased attention to gestures, in general they were still seen as a kind of appendage or add-on to speech. A more radical view is offered by Rotman (2009), who argues rather that speech is a species of gesture, perceived by auditory rather than visual means, but a gesture nevertheless. This implies that gestures are as much a part of the set of language resources used to share experiences of the world as are the components of spoken and signed languages. It also brings us back to the beginning of this section, and to Condillac's conjecture that languages emerged from a process of transforming sensations (Hewes, 1996; LeBaron & Streeck, 2000). Hence, we might argue that the valuing of visual–gestural–somatic modalities of communication goes hand in hand with recognizing the embodied nature of our sense-making experiences.

In mathematics education, as well as in the area of linguistics, it is only recently that such recognition has gained space. As a consequence, we still know very little about what it means to learn and to do mathematics using the visual–gestural modality. To a certain extent, in relation to this modality, the learners on whom we focus in this chapter represent two extremes. The visual modality is ever present for the deaf, with evidence to suggest a preference for visual reasoning amongst those who do not hear (Bull, 2008; Kelly, 2008; Monteiro & Andrade, 2005; Nunes & Moreno, 2002), while, for the blind, spatial and visual information is not seen but felt or heard. By concentrating on these learners, then, what might be learnt about the role of visual–gestural–somatic language resources in mathematics learning?

8.3 Sensory Modalities and Knowledge Mediation

It was a related question, though concerned with learning more generally, which appears to have motivated the construct of knowledge mediation in the sociocultural perspective of Vygotsky. This construct has its roots in his work with differently-abled individuals (Vygotsky, 1997). In Vygotsky's view, all higher mental functions are mediated. A mediated mental function involves an indirect action on the world, which incorporates and transforms the natural, basic mental processes, extending their range and mode of functioning. The inclusion of the tool in activity alters the course both of the activity and of all the mental processes that enter into the instrumental act (Vygotsky, 1981). Material and semiotic tools do not just enable cognitive activity; they are part of the act of thinking *and* that which is being thought about (knowledge).

For Vygotsky, the use of language as an instrument in thinking is central to the ways that learners appropriate—that is, make their own—the forms of acting and communicating which characterize the social groups to which they belong. Generally speaking the instruments and languages of culture tend to be designed for those considered "normal," for those who have all the organs of the senses and the sensorial functions intact, meaning that they may not be accessible to some. In the perspective offered by Vygotsky, the solution for the inclusion of individuals with disabilities in social (cultural) activities lies in seeking ways to substitute the traditional mediational means with others, more suited to the specific ways in which they interact with others and with culturally defining objects.

Consistent with his ideas about mediation, he believed that while enabling intellectual development, the substitution of one tool by another (for example, signed rather than spoken language for the deaf or hands as seeing tools for the blind) could lead to the emergence of different developmental paths, since, just as the inclusion of any other tool in the process of activity alters its entire structure and flow, so too the substitution of the ear or the eye by another instrument would be expected to be associated with a profound restructuration of the intellect.

Here, we should make very clear that we are not referring to a state of deficiency, but to one of difference. To better understand the deaf mathematics learner, we need to better understand what it means to practice mathematics in the medium of sign language and how those whose cognitive processes are mediated by a visual–gestural–somatic language as opposed to a sequential-auditory language come to think mathematically. Similarly, to better understand the blind mathematics learner, we need to investigate how those who process visual data through touch or sound express the mathematical properties that they feel or hear.

We now turn to our attempts to contribute to the development of such understandings. We offer two examples involving deaf learners in order to consider in more detail the challenges associated with learning mathematics when visual–gestural–somatic language resources are the dominant forms of communication. We then extend the discussion of the relationships between sensory experiences, language and mathematics learning by including an example examining gesture use by blind learners.

In the light of the previous discussion, our aim in presenting these examples is threefold: to explore the characteristics of mathematical activity privileged by the visual–gestural–somatic forms of expression; to consider the relationships between lexical terms (be they signed or spoken) and gestures; and, more generally, to seek evidence of how mathematical activity and understandings are shaped by (and shape) different ways of sensing and acting and different ways of attempting to share these experiences with others.

8.4 Language Resources of Deaf Mathematics Students

The rise, fall, and recent rise again of interest in Sign and gesture has had, and continues to have, profound implications for the education of the deaf. In the countries in which we work, Brazil and Mexico, it is only during this century that Sign has begun to be (re-)considered as a medium for teaching and learning. Currently, at least some schools (both specialized and mainstream) are beginning to adopt approaches in which Sign is considered as the deaf learner's first language (L1) and the written version of Portuguese or Spanish as a second language (L2).

While linguistic research has demonstrated that signed languages are the natural languages of deaf communities (Cruz, 2008), Sign is not a universal language. There are a large variety of different signed languages across the world, many of which have developed independently of each other. However, the largest differences between signed languages appear to be lexical in nature, with grammatical features shared in most of the signed languages studied to date. This is the case for the two signed languages that figure in the examples to follow Libras and MSL.[3] Signs in both these languages are divided into five categories:

(a) The *form or configuration* which the hand or hands adopt when performing the sign.
(b) The *place or location* in which the sign is performed.
(c) The *movement* performed by the hand, which can have a variety of internal elements, including trajectory, direction, speed, rotation, muscular tension, and vibration among others.
(d) The *orientation* of the hand's palm.
(e) The *body or facial expression*, which accompanies the sign.

Differences in any one of these five parameters can give rise to assigning distinct meanings.

Their grammars, like the grammars of many other signed languages, make similar use of locations and orientations in space, the direction, quality and speed of movements, facial expressions, and sign orders. Thus, Sign is "written" in space, the signer manipulates the space to refer to spatial, temporal, and grammatical matters and different spatial planes are used to manage the timeline, present, past, and future.

Another shared construct of signed languages, and one that has been associated with a degree of linguistic controversy, is its iconicity. Because Sign is visual–gestural–somatic and the visual properties of entities and actions are so readily accessible, they are utilized in abundance in Sign. How iconicity is used in the emergence of language in creating novel practices and in historical change are questions that are beginning to attract research (Brentari, 2010) and the first two examples bring this discussion to the area of mathematics education.

---

[3] See Brito (1995) for a discussion of Libras and Cruz (2008) for a description of MSL.

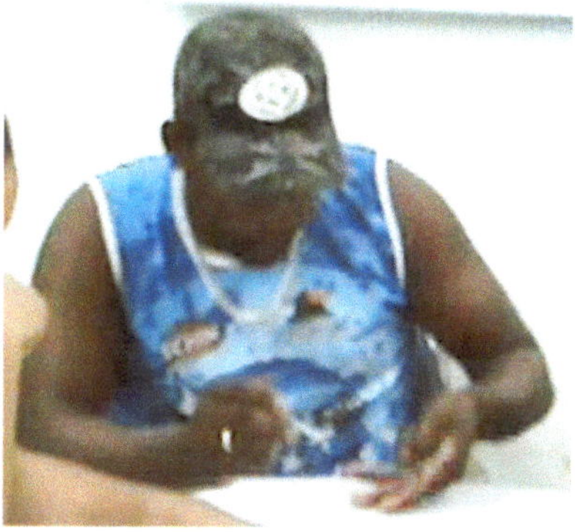

Fig. 8.1 Signing 3 and beating his hand three times on the tables

8.4.1 *Example 1: The Visual Modality in Arithmetic Calculations*

The first study we consider was carried out in the State of Bahia in the Northeast of Brazil. The data reported here were part of a larger study whose aim was to survey school students' knowledge of additive and multiplicative structures.[4] Here, we focus on the strategies of a profoundly deaf student, Rodrigo[5] (aged 24) as he carried out two calculations: 32×3 and 65×6. Rodrigo only started to learn Libras as an adolescent, having begun his education in a specialized school before the policies on bilingual education had taken hold in Brazil. We focus here on the signs, gestures, and written expressions that emerged during his calculations.

Faced with the task of multiplying 32 by 3, Rodrigo began by signing with his left hand the number 3, then beating this configuration three times on the table (Fig. 8.1). This seems to have been a way of signalling to himself the calculation he should perform. To obtain the result, he signed 3 once again, this time to referring to the digit 3 in 32. He used his right hand, which he moved in space twice (Fig. 8.2). As his right hand was moved to the second location, he simultaneously signed 6 with his left. Maintaining the three on his right hand, he then used his left to sign 7, 8, 9, arriving at the first part of his answer.

Having obtained the result associated with the digit 3 in 32, Rodrigo repeated the same procedure with the 2, moving the sign of 2 on his left hand (Fig. 8.3), to three locations in spaces while counting, in Sign, with his other hand, 3, 4 then 5, 6. He then registered his result on paper (Fig. 8.4).

---

[4]This research was funded by the Fundação de Amparo à Pesquisa do Estado da Bahia (FAPESB 2008–2010).

[5]The names of the students whose work is presented in the paper have been changed. All of them participated voluntarily in the respective studies.

Fig. 8.2 Signing 3 in different locations while counting

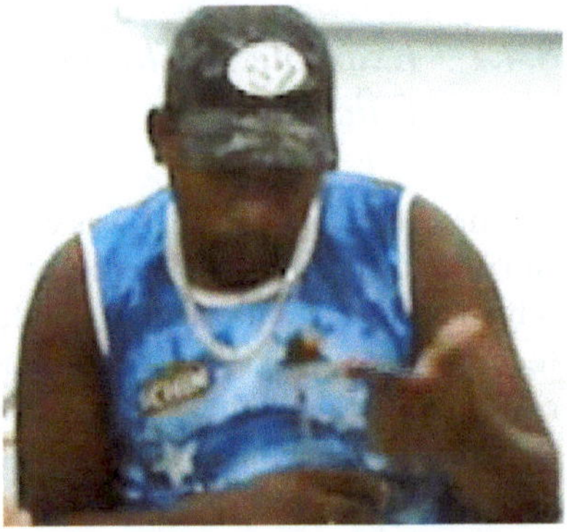

Fig. 8.3 Signing 2 and moving the sign in space

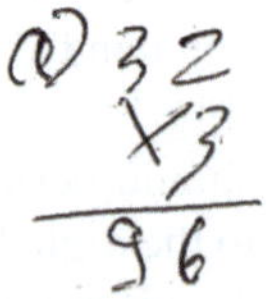

Fig. 8.4 Rodrigo's written expression

This strategy illustrates a difference between signed and spoken languages, in that using Sign, Rodrigo is able to say more than one word at the same time. What is possible, though, in spoken language is that the speech be accompanied by spatial gestures, indicating that the visual–gestural–somatic modality can bring simultaneity to expressions in both cases. Indeed, some of Rodrigo's hand movements are also better described as gestures rather than signs: the beating of the 3 on the table, for example, and the positioning of signs in space to keep track of the number of repetitions.

Although the difference in the meanings of the 3 and the 2 in 32 are not made explicit in Rodrigo's expressions, his negotiation of the calculation 65×6 suggests he was aware of this difference. This time, he started with the digit 5 related to the unit value. Once again, he combined a variety of expressions, including signs, gestures and written inscriptions. He began by signing 5 on his right hand and simultaneously

Fig. 8.5 Simultaneously signing 5 on one hand and 1, 2, 3, 4, 5, 6 on the other

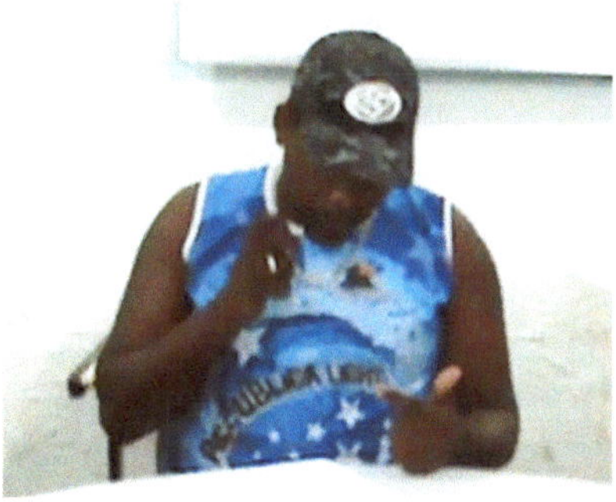

Fig. 8.6 Joining his fingers 2 by 2

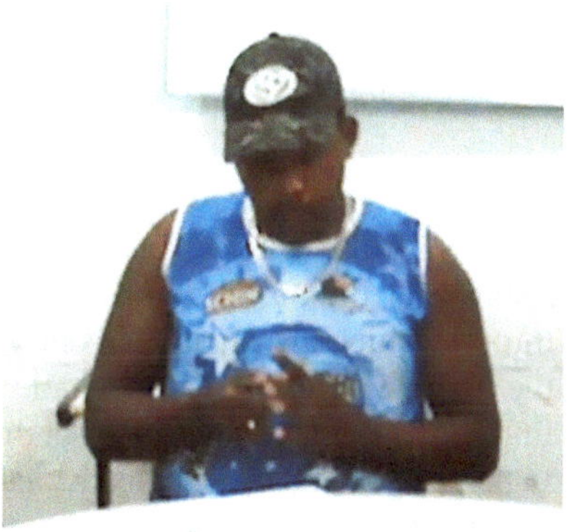

signing from 1 to 6 with his left, suggesting that he was thinking 5 once, twice, thrice up until 6 times (Fig. 8.5).[6] He then calculated this by holding out 6 fingers, each of which to represent 5 and then joining his fingers in pairs, with each pair representing 5+5 (Fig. 8.6). He repeated the strategy for the digit 6, this time also recording his method on paper (Fig. 8.7). Finally he completes the written multiplication correctly (Fig. 8.7).

In both the calculations that Rodrigo performed, he capitalized on the fact that when using a visual–gestural–somatic language it is possible to say more than one word at the same time. Nunes (2004) also observed this practice in her work with British deaf learners, describing a spontaneously developed strategy to arrive at the sum of two whole numbers by counting up with one hand while simultaneous counting

---

[6]We note here that the sign in Libras for the number 5 does not involve holding out five fingers.

In fact this is the case only for the numbers from one to four, as shown alongside.

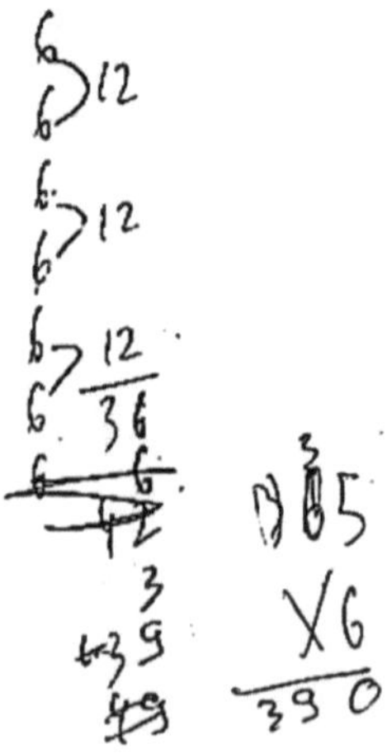

Fig. 8.7 Rodrigo's written calculations

down with the other. Considering these strategies through a Vygotskian lens, it seems that the substitution of spoken language by Sign altered the possibilities through which the calculations might be negotiated. At the very least, we argue that the use of visual–gestural–somatic communication modes brings a simultaneity to linguistic practices not present in the serial-auditory utterances of spoken language. In the next example, we consider another consequence of this substitution, and compare the dynamism of Sign to the more static nature of spoken language.

8.4.2 Example 2: Bringing Dynamism to Mathematical Discourse

The second example is drawn from a study carried out in a Mexican school for deaf students. The school adopts a bilingual educational model (written Spanish and MSL), but the particular aim of this study was to design and investigate mathematics learning scenarios in which the dynamics of the class are mediated entirely through Sign and through visual representations: that is, with no recourse to written (or oral) language. The scenario we describe here involves Pythagoras' Theorem and interactions with visual proofs of the relationships between the sides of right-angled triangles that it specifies. Four deaf students participated in the activities, two 19 year olds and two 16 year olds. Both the older students were competent users of MSL, while the younger students were still in the process of learning MSL.

To begin, the four students were presented with a series of figures (Fig. 8.8). They were asked to describe the figures, in accordance with characteristics of their choice, which might include form, area, size, or even color. They were then asked specifically to compare the areas of the different polygons in the figures.

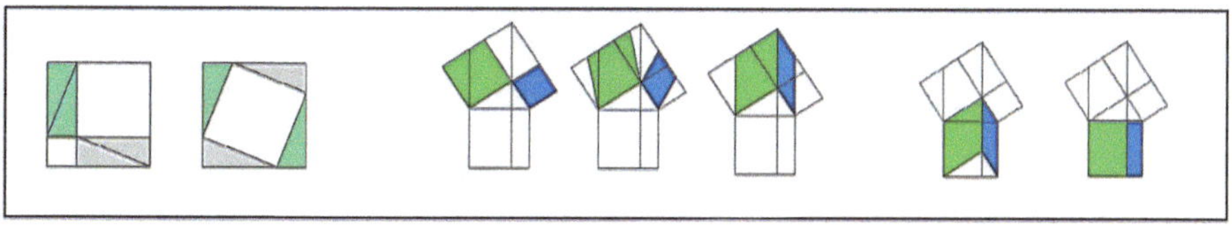

Fig. 8.8 Figures for Pythagoras' Theorem (from the book "Proofs without words," Nelsen, 1993)

Fig. 8.9 "Square" in MSL

In turn, they were shown a series of dynamic geometry applets also illustrating, this time with movement, demonstrations of Pythagoras' Theory. Following the consideration of static and dynamically presented visual proofs, the group worked on a series of examples in which they were asked to relate diagrams with the algebraic representation of the theory and to calculate the values of the sides of the triangle and the area of their squares in a number of specific cases.

A considerable difficulty of researching the interactions in scenarios involving Sign-speakers is the registering of evidence. Video-recordings help to capture the dialogue, but without several cameras filming the same scene, it is easy for signs to occur off camera, or for only part of a sign, say hand-form and movement to be captured, without the accompanying facial expression. Even where the video data is complete, literal transcriptions cannot be made as signed languages are written in space, and there is a danger that, if we rely on written interpretations of the dialogues, we may lose some of the particularities associated with doing mathematics in Sign.

Perhaps the best way to illustrate this difficulty is to present examples of the signs that emerged during the sequence of activities. Some are totally iconic, their form corresponding pictorially to the object being signed. Examples include the sign for squares, rhombi, and for some kinds of triangle. Figure 8.9 presents the MSL sign that was used in the study to represent squares.

Other signs have an iconic component, but are accompanied by some sort of movement. Examples include the signs for rectangles, isosceles, and right-angled triangles, and some of the signs used to make comparisons. In the case of the geometrical figures, the movement conveys a certain level of generality, indicating a class of objects rather than one a specific case, as well as making explicit particular properties of the shape in question. Figure 8.10, for example, presents the MSL sign for rectangle. In its initial appearance, it is the same as the square, but whereas the square sign does not involve movement, to sign "rectangle" involves keeping the hand showing three sides stationary while the finger representing the fourth side is translated in a horizontal way, as if the two horizontal sides were being gradually extended.

Fig. 8.10 Signing "rectangle" in MSL

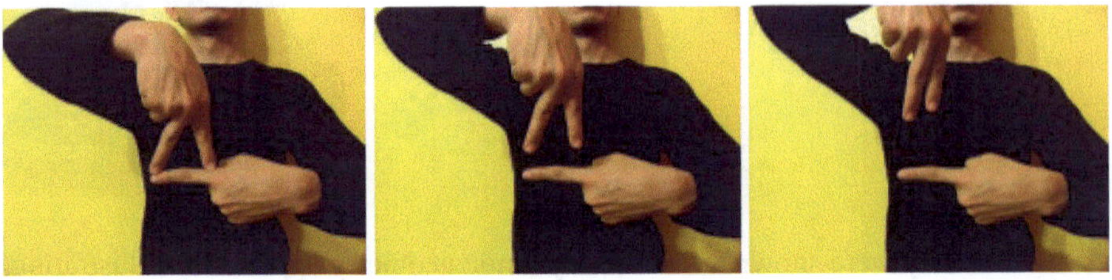

Fig. 8.11 Signing "isosceles triangle" in MSL

Fig. 8.12 Signing "more than…"

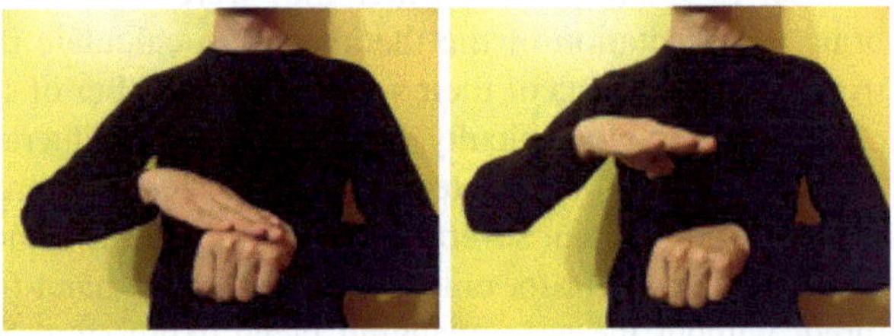

Fig. 8.13 :… and "less than"

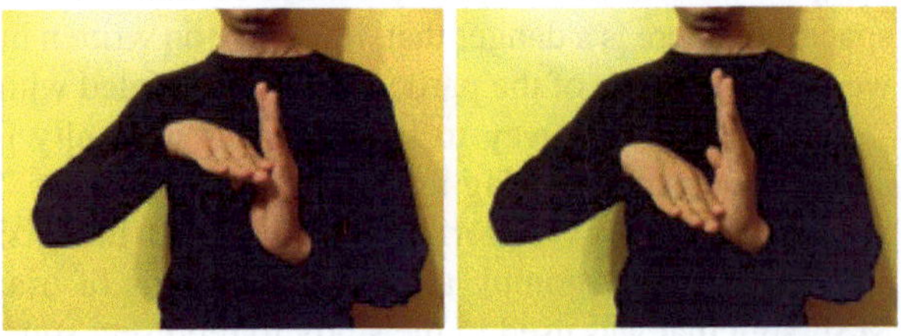

The sign for "isosceles triangle" is similarly dynamic and involves a movement which implies maintaining two sides congruent (Fig. 8.11).

These signs are interesting in that they differ both from the spoken words and from the visual drawings that are usually used to represent geometrical objects. In the dynamic sign, the generic nature of the object being denoted is much more explicit than in a paper and pencil drawing or perhaps even in a spoken word. MSL, then, brings a dynamism to mathematical discourse. This dynamism is not only evident in the signs representing objects, but also in those which denote relationships. Figures 8.12 and 8.13 show the movements associated with the MSL signs for comparing "more than" and "less than."

Through the use of signs like these, the four students were able to argue that the sums of the area of the smaller squares in each of the diagrams shown in Fig. 8.8 were equal to the area of the larger square, and to describe the transformations by which they determined this. They were also able to successfully relate the visual representation of the Pythagorean Theorem with its algebraic representation $c^2 = a^2 + b^2$. This indicates that MSL allows for the abstraction of the algebraic representation from a graphic one. Thus, in spite of the fact that the way of expressing their ideas is very iconic, and the ideas themselves based also on the visual material at their disposal, the students could make sense of a less visual means of expressing the mathematical ideas they worked with.

We might even speculate that the ways in which movement is incorporated into the signs expressed by deaf mathematics learners embodies variation in a way that is different to algebraic symbolism but perhaps serves as an effective means of enacting meanings for it: the dynamic movement of, say, the opposite sides of a rectangle in its sign already indicating that the measure of these sides is a value that varies, while their relationship to each other does not. Of course, it may also be the case that the signs constrain as well as afford the interpretations that come to be associated with the objects they are intended to represent. The signs for both "rectangle" and "isosceles triangle" privilege the prototypical orientations of the objects, presenting them horizontally. Based on these observations, we would suggest a need for more research investigating the signs for mathematical terms in different signed languages, so we might better understand how the signs were created and the properties that they appear to privilege or constrain in use.

8.4.3 Traces of Enactments in the Signs and Gestures of Deaf Mathematics Learners

Although the mathematical content, demands and structuring of the activities were rather different in these two examples, there is evidence in both that language resources of a visual–gestural–somatic nature not only enable the expression of mathematical objects and properties, but also shape all the aspects of the learners' activities with them.

The combination of signs and gestures used by Rodrigo as he multiplied were made not so much in order to communicate his ideas to others, they were integral to the processes of thinking that occurred. We might say that the gestures and signs served as visually expressed enactments of imagined activity: Rodrigo's holding up of six fingers, each one of which represented five objects, suggests he was imagining a physical process of combining objects. In a similar way, the signs used to refer to mathematical objects in Mexican Sign Language might also be seen as enactments or re-enactments of the activities involved in producing and exploring such objects. Yet, although firmly connected to enactments of physical doings, the visual–gestural–somatic expressions also bring evidence of processes of abstraction, at least if we define abstraction as some conscious appreciation by learners of the generalized relationships implied in their expressions (as do Mason, 1989; Noss & Hoyles, 1996).

This is especially so in relation to the signs described in the second example: creating a set of isosceles triangles by moving the hand in a way that preserves the congruency of two sides, for instance, a form of explicitly articulating this general property.

In the context of the visual–gestural–somatic expressions of generality made by the deaf learners, abstraction does not seem to involve the detachment of objects, properties, and relationships from the settings in which they were encountered: that is, their mathematics learning does not seem to involve a process of de-contextualization. They continue to sense, to feel, the multimodal experiences involved in the process of identifying generalities: thinking, speaking or signing and gesturing, are accompanied by feeling. Indeed, we offer the gestures and signs described in the examples as one source of evidence to support a premise of embodied cognition: that thinking involves re-enacting and hence re-feeling previously experienced activities. Re-enactment, though, should not necessarily be seen simply as a replay. It is a new performance, that Nemirovsky et al. (2013) describe as a "social-interactive experience of bringing to presence something which is absent in the current surroundings of the participants" (p. 3). In its virtual form, this something can be acted upon in new ways, providing new forms of experiencing its potential in new kinds of activity.

8.5 Visual–Gestural Expressions of Blind Mathematics Students

While it is not surprising that the visual–gestural–somatic modality is central in the case of deaf learners, what would we expect in the language activities of blind mathematics learners? The following example suggests that this modality represents as integral a part of the language activities of the blind as it does for those who see with their eyes, with even students who have never seen (visually) the gestures of others spontaneously producing gestures in the course of their mathematical explorations and explanations.

8.5.1 Example 3: Embodied Abstractions in the Gestures of a Blind Student

This last example comes from an ongoing programme of research[7] in which we are exploring relationships between sensory experience and mathematical knowledge. Our research activities have included attending to how the use of hands to substitute eyes, and touch to substitute sight, in the sense proposed by Vygotsky, impacts upon the mathematical practices of blind learners. Focusing on the hands of these learners as they explore material–tactile–representations of geometrical objects, is

[7]We are grateful to the funding we have received from FAPESP (Project no. 2004/15109-9) and CAPES (Project no. 23038.019444/2009-33) in the course of this programme of research.

illuminating both the extensive use of visual–gestural–somatic language resources by blind students, as well as the embodied nature of the mathematical interpretations that these expressions imply.

We have chosen an episode from a sequence of learning situations undertaken with a group of four learners who attended a mainstream school in São Paulo, Brazil. The learners, whose ages were between 14 and 18 years, were first-year high-school students. Marcos (18 years old), Fabio (16 years old), and Caio (17 years old) were all born with different kinds of congenital blindness, while Leandro (14 years old) lost his sight at the age of 2 in an accident that resulted in the severing of his optic nerve. The sequence involved a series of tasks associated with the study of volume, area, and perimeter, implemented over four research sessions, each of approximately 90 min. In the first two sessions, the activities centered on the area and perimeter of plane figures, the third initiated work on volume and in the final session, the students worked on a task which involved determining the most economical amongst a range of boxes and other rectangular prisms. All the research sessions were videotaped and transcribed for analysis. Two researchers and the schools' special needs teacher also participated in the sessions.

In the first session, the students worked with a wooden board containing the impressions of four different rectangles, which could be filled either with wooden unit cubes or with rectangular and triangular shapes in foam rubber. In this session, the students explored the perimeters and areas of rectangles and right-angled triangles. In particular, they experienced how the area of a triangle could be perceived as half the area of a rectangle with the same height and base (for a more detailed analysis of the activities during this session, see Fernandes & Healy, 2010).

The second session was dedicated to determining the areas of the plane figures represented in foldable cardboard. During this session the students worked with adapted rulers in which the number marks were raised so they could be read tactilely. The episode we present here occurred during the explorations of one of the first cardboard figures analyzed by the students, a right-angled triangle with sides of 5, 12, and 13 cm. Each of the students received a cardboard representation of the same triangle, the idea being they would first determine its area and perimeter individually and then share and agree upon their results.

This example centers upon the strategies of one of the students, Leandro, as he attempted to determine the triangle's area and perimeter. Reminded by Caio and Fabio that its area would be half the area of a rectangle of the same height and width, he calculated it to be 30 cm, and explained his thinking to one of the researchers.

Leandro: The area, the area, I understand it how they said. The rectangle would be 60 and dividing would give 30.

Researcher 1: And why would the area of the rectangle be 60?

Leandro: Because it has one side of 12 (*he traces the side of the triangle which measured 12 cm, Fig.* 8.14a), and the other would be 12 (*traces an imaginary segment in space parallel to the side of triangle he had previously indicated, as shown in Fig.* 8.14b). And then 5 (*again tracing the side of measure 5 cm, Fig.* 8.14c) and 5 (*and an imaginary parallel, Fig.* 8.14d).

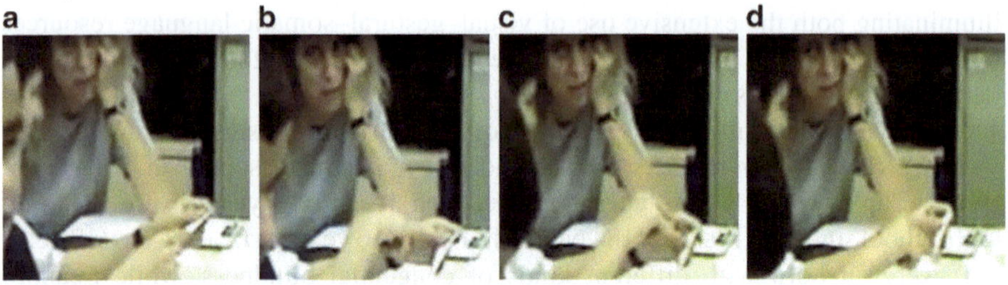

Fig. 8.14 (**a**) "one side of 12" (*Leandro traces the 13 cm side…*). (**b**) "the other would be 12" (*… and indicates a parallel*). (**c**) "And then 5" (*He traces the 5 cm side…*). (**d**) "and 5" (*… and indicates its parallel*)

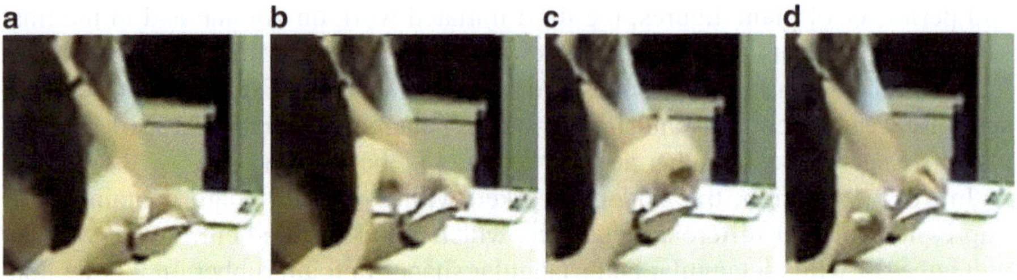

Fig. 8.15 (**a**)"I thought 12," (**b**) "plus 12 on the other side," (**c**) "5," (**d**) "plus 5"

Researcher 1: OK
Leandro: So it would be 30.

The gestures shown in Figs. 8.14a–d were an integral part of Leandro's explanation. We cannot be sure whether they were part of an intentional act to communicate to the researcher (although Leandro could not see his gestures, he knew that the researcher could) or whether they were for himself, as much a part of his thinking as the spoken words. Our conjecture is that, in practice, they served both these functions. Moreover, just as the sign for "rectangle" in MSL emphasizes defining properties of the shape, so too did Leandro's gesturing. Gestures continued to be fundamental as Leandro attempted to determine the perimeter. His first conjecture involved first calculating the perimeter of a 12 by 5 rectangle and then halving the value.

Leandro: I think the perimeter is 17.
Researcher 1: The perimeter of this is 17? Explain why.
Leandro: I thought 12 plus 12 on the other side (*once again he traces two sides of an imaginary rectangle, Fig.* 8.15a, b). 5 plus 5 (*tracing the other two opposite sides Fig.* 8.15c, d). That would give 34 and I divided by 2.17.

This time evoking, through the same set of gestures, the rectangle that had enabled a correct calculation of the triangle's area, led Leandro to overgeneralize the strategy of dividing by 2. Without commenting on the strategy, the researcher suggested that Leandro share his thinking with the other students.

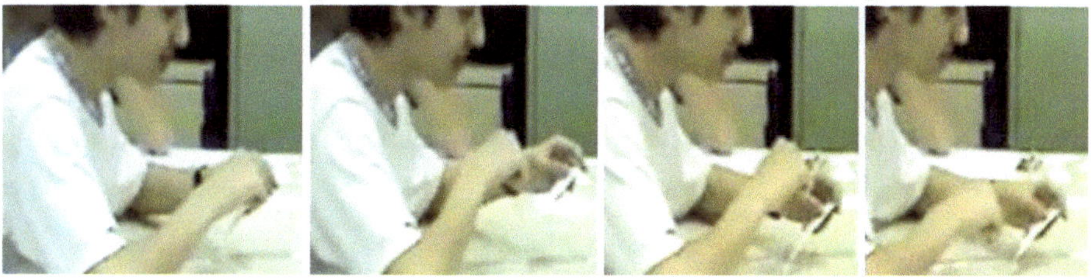

Fig. 8.16 (**a**) "12," (**b**) "plus 12," (**c**) "5," (**d**) "plus 5"

Researcher 1: Leandro has another conjecture about the perimeter. He says that it is 17.

Caio: The right-angled triangle? (*Caio picks up the triangles*)

Leandro: Yes, this here (*holding up his triangle*)

Caio: I don't think so, because if you add the 12 that is here (*runs his finger on the side measuring 12*), the 5 (*traces the 5 cm side*) and the hypoteneuse, you get 30.

This exchange is interesting because both Leandro and Caio use gestures and the deictic reference "here," which the other cannot see. In order to determine that they are discussing the same shape, they hold up in turn the triangle and then, as Caio explains how he calculated the perimeter by adding the measures of the three sides, he traces his finger along each side as he mentions it. In this case, we imagine the gestures were more for themselves than for the other, since, as the dialogue continues, it seems that Leandro has not yet appropriated Caio's method and wants to explain his own,

Leandro: What I did was this Caio. I calculated as if I had the rectangle in my hand. So I went 12 (Fig. 8.16a) plus 12 (Fig. 8.16b). 24. 5 (Fig. 8.16c) plus 5 (Fig. 8.16d). 34. And then I divided.

As he explained to Caio, Leandro repeated, for a third time, the same four gestures in which he traces the two parallel sides of measure 12 cm (one along a side of the triangle and the other in the air) then the two parallel sides of measure 5 cm (again one along a side of the triangle and the other in the air). Caio could not see these gestures, so perhaps it was the verbal reference to the rectangle or the way in which the measures of its sides were recited in congruent pairs that enabled him, and Fabio who had also begun to pay attention to Leandro's explanation, to understand that Leandro's method was incorrect.

Caio: Ah, but the sides of the triangle are not equal (traces the perimeter with his fingers).

Fabio: Not to mention that the hypotenuse is irregular and longer than the others.

In the light of these comments, Leandro revised his method, with Caio's approval.

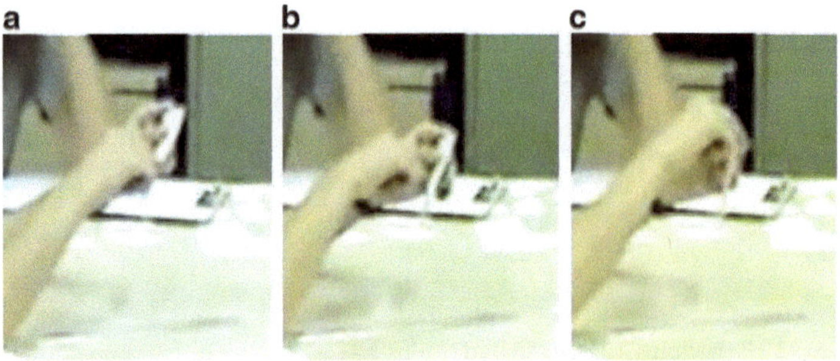

Fig. 8.17 (**a**) "Perimeter is this here," (**b**) "plus this here," (**c**) "Plus this here"

Leandro: Now I get it, So, if you measure the hypotenuse (*he measures the hypotenuse*), the hypotenuse is 13. To find the perimeter of this (*hits the triangle twice*), I calculate 12 plus 13 plus 5?

Caio: Yes, yes.

To check that Leandro had really understood, the second researcher asks Leandro what a perimeter is: his reply, completely correct, once again relied as much on the visual–gestural–somatic modality as on the verbal.

Researcher 2: What is perimeter Leandro?

Leandro: Perimeter is this here (Fig. 8.17a) plus this here (Fig. 8.17b) and this here (Fig. 8.17c).

Leandro's gesture communicates unambiguously to the researcher that he knows what a perimeter is. Because he has a particular figure in hand, it could be argued that his answer is specific rather than general, but in the context of the complete set of gestures that accompanied the dialogues in this episode, we are convinced that he had appropriated a general sense of perimeter and the fact that he chose to define perimeter using a visual–gestural–somatic expression provides evidence for our claim that this modality serves as a language resource as much for those who feel gestures as for those who see them.

The data from this episode also strengthens the claim that gestures emerge as a consequence of imagined re-enactments: in this case as Leandro imagines a rectangle that is not physically there, his hand movements indicating it is almost as if he is feeling the nonpresent shape. The gestures were far from arbitrary hand-waving, with the repeated use of the exact same sequence (Figs. 8.14, 8.15, and 8.16) suggesting that they acted as a kind of embodied abstraction, and a representation of rectangle incorporating aspects of its meaning for Leandro. A striking feature of his gestures, and one that makes them, like the signs of the deaf learners, difficult to capture using the paper and pencil medium, is their dynamic nature. Leandro's fingers were constantly moving as he traced out line segments in ways which preserved their relationship to others in space. Such dynamic gestures in which blind students abstract and express mathematical relationships appear to be characteristic of their interactions with geometrical objects (other examples are available in Healy & Fernandes, 2011 and Healy, 2012).

Just as in the case of deaf mathematics learners, then, it seems that for blind learners too the visual–gestural–somatic modality brings dynamism to mathematical discussions and practices and the ubiquity of such expressions, even amongst those who do not see them, suggest that they are an integral part of the process of learning and doing mathematics.

8.6 Mathematics in Our Hands

The examples presented in this chapter have attempted to explore the role of visual–gestural–somatic expressions in the mathematical practices of deaf learners and blind learners. We began by outlining how, for a considerable period in history, such communication forms were devalued in relation to spoken languages, with, in the extreme case of educating deaf learners, their use prohibited in the classrooms of schools in many countries across the world. One result is that it is only relatively recently that attempts to investigate the role of visual–gestural–somatic expressions in doing and learning mathematics are beginning to emerge. There is, we believe, a factor that we have not yet mentioned that is also contributing to this growing interest. Previously, much research into language use, in mathematics education and beyond, relied on analyses of written transcripts. It is now much more common for video data to form the bases of such analyses. Video-recordings of visual–gestural–somatic expressions, whether in conjunction with spoken language or not, provide a way of recording such utterances for future reference, in a way that writing has traditionally served for recording spoken words.

Not only does this technology make it possible for detailed analyses of Sign and gesture to be undertaken, it also offers the possibility of bringing new language resources to the teaching of mathematics. Neither Libras nor MSL has a widely-used official written representation. This has meant that to access most teaching materials, deaf learners have had to work mainly in their second language. This no longer needs necessarily to be the case, as it would be possible to develop digital resources in which activities are presented in Sign and even in which students' solutions are also recorded in the visual–gestural–somatic modality in which they are produced. Such a scenario is still distant from the realities of the mathematics classrooms in our countries, but we see it as one promising area for future research.

Returning to the present, and to the examples presented in this paper, we have made three main claims. The first is that the visual–gestural–somatic modality is amply used by both deaf students and blind students, in both cases bringing a dynamism and a simultaneity to mathematical discourse less easily expressed in spoken or written representations. These dynamic expressions incorporate properties of the mathematical objects, relations, and operations that they represent and are hence central constituents of the conceptual meanings that the students use in practice.

The second claim concerns the relationships between gestures and Sign and between gestures and speech. Looking at gestures in multiple sensory modalities is central to understanding signed languages, since it is not always easy, even with a competent management of Sign, to distinguish between conventional signs that have a priori meaning and natural gestures that emerge in the discourse. Our data suggest that understanding the meanings intended in the speech and gestures of our blind participants similarly involved considering their utterances as a combination of words and gestures. The deaf student, Rodrigo, and the blind student, Leandro, both used gestures in coordination with the other language resources as tools for thinking. In fact, for Rodrigo both gestures and signs seemed to be directed to organize his own strategies, rather than to communicate with an external interlocutor, while Leandro's words and gestures simultaneously served both roles. Vygotsky (1962) posited word meaning "as a unit of both generalizing thought and social interchange" (p. 9). Our examples suggest that gestures can be similarly conceived as a union of generalization and communication. Words and gestures or signs and gestures were used as simultaneous mediational resources throughout the learners activities, and, as co-temporal simultaneous productions (Goldin-Meadow, 2003), their roles in thinking and communicating are difficult to separate.

It is in this sense that we are attracted to Rotman's view of speech (and by analogy also Sign) as a species of gesture (Rotman, 2009), rather than a position that treats gesture as an appendage or add-on to the "official" language. Indeed, our examples suggest that gestures emerge when no word or sign is available that would communicate the meaning that the students wish to stress. Looking at the relationship between gestures and official languages hence offers a form of reflecting upon the origin and formation of languages.[8]

Our third claim is that the visual–gestural–somatic expressions that emerged in all three examples evince the embodied nature of mathematical cognition perhaps more clearly than the verbal-auditory mode. That is not to say that Sign and gestures are bodily things while spoken and written languages are not; it was this very thinking that led us to neglect the visual–gestural–somatic modality for so long. Our view is that words, signs, and gestures, all forms of what Vygotsky termed symbolic language, are constructions with their roots in the sensory experiences of the learners who produce them. To understand their meanings, we should not try to strip them of the connections with the senses and with feelings, instead we should seek to illuminate these connections so we can better feel, hear, and express the mathematics of all our students.

[8]It has even been argued that the origin of language itself, be it aural or manual, can be plausibly traced back to gesturing (Armstrong & Wilcox, 2003).

References

Armstrong, D. F., & Wilcox, S. (2003). Origins of signed languages. In M. Marschark & P. Spencer (Eds.), *Oxford handbook of deaf studies, language, and education* (pp. 305–318). Oxford, England: Oxford University Press.

Brentari, D. (Ed.). (2010). *Sign languages: A Cambridge language survey*. Cambridge, England: Cambridge University Press.

Brito, L. F. (1995). *Por uma gramática de Língua de Sinais*. Rio de Janeiro, Brazil: Tempo Brasileiro.

Bull, S. (2008). Deafness, numerical cognition and mathematics. In M. Marschark & P. C. Hauser (Eds.), *Deaf cognition: Foundations and outcomes* (pp. 170–200). New York: Oxford University Press.

Cruz, M. (2008). *Gramática de la lengua de señas mexicana*. Doctoral dissertation, Colegio de México, México.

Edwards, L., Ferrera, F., & Russo-Moore, D. (Eds.). (2014). *Emerging perspectives on gesture and embodiment in mathematics*. Charlotte, NC: Information Age.

Fernandes, S. H. A. A., & Healy, L. (2010). A inclusão de alunos cegos nas aulas de matemática: Explorando área, perímetro e volume através do tato. *Bolema: Boletim de Educação Matemática, Rio Claro (SP), 23*(37), 1111–1135.

Gee, J. P. (2014). *An introduction to discourse analysis: Theory and method* (4th ed.). Abingdon, England: Routledge.

Goldin-Meadow, S. (2003). *Hearing gesture: How our hands help us think*. Cambridge, MA: Harvard University Press.

Healy, L. (2012). Mathematical opportunities for students with disabilities. In T. Y. Tso (Ed.), *Proceedings of the 36th Conference of the International Group for the Psychology of Mathematics Education* (Vol. 1, pp. 100–106). Taipei, Taiwan: PME.

Healy, L., & Fernandes, S. H. A. A. (2011). The role of gestures in the mathematical practices of those who do not see with their eyes. *Educational Studies in Mathematics, 77*, 157–174.

Healy, L., & Fernandes, S. H. A. A. (2014). The gestures of blind mathematics learners. In L. Edwards, F. Ferrera, & D. Russo-Moore (Eds.), *Emerging perspectives on gesture and embodiment in mathematics* (pp. 125–150). Charlotte, NC: Information Age.

Hewes, G. W. (1996). A history of the study of language origins and the gestural primacy hypothesis. In A. Lock & C. R. Peters (Eds.), *Handbook of human symbolic evolution* (pp. 571–595). New York: Oxford University Press.

Iverson, J. M., & Goldin-Meadow, S. (1998). Why people gesture when they speak. *Nature, 396*(6708), 228.

Kelly, R. R. (2008). Deaf learners and mathematical problem solving. In M. Marschark & P. C. Hauser (Eds.), *Deaf cognition: Foundations and outcomes* (pp. 226–249). New York: Oxford University Press.

Kendon, A. (2008). Some reflections on the relationship between 'gesture' and 'sign'. *Gesture, 8*(3), 348–366.

Lane, H. (1984). *When the mind hears: A history of the deaf*. New York: Random House.

Lang, H. G. (2003). Perspectives on the history of deaf education. In M. Marschark & P. E. Spencer (Eds.), *Oxford handbook of deaf studies, language, and education* (pp. 9–20). New York: Oxford University Press.

LeBaron, C., & Streeck, J. (2000). Gestures, knowledge, and the world. In D. McNeill (Ed.), *Language and gesture* (pp. 118–138). Cambridge, England: Cambridge University Press.

Mason, J. (1989). Mathematical abstraction as the result of a delicate shift of attention. *For the Learning of Mathematics, 9*(2), 2–8.

McNeill, D. (1985). So you think gestures are nonverbal? *Psychological Review, 92*, 350–371.

Monteiro, J. K., & Andrade, C. G. (2005). Avaliação do raciocínio abstrato, numérico e espacial em adolescentes surdos. *Aletheia, 21*, 93–99.

Moura, M. C. (2000). *O Surdo, caminhos para uma nova identidade*. Rio de Janeiro, Brazil: Editora Revinter.
Nelsen, R. (1993). *Proofs without words: Exercises in visual thinking*. Washington, DC: The Mathematical Association of America.
Nemirovsky, R., Kelton, M. L., & Rhodehamel, B. (2013). Playing mathematical instruments: Emerging perceptuomotor integration with an interactive mathematics exhibit. *Journal for Research in Mathematics Education, 44*(2), 372–415.
Noss, R., & Hoyles, C. (1996). *Windows on mathematical meaning: Learning cultures and computers*. Dordrecht, The Netherlands: Kluwer.
Nunes, T. (2004). *Teaching mathematics to deaf children*. London: Whurr.
Nunes, T., & Moreno, C. (2002). An intervention program for promoting deaf pupils' achievement in mathematics. *Journal of Deaf Studies and Deaf Education, 7*(2), 120–133.
Radford, L. (2009). Why do gestures matter? Sensuous cognition and the palpability of mathematical meanings. *Educational Studies in Mathematics, 70*(2), 111–126.
Radford, L., Edwards, L., & Arzarello, F. (2009). Introduction: Beyond words. *Educational Studies in Mathematics, 70*(2), 91–95.
Roth, W. M. (2010). Incarnation: Radicalizing the embodiment of mathematics. *For the Learning of Mathematics, 30*(2), 8–17.
Roth, W. M., & Thom, J. S. (2009). Bodily experience and mathematical conceptions: From classical views to a phenomenological reconceptualization. *Educational Studies in Mathematics, 70*(2), 175–189.
Rotman, B. (2009). Gesture, or the body without organs of speech. *Semiotix. A Global Information Bulletin*, (15). Retrieved April 28, 2011, from http://www.semioticon.com/semiotix/semiotix15/sem-15-02.html.
Sacks, O. (2000). *Seeing voices*. New York: Vintage Books.
Stokoe, W. C. (1960/2005). Sign language structure: An outline of the visual communication system of the American deaf. *Journal of Deaf Studies and Deaf Education, 10*, 3–37.
Vygotsky, L. S. (1962). *Thought and language*. Cambridge, MA: MIT Press.
Vygotsky, L. S. (1981). The instrumental method in psychology. In J. V. Wertsch (Ed.), *The concept of activity in Soviet psychology* (pp. 134–143). Armonk: M.E. Sharpe.
Vygotsky, L.S. (1997). *Obras escogidas V–Fundamentos da defectología* [The fundamentals of defectology] (Julio Guillermo Blank, Trans.). Madrid, Spain: Visor.
Wilcox, S., & Morford, J. P. (2007). Empirical methods in signed language research. In M. Gonzalez-Marquez, I. Mittelberg, S. Coulson, & M. J. Spivey (Eds.), *Methods in cognitive linguistics* (pp. 173–202). Amsterdam, The Netherlands: John Benjamins.

Chapter 9
Student Agency and Counter-Narratives in Diverse Multilingual Mathematics Classrooms: Challenging Deficit Perspectives

Jennifer M. Langer-Osuna, Judit Moschkovich, Eva Norén, Arthur B. Powell, and Sumaia Vazquez

9.1 Introduction

With increased global migration, many mathematics classrooms now serve students who are also learning the dominant language of instruction. The forms of participation in mathematics classroom activity of emergent[1] bilinguals or multilinguals have often been examined from deficit perspectives (see, for instance, Orr, 1987) that focused on the obstacles for vocabulary, oral fluency, and comprehension in the new

[1] We use the term "emergent" bilingual or multilingual to highlight that English learners are not only learning English but also becoming bilingual/multilingual along a continuum of different types of bilinguals (see Fig. 1 in Valdes, 2005).

J.M. Langer-Osuna
Stanford University, Stanford, CA, USA
e-mail: jmlo@stanford.edu

J. Moschkovich (✉)
University of California, Santa Cruz, 1156 High Street, Santa Cruz, CA 95064, USA
e-mail: jmoschko@ucsc.edu

E. Norén
Stockholm University, Stockholm, Sweden
e-mail: eva.noren@mnd.su.se

A.B. Powell
Rutgers University, Newark, NJ, USA
e-mail: PowellAB@andromeda.rutgers.edu

S. Vazquez
Universidade Federal de Vitoria, Vitória, Brazil
e-mail: sumaia_ns@yahoo.com

R. Barwell et al. (eds.), *Mathematics Education and Language Diversity*, New ICMI Study Series, https://doi.org/10.1007/978-3-319-14511-2_9

language rather than the resources they bring to learn mathematics[2] (see Cuevas, 1983, 1984; Moschkovich, 2000, 2002; Spanos & Crandall, 1990 for criticisms). Recent studies focused on language learning and engagement in mathematics classroom activity from sociocultural perspectives have examined the construction of mathematical knowledge in social context, focusing not only on the challenges that emergent multilinguals face in the classroom, but also on the strengths of their linguistic resources (Adler, 1998; Brenner, 1994; Khisty, 1995, 2002; Moschkovich, 1999, 2002; Powell & Frankenstein, 1997; Rosebery, Warren, & Conant, 1992). Such work, which this chapter builds on, provides counter-narratives to deficit perspectives and offers what we argue is a more productive frame for understanding issues of mathematics teaching and learning in multilingual classrooms.

Deficit perspectives arise from and are fed by broader sociopolitical contexts that often position emergent multilinguals learning the language of instruction as lacking in cognitive resources necessary for competence (Healy & Powell, 2013; Morgan, 2006; O'Halloran, 2004). Deficit perspectives also derive from existing literature on language learning and development framed with a monolingual bias, obscuring not only competence in children's native languages, but also how children can productively draw on native and emergent languages to support communication in the language of instruction (Gort, 2006; Moschkovich, 2002). This bias is problematic not only because it assumes the monolingual student as the norm (Moschkovich, 2010), but also because it distorts the reality of emergent multilinguals' learning trajectories (Gort, 2006). Emergent bilinguals have a variety of unique linguistic resources at their disposal that they utilize productively to learn in their second language. When bilingual learners are treated as if they were imperfect monolinguals, these competencies are ignored, downplayed, or disallowed in educational environments structured around policies based on a monolingual bias (Gort, 2005). For example, an adolescent who is developing bilingual competence in English will typically use reading comprehension strategies developed for reading in her first language to understand a text in English. In contrast, when using a monolingual bias, such students are treated as if they are learning to read from scratch.

Research that illuminates the ways in which multilingual learners positively utilize available resources, both linguistic and otherwise, during mathematical activity is greatly needed (Bialystok, 2001; Gutierrez, Sengupta-Irving, & Dieckmann, 2010; Moschkovich, 2002, 2010). Such work begins to illuminate the productive use of resources that educators could strategically build on during instruction.

[2] It is important to acknowledge that some students are learning not a second language but their third or *n*th language. For example, in some schools in the United States there are African and Haitian students who already speak two languages (French and a mother tongue) and are now learning mathematics in English, their third language. It is critical that these students be recognized in the literature, especially since they provide evidence that multilingual students have already demonstrated that they are capable of functioning effectively in more than one language and that some of the difficulties they encounter in school mathematics—taught in a new language of instruction they are yet to master—result not from deficient cognitive resources but from sociopolitical context of instructional practices.

In this chapter, we focus on one particular resource: multilingual learners' agency in positioning themselves and one another with competence in ways that initiate, support, or further productive mathematics classroom interactions.

Practice theories of identity and learning offer researchers ways to highlight the individual agency of emergent multilinguals' interactions in mathematical activity, allowing for the emergence of analytical counter-narratives that challenge deficit perspectives. Holland, Lachicotte, Skinner, and Cain's (2001) figured worlds-based theory of identity considers both the sociopolitical forces that shape the positional identities made available to particular persons and the role of individual agency that can accept, reject, or alter such positions. Figured worlds are "socially produced, culturally constituted activities" (1998, pp. 40–41) where people come to conceptually (cognitively) and materially/procedurally produce (perform) new self-understandings (identities). Identity making is defined as "producing, from the cultural resources available to them, understandings of themselves" (2001, p. 4). For example, Holland et al. (2001) describe the figured world of the caste system in Nepal, where actions are interpreted from the perspective of a narrative that includes such rules as persons from a lower caste cannot enter the homes of persons from a higher caste. Thus a woman from a high caste who enters the home of another woman from a high caste is interpreted as being appropriate, while a woman from a lower caste who engages in the same act is interpreted as inappropriate. Such "characters" and "story lines" in particular figured worlds enable and constrain possible acts. Yet, even these constraints are still subject to agentive acts. Holland et al. (2001) describe, for instance, an encounter with a woman from a lower caste who was invited to enter the researcher's (accorded a higher caste status) apartment for an interview. As the researcher walked down to greet her, the low-caste woman instead scaled the outside of the building to the second floor balcony for the interview as a way to participate in the interview without entering the researcher's building. This act, profoundly shaped by how the narrative of caste positioned the woman, was nonetheless an agentive act; no other community members climbed walls to reach second stories of buildings.

A framework centered on human agency within dominant sociopolitical narratives and discourses is a potentially useful way to examine how emergent multilinguals might utilize the resources available to them in creative and competent ways as they navigate interactions during mathematical activity. In the remainder of this chapter, we draw on these ideas to examine several vignettes that highlight creative and productive uses of emergent multilinguals' available resources. In doing so, we hope to offer counter-narratives that both challenge deficit perspectives and offer a promising approach to move forward research in the teaching and learning of mathematics in linguistically heterogeneous classrooms that capture the full range of students' competencies. The vignettes are drawn from a variety of mathematics classroom contexts: Spanish-dominant emergent bilinguals from different states in the United States, and multilinguals in Sweden using various languages, most commonly Arabic and Swedish. The vignettes show students as creative users of linguistic resources in ways that serve a variety of functions in mathematical activity.

9.2 Vignettes of Counter-Narratives in the Experiences of Emergent Bilinguals in Mathematical Classrooms

The three vignettes in this section draw from a variety of mathematics classroom contexts. Vignette 1 focuses on how a Spanish-dominant bilingual fifth grader in the Southeastern United States negotiated mathematical ideas and her position as an English language learner as she collaborated with an English proficient peer. Vignette 2 focuses on multilingual students in Sweden who successfully re-directed their teacher's planned lesson toward Swedish counting words, an aspect of mathematical language they were developing. Vignette 3, set in the Northeastern United States classroom, evidences students' use of linguistic and representational resources to interpret open-ended problems in combinatorics.

9.2.1 Vignette 1: Mispronunciation as Wordplay

Based on a fifth grade mathematics classroom in the Southeastern United States, this first vignette highlights the ways in which language learners can positively alter potentially negative positionings of their struggles to simultaneously learn both mathematics and the language of instruction. The vignette focuses on a dyad collaboratively solving a mathematics word problem. As they take turns reading the problem aloud, one student, an English language learner, struggles with the pronunciation of particular words, in particular "cello." The student follows her own mispronunciation of the word "cello" with immediate subsequent wordplay. In doing so, she repositions her error as creative play. Her partner takes up this alternate, positive positioning and the two students move forward into problem solving.

Fifth graders Amanda and Roger collaborated on a worksheet with several mathematics word problems on the topic of sets. Amanda, an English language learner, agrees to read the second problem that included finding the set of students who play violin, the set of students who play the cello, and the set of students who play both. While reading, Amanda repeatedly gets stuck on the pronunciation of the word "cello," alternating the correct pronunciation "CHE-loh" with the incorrect pronunciations "CHAY-loh" and "SELL-oh."

Immediately, Amanda turns to word play to make light of her mispronunciation, creating humorous rhymes and joking:

[in sing-songy, dramatic voice] che-LO::H! che-LO::H! Or jell-o! *[giggles]* You see what I just did that? Che-lo::::h? *[mimicking Spanish accented form of saying hello]*

(continued)

Instead of responding to her mispronunciations, Roger, an English proficient peer, responds to her word play by smiling and giggling, which is positioned positively. Directly after her wordplay, Amanda resumes working on the problem by stating, "Okay. So eight play cello and eight play both. Your turn." Roger pulls the worksheet closer to him and takes up Amanda's solution by stating, "Eight play both." Amanda follows by stating, "And eight play cello, or jell-o." Both students smile and move on to the third problem on the worksheet.

Language learners like Amanda are likely to show errors in pronunciation (Brown, 1973). The social positioning of these errors, however, is constructed in interaction by the speaker and his or her audience (Holland et al., 2001). Typically, a speaker who makes a pronunciation error is positioned as lacking competence. Here, Amanda drew on other linguistic and cultural resources—her knowledge of rhyme, humor, popular culture, and even accented speech—to position her pronunciation of the word "cello" not as lack of competence, but as creative wordplay. This bid in positioning is taken up by her student partner Roger, who joins Amanda in enjoying the word play and also subsequently takes up her mathematical ideas. Such subsequent uptake of her ideas provides further evidence that the mispronunciation did not downgrade her perceived academic competence. Rather, Amanda is positioned with both social and intellectual authority. This is further highlighted at the end of her turn solving the problem, where she concluded, "And eight play cello, or jell-o." Her concluding statement was both mathematically correct and socially sophisticated and, importantly, was taken up by Roger as such. Although the resources that Amanda used are not mathematical by themselves, they provided her with access to participation in the mathematical activity as well as a position of intellectual authority in the dyad.

Research on cooperative groups made up of culturally heterogeneous students suggests that minority students', including language learners, opportunities for engagement in mathematical activity are constrained by the negative ways in which their ideas and identities are positioned in interaction (Kurth, Anderson, & Palinscar, 2002; Langer-Osuna, 2011). This vignette offers a counter-example, where a language learner successfully navigated social dynamics in ways that "saved face" (Goffman, 1967/2005) and supported continued productive engagement. This counter-example points to the fruitfulness of research that examines the interactional strategies that language learners utilize, such as humor and play, to negotiate mathematical ideas and positional identities during collaborative student-led problem solving.

9.2.2 *Vignette 2: Teenagers and Counting Words*

Based on a first grade mathematics classroom in a suburb of Stockholm, this second vignette illustrates the ways in which first grade second language and mathematics learners were able to shift the direction of their teacher's planned mathematics lesson to one of their own choosing. Further, the direction of their choosing was one that addressed mathematical language. The vignette highlights important shifts in power relations, where students shaped classroom discussions around their learning needs.

Students Amina and Liljana ask their teacher, Ms. Anna, whether everybody has to become "cheeky" when they become teenagers. After a few minutes of interaction with the girls and the rest of the class, the teacher acknowledges that many of her students do not understand the connection between the age range of being a "teenager," and the ending of the counting words between 13 and 19.[3]

Ms. Anna then poses the question, "How old are they then? How old are you when you are a teenager? What is a teenager?"

Ms. Anna changed the direction of the mathematical discussion based on the girls' interests in teenagers, a change that the students initiated. Here we see student agency in their collective ability to take up positions that suit them (Burr, 2003; Foucault, 1982). After this re-direction, students and teachers discussed the counting words from 11 to 19, for about 45 min, rather than the originally planned topic. Liljana's and Amina's out-of-school identities and experiences were thus legitimized and valued as part of the mathematical learning experience.

Another example occurred moments later in the same lesson, when one of the boys, Tony, decided to pick up a pointer to point to all the -teen words (thirteen, fourteen, … nineteen) written on the whiteboard, positioning himself as teacher-like:

Tony: There it says -teen, there it says -teen, /…/there it says -teen, it says -teen the whole row.

[3] This is not the case for all languages. In some languages, the word **teen**ager is related to the counting words for thir**teen**, four**teen**, …, nine**teen**. In Swedish, teenager is **ton**åring and the counting words from 13 to 19 are tret**ton**, fjor**ton**, …, and nit**ton**. This example works in English and Swedish (as well as in some other languages) but not, for example, in Spanish, Arabic, Japanese, or Korean.

After about 45 min another boy, Melvin, initiated a change of topic once again, asking:

Melvin: But ... this thing about **ton,**[4] two tons, three tons ... what is that? It is like ... [pointing at a desk] This weighs a ton!

Ms. Anna shifted yet again in the mathematical discussion, based on Melvin's question. The shift provided the opportunity for the students to get involved and tell their stories about heavy tables and elephants. The open nature of the classroom discussion determined students' possible ways of acting (Cotton & Hardy, 2004). In this classroom, students were very well disciplined into a mainstream school mathematics discourse where the teacher was the one with the "right" answers. Yet, in this discussion, Ms. Anna supported students' agency, as when Melvin changed the focus from ton as a counting word to ton as a measuring unit. Students' language and out-of-school experiences were supported when the teacher and students engaged in discussion that bridged mathematics and (Swedish) language issues. These young students positioned themselves as active mathematics learners by coordinating their interests in both linguistic and mathematics knowledge.

9.2.3 Vignette 3: Open-Ended Tasks

This final vignette is based on data from a larger study of Spanish-dominant, emergent multilingual third graders in a poor urban community in the Northeastern United States (Vazquez, 2009). The study investigates how students use their linguistic resources to develop mathematical and heuristic ideas as they solve a series of open-ended combinatorial tasks. The vignette below illustrates the work of a group of students who received mathematics instruction in Spanish and English. In the classroom, the students were allowed to use either language and tended to move fluidly between English and Spanish. To analyze these data, Vazquez and Powell focus particularly on the discursive interactions as students within a group justified and attempted to persuade each other of their understandings and resolutions.

To understand the students' oral and inscriptive mathematical interactions and representations, they use the analytic lens of agency and cognitive elasticity. Agency is understood as individuals' intentional, self-initiated action directed toward accomplishing an explicit or implicit goal. In the context of the third vignette, students enact agency when they author their own procedures and heuristics to resolve mathematical tasks or aspects of them or when they pose their own questions upon which to work. This conceptualization highlights students' interdependent, independent, and autonomous mathematical performances through student-to-student discourse. As Powell (2004) notes, "understanding agency is particularly important since both failure and success can be located within the same set of social, economic, and

[4] Note that In Swedish **teen-**ager is **ton**-åring, and **ton** is **ton,** as in the measuring unit. This works in Swedish but not in English or other languages.

school conditions that usually is described as only producing failure" (p. 42). Examining how students exercise agency in their use of linguistic resources as they interact jointly to meet challenges of a mathematical task can provide counter-narratives to deficit perspectives. By cognitive elasticity, we refer to students' ability to coordinate discursively between abstract and concrete representations. Discursive manifestations of cognitive elasticity can occur orally, inscriptively, or gesturally and provide evidence of emergent multilinguals' use of their linguistic resources.

In the excerpt below, a group of emergent multilingual students, whose mathematics class was in English, work together to understand and resolve this open-ended combinatorial task:

> Stephen has a white shirt, a blue and a yellow. He has a blue pair of jeans and a white pair of jeans. How many different combinations can he do?

The analytic goal is to understand how students enacted agency and cognitive elasticity to mobilize their linguistic resources to understand and develop solutions to combinatorial mathematics problems. In the vignette, a group of three students discuss their understanding and solution of the task presented above:

Ciara: We can't use the yellow shirt because you don't have pants that go with it. Because it says that also has a pair of blue jeans and white jeans and has no yellow pants, so we can't use yellow shirt

Parniery: The color doesn't matter!

Kayla: You can put blue, white, yellow and blue, and white and yellow. Three of each is six. Six ways to combine!

Ciara: How?

Kayla: See [pointing to Ciara's diagram] white with yellow, yellow and blue, and blue and white, yellow and white, and blue and white (…) all colors of with each pants, each group.

Ciara: One, two, three, four, five, six … we have six ways to combine.

The vignette shows students' agency as well as their cognitive elasticity. The exercise of students' agency is a hallmark of the classroom's linguistic and collaborative atmosphere which provides space for students to be comfortable discussing and challenging each other's ideas. Discursively, the group challenges Ciara's idea that colors matter and, thereby, convinces her that there are six different combinations of shirts and pants Stephen can use as outfits. Inscriptively, they use their own strategic representations to solve the problem. In their discursive interactions, they reveal the importance of the interaction between members of the group that are at a high level of mathematical understanding (Fig. 9.1). This excerpt also illustrates Vygotsky's theory of the Zone of Proximal Development (ZPD), where the assistance of a more experienced interlocutor contributes to the activity of another. With the help of the group, Ciara uses agency and cognitive elasticity to develop the diagram and represent her ideas (Fig. 9.2). She and other members of her group exhibit cognitive elasticity as they move back and forth between concrete and more abstract representational forms. The emergent multilingual students in this vignette evidenced ease in communication and construction of mathematical representations as well enactment of agency and cognitive elasticity.

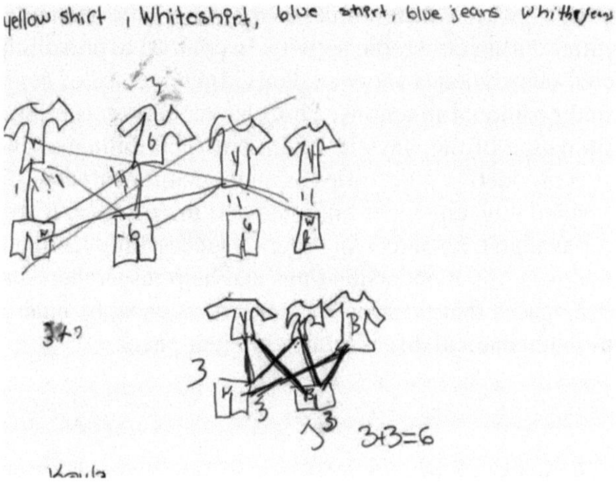

Fig. 9.1 Kayla's inscription

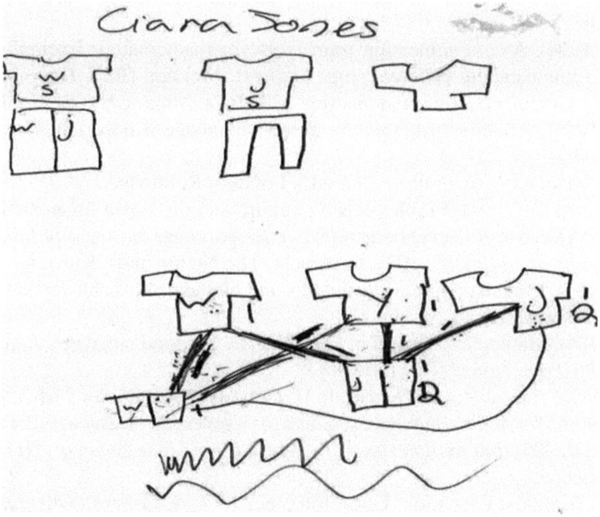

Fig. 9.2 Ciara's inscription

9.3 Concluding Thoughts

The purpose of this chapter is to broaden the conversation around language learners' competence in negotiating interactions in mathematics classrooms. Our three excerpts highlight how language learners draw creatively on humor, personal interests, and bilingual competencies to position themselves productively as learners, shift the nature of mathematical discussions, and engage in complex debates of mathematical ideas.

Learners draw on a variety of resources as they negotiate mathematical ideas and positional identities during classroom activity. In contrast to prevailing emphases on linguistic deficits, our vignettes show students' agentive use of resources for both mathematical and positional functions. This chapter serves as a jumping off point for further examination of the ways in which emergent bilingual and multilingual students construct productive trajectories as mathematics learners. More research is needed to understand how emergent bilingual and multilingual learners utilize linguistic and other available resources in order to successfully navigate mathematics classroom interactions. Such understandings can help researchers and educators to design classroom spaces that position all learners as capable, enabling their entry into meaningful mathematical discussions with their peers.

References

Adler, J. (1998). A language of teaching dilemmas: Unlocking the complex multilingual secondary mathematics classroom. *For the Learning of Mathematics, 18*(1), 24–33.

Bialystok, E. (2001). *Bilingualism in development: Language, literacy, and cognition*. New York: Cambridge University Press.

Brenner, M. E. (1994). A communication framework for mathematics: Exemplary instruction of culturally and linguistically different students. In B. McLeod (Ed.), *Language and learning: Educating linguistically diverse students* (pp. 233–267). Albany, NY: SUNY Press.

Brown, H. D. (1973). Affective variables in second language acquisition. *Language Learning, 23*(2), 231–244.

Burr, V. (2003). *Social constructionism* (2nd ed.). London: Routledge.

Cotton, T., & Hardy, T. (2004). Problematising culture and discourse for mathematics education research. In P. Valero & R. Zevenbergen (Eds.), *Researching the socio-political dimensions of mathematics education* (pp. 85–103). Dordrecht, The Netherlands: Kluwer.

Cuevas, G. J. (1983). Linguistic minority students and mathematics achievement. *Zentralblatt für Didaktik der Mathematik, 6*, 24–25.

Cuevas, G. J. (1984). Mathematics learning in English as a second language. *Journal for Research in Mathematics Education, 15*(2), 134–144.

Foucault, M. (1982). The subject and power. In H. Dreyfus & P. Rabinow (Eds.), *Michel Foucault: Beyond structuralism and hermeneutics*. Chicago: University of Chicago Press.

Goffman, E. (1967/2005). *Interaction ritual: Essays in face to face behavior*. New Brunswick, NJ: Aldine Transaction.

Gort, M. (2005). Bilingual education: Good for U.S.? In T. A. Osborn (Ed.), *Language and cultural diversity in education*. Westport, CT: Praeger.

Gort, M. (2006). Strategic code-switching, interliteracy, and other phenomena of emergent bilingual writing: Lessons from first grade dual language classrooms. *Journal of Early Childhood Literacy, 6*(3), 323–354.

Gutierrez, K., Sengupta-Irving, T., & Dieckmann, J. (2010). Developing a mathematical vision: Mathematics as a discursive and embodied practice. In J. N. Moschkovich (Ed.), *Language and mathematics education: Multiple perspectives and directions for research* (pp. 29–71). Charlotte, NC: Information Age.

Healy, L., & Powell, A. B. (2013). Understanding and overcoming "disadvantage" in learning mathematics. In M. A. K. Clements, A. J. Bishop, C. Keitel, J. Kilpatrick, & F. K. S. Leung (Eds.), *Third international handbook on mathematics education* (pp. 69–100). New York: Springer.

Holland, D., Lachicotte, W., Jr., Skinner, D., & Cain, C. (2001). *Identity and agency in cultural worlds*. Cambridge, MA: Harvard University Press.

Khisty, L. L. (1995). Making inequality: Issues in language and meanings in mathematics teaching with Hispanic students. In W. G. Secada, E. Fennema, & L. B. Adajian (Eds.), *New directions for equity in mathematics instruction* (pp. 279–297). Cambridge, England: Cambridge University Press.

Khisty, L. L. (2002). Mathematics learning and the Latino student: Suggestions from research for classroom practice. *Teaching Children Mathematics, 9*(1), 2–35.

Kurth, L. A., Anderson, C. W., & Palinscar, A. S. (2002). The case of Carla: Dilemmas of helping all students to understand science. *Science Education, 86*(3), 287–313.

Langer-Osuna, J. (2011). How Brianna became bossy and Kofi came out smart. *Canadian Journal for Research in Mathematics, Science, and Technology, 11*(3), 207–225.

Morgan, C. (2006). What does social semiotics have to offer mathematics education research? *Educational Studies in Mathematics, 61*, 219–245.

Moschkovich, J. N. (1999). Supporting the participation of English language learners in mathematical discussions. *For the Learning of Mathematics, 19*(1), 11–19.

Moschkovich, J. N. (2000). Learning mathematics in two languages: Moving from obstacles to resources. In W. Secada (Ed.), *Changing the faces of mathematics: Perspectives on multiculturalism and gender equity* (pp. 85–93). Reston, VA: NCTM.

Moschkovich, J. N. (2002). A situated and sociocultural perspective on bilingual mathematics learners. *Mathematical Thinking and Learning, 4*(2 & 3), 189–212.

Moschkovich, J. N. (2010). Language(s) and learning mathematics: Resources, challenges, and issues for research. In J. N. Moschkovich (Ed.), *Language and mathematics education: Multiple perspectives and directions for research* (pp. 1–28). Charlotte, NC: Information Age.

O'Halloran, K. L. (2004). *Multimodal discourse analysis*. London: Continuum.

Orr, E. W. (1987). *Twice as less: Black English and the performance of Black students in mathematics and science*. New York: W. W. Norton.

Powell, A. B. (2004). The diversity backlash and the mathematical agency of students of color. In M. J. Høines & A. B. Fuglestad (Eds.), *Proceedings of the Twenty-eighth Conference of the International Group for the Psychology of Mathematics Education* (Vol. 2, pp. 37–54). Bergen, Norway: PME.

Powell, A. B., & Frankenstein, M. (1997). Ethnomathematical praxis in the curriculum. In A. B. Powell & M. Frankenstein (Eds.), *Ethnomathematics: Challenging eurocentrism in mathematics education* (pp. 249–259). Albany, NY: State University of New York.

Rosebery, A. S., Warren, B., & Conant, F. R. (1992). Appropriating scientific discourse: Findings from language minority classrooms. *Journal of the Learning Sciences, 2*(1), 61–94.

Spanos, G., & Crandall, J. (1990). Language and problem solving: Some examples from math and science. In A. M. Padilla, H. H. Fairchild, & C. M. Valadez (Eds.), *Bilingual education: Issues and strategies* (pp. 157–170). Beverly Hill, CA: Sage.

Vazquez, S. C. (2009). Combina ou não combina? Um estudo de caso com alunos bilíngües e não bilíngües nos EUA. *BOLEMA: O Boletim de Educação Matemática, 55*, 41–64.

Chapter 10
Tensions in Teaching Mathematics in Contexts of Language Diversity

Richard Barwell, Lim Chapsam, Thulisile Nkambule, and Mamokgethi Setati Phakeng

10.1 Introduction

It is apparent from existing research that teaching and learning mathematics in contexts of language diversity presents learners, teachers, and policymakers with challenges. These challenges take many forms, but all can be thought of in terms of tensions. By "tension," we mean a challenging situation with no clear-cut solution, a situation in which competing influences or forces suggest different, often opposing courses of action. In this chapter, we present four case studies of such tensions arising in mathematics classrooms in Canada, Malaysia, and South Africa. One of our aims is to show how similar tensions arise in quite different contexts. A second aim is to examine the role of teachers in mediating these tensions. To do so, we draw on aspects of the theory of language developed by Bakhtin (1981). From this perspective, language use always results in particular kinds of tension, as we explain later in the chapter.

The chapter is organised into four main sections. First, we summarise some of the tensions that have already been reported in the literature on teaching and learning mathematics in contexts of language diversity. Next, we introduce the theoretical perspective on language, drawing on Bakhtin's work. We then present our four case studies. Finally, we discuss the similarities and differences arising across the cases and consider some of the implications of this work.

R. Barwell (✉)
Faculty of Education, University of Ottawa, 145 J-J-Lussier, Ottawa, ON, Canada K1N 6N5
e-mail: richard.barwell@uottawa.ca

L. Chapsam
Universiti Sains Malaysia, Penang, Malaysia
e-mail: cslim@usm.my

T. Nkambule • M. Setati Phakeng
University of South Africa, Pretoria, South Africa
e-mail: nkanbt@unisa.ac.za; phakerm@unisa.ac.za

R. Barwell et al. (eds.), *Mathematics Education and Language Diversity*, New ICMI Study Series, https://doi.org/10.1007/978-3-319-14511-2_10

10.2 Tensions in Mathematics Classrooms in Contexts of Language Diversity

While the term "tension" has not been widely used, similar words arise frequently in the literature (see Barwell, 2009, 2012). Most notably, Adler (2001) wrote of the different 'teaching dilemmas' facing mathematics teachers in multilingual South Africa. For Adler, teaching dilemmas were situations in which teachers perceived a choice of actions, and each one presented both advantages and disadvantages. For example, Adler referred to the dilemma of transparency to describe a situation described by teachers in which they felt that they had to decide whether mathematical language should be made visible through explicit attention, or whether it should be left as a transparent medium for mathematical discussion. A similar tension between informal language and mathematical language has been observed in different forms in several other studies in different parts of the world, including in the United States (e.g., Khisty, 1995; Moschkovich 2008), South Africa (Setati & Adler, 2000), Canada (Barwell, 2014), and Australia (Clarkson, 2009). As implied by our definition of tension, there is no neat resolution to the tension between informal and mathematical language. Teaching and insisting on highly formalised mathematical language in contexts of language diversity will simply disenfranchise many learners. On the other hand, not to provide them with the opportunity to learn more formal mathematical language will, in the long run, also disenfranchise them, even if they have a good understanding of mathematics. Again, teachers' responses to these kinds of situations will have an impact on students' participation in mathematics.

A second widespread tension arises between the language of instruction for mathematics in school, and the languages used by students outside of school. This tension is apparent in many other studies in different parts of the world, including Australia (Clarkson, 2007; Clarkson & Dawe, 1997; Parvanehnezhad & Clarkson, 2008), Malawi (Chitera, 2009), Malta (Farrugia, 2009), Pakistan (Halai, 2009), Swaziland (Dlamini, 2008), Catalonia, Spain (Planas & Setati, 2009), and Malaysia (Lim & Ellerton, 2009). It is also worth remembering that in many contexts, code-switching is *not* used, despite at least some students using two or more languages in their daily life outside school (Setati & Barwell, 2006). In each of these studies, there is a tension between the languages students use in their daily life and the language used to teach and learn mathematics. There is no simple way out of this tension. Students in most parts of the world live with language diversity: many languages are spoken outside of school, even if a single language is used for teaching and learning mathematics. It is clear that teachers play a key role in navigating this tension.

Finally, Setati's (e.g. 2008) work has highlighted a third significant tension arising in language-diverse mathematics classrooms. This tension derives from the status that different languages or ways of talking have in wider society. Specifically, there is a tension between the desirability of English in South African society and the facility with which learners are able to make sense of mathematics taught in English. Similarly, in work conducted in the United States and in Catalonia, Spain, Planas and Civil (2013) show how the societal value of different languages plays an important mediating role in mathematics classroom participation. They make the point that this mediation can undermine approaches to teaching mathematics that in other respects are open to the various linguistic resources that students may bring.

Our summary highlights some common tensions that have been reported in different mathematics classrooms around the world. We offer three observations about this work. First, the fact that each of them has been reported in several contexts suggests that they are likely to be widespread. There is no reason to suppose that a tension around the use of formal and informal language in multilingual mathematics classrooms is restricted to the places where research has been conducted.

Second, it is also apparent that the above tensions are all interrelated. The use of code-switching in mathematics classrooms, for example, is related to the policy and preference for using high-status languages rather than languages that students use at home. Similarly, the tension around the use of informal mathematical language is accentuated in contexts of language diversity, when students have less confidence or proficiency in using the official language of schooling. It is also clear that teachers find themselves in a pivotal position in dealing with these tensions in their practice. The choices teachers make and the strategies teachers use will have a potentially significant impact on their students' opportunities to engage in mathematical thinking and to learn mathematics. A teacher who enforces a one-language-only policy, for example, will empower some students in her class (those with the cultural resources to support use of that language) and disempower others. A teacher who encourages multiple language use in mathematics will also empower some students and disempower others, but not in the same way as in the first situation.

Third, the research we have summarised has been valuable in describing these tensions but has not sufficiently developed a theoretical perspective that accounts for or explains them. This is not to say that this work has not drawn on robust theoretical frameworks. Adler (2001) drew on a Vygostkian perspective to show how competing pedagogical priorities lead to teachers experiencing dilemmas. Setati's (2005) work is framed by Gee's (1999) theory of discourse and cultural models, through which she was able to show how individuals often held competing cultural models about language or mathematics. What this work has done less well, however, is to develop a theoretical account that explains how these tensions arise. Such a perspective would allow for a deeper understanding of the nature of the many tensions that arise in different contexts and would suggest new ways to tackle them. In the next section, we show how Bakhtin's theory of language provides such a perspective.

10.3 Bakhtin's Theory of Language

Bakhtin was predominantly a literary theorist with an interest in, among other things, the nature of language used in novels or in poetry. Embedded in his work, however, is a theory of language that challenges many common assumptions. In particular, Bakhtin's work challenges the ideas that language can be codified and categorised in a simple way, or that words or definitions are the basic units of a language. For Bakhtin (1981), language lives in interaction; he argued that the basic unit of language is, therefore, the utterance, situated within interaction. Moreover, the words that we all use when we speak or write are always derived from our prior experience of these words and of language. We are always in some sense borrowing or recycling words we first heard or read in other people's utterances. A key feature of Bakhtin's view of language is that he does not privilege a formal model of language purity, nor does he privilege a use-based model of language diversity. Instead, he proposes that both are present in the form of opposing forces.

Bakhtin (1981) used the term unitary language to refer to a view of language as coherent and unified. Unitary language "gives expression to forces working toward concrete verbal and ideological unification and centralization, which develop in vital connection with the processes of sociopolitical and cultural centralization" (p. 271). Implicit in this idea is the link between unitary language and political projects of nationalism or colonialism, as well as ideas about language purity or correctness. Bakhtin argues that the idea of unitary language exerts a constant pressure on language use, for which he uses the metaphor of centripetal force. Centripetal forces seek to impose or coerce language and its speakers into idealised standard ways of talking. Such forces are familiar in education: the preference for a single language of instruction; the teaching of the rules of grammar; the correction or suppression of non-standard ways of speaking or writing.

Bakhtin (1981) also stresses the diversity apparent in language in use, highlighting "the social diversity of speech types" (p. 263), translated as heteroglossia, to encompass the different ways of talking apparent in different social groups, political groups, regional dialects, and accents, even of families or individuals or particular times. Heteroglossia is apparent in any classroom, in which many different intersecting social languages can be identified: the languages of curriculum, of subject matter, of pedagogy, of teachers, of students, of school, of home, of working class or middle class, of neighbourhoods, and of age groups, and so on. Bakhtin uses the metaphor of centrifugal force to characterise the pressure that this diversity exerts.

These two ideas of language—unitary language and heteroglossia—are in a constant struggle analogous to the tension between centripetal and centrifugal forces. Centripetal forces seek to impose or coerce language and its speakers into idealised standard ways of talking, while centrifugal forces are apparent in the multivocal, multilingual, hybrid ways in which language is actually used. The tension between the centripetal force of unitary language and the centrifugal force of heteroglossia is present each time we speak and shapes what we say. Moreover, this tension is inherent in language. Without a degree of uniformity, language would be meaningless; language needs patterns and rules to make communication possible. At the same time, without diversity self-expression would be impossible. As Duranti (1998) points out, this tension is related to issues of power and marginalisation. The unitary forces tend to promote versions of language that correspond to the language of the powerful.

It is apparent that the tensions that arise in teaching and learning mathematics in contexts of language diversity can all be understood as reflecting the tension between centripetal and centrifugal forces. The tension between formal and informal language in mathematics classrooms is precisely a tension between more unified standardised forms of mathematical expression (i.e. unitary language) and more diverse, idiosyncratic expressions of mathematical meaning. The tensions around code-switching arise from a centralising tendency to demand that a single language be used for learning and teaching mathematics and the need to use a wider range of forms of expression to make mathematics meaningful. And Setati's (2008) research highlights the tension between the unifying movement to prefer English as a high-status international language and the multilingual reality of South African society. These forces, moreover, are not a simple case of government policy vs. the people. The pressure for English in South African schools comes from parents; official policy permits instruction in any of South Africa's official languages.

This theoretical perspective has the potential to extend the work summarised in the previous section. In particular, Bakhtin's theory of tensions explains the widespread prevalence of various specific tensions in the literature on mathematics teaching and learning in contexts of language diversity. They are widespread because, in a general sense, language is always filled with such tensions. We can thus clarify the nature of the teaching dilemmas described by Adler (2001): these dilemmas arise when the tensions that are present in language lead to teaching situations in which teachers are faced with competing or conflicting choices. Similarly, the competing cultural models in Setati's (2005) work reflect the tension between centripetal forces (English is valuable) and centrifugal forces (multilingualism is valuable).

To explore the role of centripetal and centrifugal forces in mathematics classrooms in contexts of language diversity, in the rest of the chapter we focus on four cases drawn from different settings in different parts of the world. Our particular focus is on the role of mathematics teachers in mediating the tensions that arise in each case. The four cases comprise: (1) a multilingual mathematics classroom in South Africa, (2) the use of an international language to teach mathematics in Malaysia, (3) teaching and learning mathematics in a class for new immigrants in Canada, and (4) immigrant students and teacher in a mathematics classroom in South Africa.

10.4 Case 1: Mathematics in a Multilingual Mathematics Classroom in South Africa

Language diversity is one of the distinctive features of South Africa. The country's eleven official languages include nine African languages. The policy environment allows schools and learners to select their language of instruction and encourages multilingualism and practices such as code-switching. Despite this seemingly progressive policy environment, research shows that schools in South Africa are not opting to use learners' home languages as the language of instruction, in both policy and practice and thus English remains the preferred language of instruction (see, for example Setati, 2008; Setati & Adler, 2000; Setati, Adler, Reed, & Bapoo, 2002). While African languages are spoken widely, using them as language of instruction is still associated with apartheid education and hence with inferior education.

This case[1] involves a multilingual Grade 4 mathematics class of 8–14 years old in a school located in an African township, west of Johannesburg. The learners in the school are all African. The chosen language of instruction in the school is English. The learners share a main language, which is Setswana and most of them can speak, read, write or understand another two or more African languages. The learners also study three languages as subjects—Setswana, English, and Afrikaans. At the time of data collection the teacher could speak six languages and shared a main language, Setswana, with her learners.

During an interview with the teacher, some of her remarks gave a sense of the centripetal forces that she experienced:

Teacher: If we changed our [mathematics] textbooks into Setswana and set our exams in Setswana, then my school will be empty because our parents now believe in English.

It became clear that she was struggling to emphasise the unitary role of language despite the local realities that have to do with the learners' level of proficiency in English. On the one hand, she expressed a view that learners should be taught in their home languages so that they can understand the mathematics that they are being taught. On the other hand, she expressed the need for her learners to develop fluency in English so that they are able to communicate with people from "other schools or cultures":

Teacher: The child must learn with the language that he or she will understand … the child must understand what you are teaching him … But still you must not confine a child to a situation where she won't be able to understand other people when she meets or where he meets other people from other schools or other cultures … it is said that English is an international language …

The teacher is attempting to ensure that her learners are initiated into the idealised way of communicating in English, the language of power. While the teacher wants her learners to understand mathematics, she also wants them to learn and practice speaking English so that they can fit into the idealised standard ways of talking in a world in which English dominates. While she indicated that during teaching she uses both English and Setswana, it seems that she uses Setswana to support not only learner access to mathematics, but also improve their fluency in English. There are pragmatic reasons for encouraging English. In the following extract the teacher explains them:

Teacher: I encourage them to use English … The textbooks are written in English the question papers are in English. So [if you use their language] you find that the child doesn't understand what is written there because all the time you encourage them to speak in Setswana and then you give him the question that has been written in English … Like for instance let me make an example last year during exam time we had a problem children asking

[1] A detailed analysis of this example can be found in Planas and Setati Phakeng (2014).

raising their hands asking the invigilator all the questions that they do not understand and the problem was the language not the question itself.

The teacher's emphasis on the use of English conforms to a unitary language perspective. While she seems to recognise its limitations she seems to be constrained by the centripetal forces related to the idea of English as an international language. All she can do is to prepare the learners for participation in this international world, and teaching mathematics in English is an important part of this preparation.

Observation of the teacher during a lesson on fractions (halves and quarters) underlined the role of centrifugal forces. The teacher introduced the lesson using paper cutting and then moved on to using pictures and stories. She did not use the words denominator and numerator; she spoke about the number on top and the number below. She focussed more on getting children to understand what the numbers represent rather than memorising the terms numerator and denominator. In most of the exercises given, learners had to make a pictorial representation of given fractions. Heteroglossia is evident in the switching between languages (Setswana and English), discourses ('academic' and 'everyday' mathematical discourses) and mathematical representations (diagrammatic and symbolic). In the extract below the teacher challenged her learners to imagine themselves at Mr. Nkomo's store buying bread.

Teacher: I want you to think. Close your eyes, and think about the shopkeeper. What will Mr. Nkomo do when you say, "Mr. Nkomo I want three halves." Anyone? (One of the learners goes to the board to draw the bread)

Teacher: So Bernard is showing us how many halves.

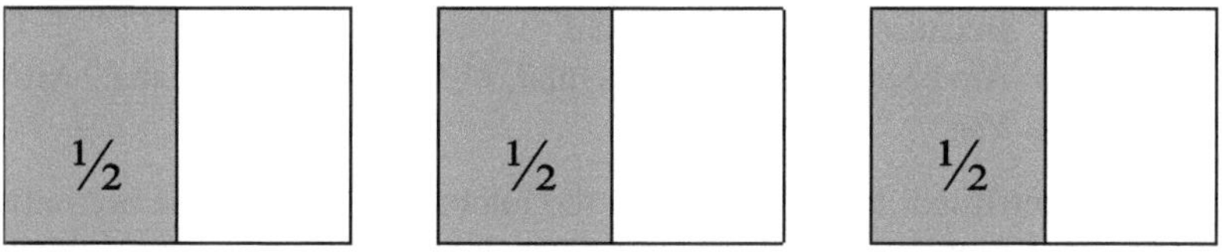

Learners: Three halves.

Teacher: Bernard is showing us three halves. (Pointing at the each of the pieces labelled ½) So this is one half, one half and another one half, neh!

Learners: Yes.

While Bernard's diagrammatic representation is mathematically correct, it seems that the teacher was not satisfied with it and therefore engaged learners in a discussion about a realistic situation of a shopkeeper selling bread. She then challenged the learners by asking them to think about how Mr. Nkomo should handle this problem in an everyday context and continued to give her representation of 3/2:

Teacher: When you go to the shop and you say you want three halves, why does not Mr. Nkomo just take one bread and say hey I don't want to waste my bread. I will just give you one bread and I will cut it into three equal parts. Why does he not do that, Molefe? Why Mr. Nkomo a sa nke borotho bo boiwane, a be a bokgaola dipiece tse three [why doesn't Mr. Nkomo take one loaf and cut it into three parts] and say I can't take another bread because you are going to waste my bread. Why a sa etsa yalo? Why does not he do that? Why does he not take one bread and cut it into two equal halves and another into three equal pieces, S'bongile?

S'bongile: Because they are three two.

S'bongile's response is based on the symbolic representation of three halves, which is 3/2, hence she says "three two". The denominator "2" indicates the number of pieces the whole is divided into. The teacher, however, missed this subtle switch in mathematical representation and continued to request for more responses.

Teacher: Why a sa nke borotho boiwane a be a bo kgaola piece tse three tse lekanang? [Why doesn't he take one loaf and cut it into three equal parts?] Let's say Mr. Nkomo takes one bread and he cuts it into two equal parts. (She draws the diagrams below on the board.) He gives the first child the one half and the second child one half. But there were three children. Mr. Nkomo then takes another bread and cuts it into two equal parts and gives another child his half. Why does he not do that, Victor?

Victor: Because they are not equal.

Teacher: Are because they are not equal. Are these parts not equal? Are they equal?

Victor: Yes.

The centrifugal forces evident in the interaction above arise not only as a result of the use of multiple representations but also multiple discourses. While the learners provided the teacher with a correct mathematical representation of three halves, the teacher was expecting them to think of the problem as it may be dealt with in real life. The teacher thus expected the learners to think about the fact that Mr. Nkomo does not want to have three half loaves of bread left over because that can be wasteful if nobody else buys them. So while the learners were engaging in academic mathematical discourse, the teacher was also expecting them to also engage in everyday mathematical discourse. The teacher mediates between these discourses, in this case preferring an everyday discourse.

10.5 Case 2: Using an International Language to Teach Mathematics in Malaysia

In Malaysia, English is not the first language of the majority of Malaysian teachers and pupils. However, in 2003, the Malaysian Ministry of Education took the step of implementing the policy of teaching mathematics and science in English. In fact, almost all teachers who are presently below 45 years old have undergone their entire education with languages other than English as the language of instruction. English was only taught as a compulsory subject before 2003. Teaching mathematics in English poses great challenges and tensions, particularly for younger mathematics teachers. After much controversy the policy was reversed in 2012. However, the debates about which language to use continue.

For this case, we draw on data from one Malaysian Chinese primary school (reported in Lim & Presmeg, 2010), from the period in which the above-mentioned language policy was still being implemented. There are three types of primary schools in Malaysia, differentiated by the medium of instruction: national primary schools [SK] (Malay as medium of instruction); Chinese primary schools [SJKC] (Mandarin) and Tamil primary schools [SJKT] (Tamil). In the Chinese primary school described in this case, Mandarin was the most used language both in school and at home for the majority of the teachers and students. Mathematics was taught in Mandarin from Grade 1. However, due to the complex demands of the Malaysian Chinese community mathematics was actually taught bilingually in Mandarin (the pupils' mother tongue) and English.

As in the previous case, in Malaysia, English is highly valued as the language of power, particularly in meeting the challenges of globalisation, and particularly in mathematics, science and technology. In view of the reported decline in Malaysian students' English language proficiency, and the belief that "teaching the subjects in the science disciplines in English would expedite acquisition of scientific knowledge in order to develop a scientifically literate nation by the year 2020" the government introduced the new language policy for mathematics and science education (Choong, 2004, p. 2). However, the implementation of this policy drew much criticism and debate. Those who were against the policy argued that the teaching of mathematics and science in English would not help to rescue the deteriorating standard of English, whereas the proponents claimed that making English a tool of learning is the most effective way of ensuring students are proficient in English as well as upgrading students' achievement in mathematics and science.

For the Chinese school community, strong opposition to the policy was apparent from the beginning, particularly from influential Chinese education groups in Malaysia. These groups opposed the teaching of mathematics and science in English for fear of increasing the burden on school children and changing the distinctive character of Chinese schools. They also argued that mathematics is best taught in students' mother tongue, and were proud that research had shown that students in Chinese schools (SJKC) consistently achieved better than their counterparts in SK and SJKT schools. Hence, after much political negotiation, mathematics was taught as two subjects in Chinese primary schools: mathematics in Mandarin (six periods) and mathematics in English (two periods). These policy initiatives are characteristic of centripetal forces since they seek to promote a particular language (English). In this case, the centralising tendencies of language policy led to a clash between two relatively powerful languages, English and Mandarin.

Analysis of mathematics classroom interaction revealed a difference in the percentage of English language used in high-attaining classes compared with low-attaining classes. English was used in more than 97 % of interaction in the high-attaining classes, but in less than 52 % of interaction in low-attaining classes. Where code-switching occurred, it was dominated by the following patterns of interaction, as shown in Table 10.1.

These findings suggest that there is greater language diversity in interaction in the low-attaining classes, at least in terms of the use of Mandarin and English. Moreover, as highlighted in Lim and Presmeg (2010), both the experienced and novice teachers were not able to teach mathematics entirely in English for both high- and low-attaining classes. Particularly in low-attaining classes, teachers said that they used extensive code-switching, in order to allow pupils who are weak in English to catch up with the peers. This way of teaching mathematics in two languages has become a teaching dilemma for the teachers. Ideally, teachers would like to teach mathematics monolingually (in English) as instructed by the newly implemented language policy. Teachers are also worried, however, about whether their students can understand their mathematics lesson fully if it is taught totally in English. The latter concern is greatly related to their confidence in their students' English language proficiency. A substantial amount of teaching time was devoted to translation, especially of the terminology of mathematics.

However, to expedite the teaching, these teachers sometimes opted to teach in the pupils' mother tongue (Mandarin) only. For example, an experienced teacher in the study, Mrs. L, disclosed that when she explained difficult concepts, even for a good class, she opted to use Mandarin sometimes. This is because "我要看到很快的效果的话，我就用华语" [If I want to see the effect quickly, I will use

Table 10.1 Types of code-switching in Malaysian mathematics classrooms

Types of code-switching	Examples
1. Translation	"Width, 宽度"; "What is the unit you use, to measure the mass of heavy object? 你们用什么单位来测量，比较重的东西?
2. Translation involving terminology	Teacher: 华语叫什么meter? [What is meter in Mandarin?] Student: 米 [meter]
3. English is used for units and certain common words, such as pencil	"一个cm等于十个mm" [1 cm equal to 10 mm] or "那边量起。量给老师看。来，第一组，来，你去写pencil,你量到多少，出去写，快点，玲伊。" [Measure from that side. Measure to show teacher. Come, first group, come, you go and write pencil, how many have you measured, go and write, quick, Linyi.]

Mandarin]. In order to speed up the lesson and to overcome time constraints, the teacher opted to use Mandarin because pupils were more familiar with this language. This is an example of how teachers mediate the tensions between centripetal and centrifugal forces, in this case between a language policy and the needs of students. It is clear in this case that the mediation of these tensions is intimately linked to concerns about students' mathematical learning (what Setati, 2008, calls 'epsitemological access').

10.6 Case 3: Immigrant Students Learning Mathematics in a Sheltered Language Program

The first two cases concerns classrooms in which all students are assumed to be growing up multilingual. In this third case, from Québec, Canada, the school system assumes that most children will grow up monolingual. Newcomers must adapt to this situation. In Québec, the children of immigrants are required to join the French-medium school system. Many such children do not, on arrival, speak French. These children are placed in a special class, called a *classe d'accueil*, the main goal of which is for the students to learn enough French to join mainstream classes. Students stay in the *classe d'accueil* for up to a year, by which point they are expected to be proficient enough to survive in the mainstream school system.

This case concerns a class of 9–10-year-old students in a *classe d'accueil* in a medium sized city in Quebec. There were 18 students in the class who had come from a variety of backgrounds, including South America, West Africa, India, and the Middle East. Some of the students have lived in a third country before moving to Canada. About half the students in the class speak Spanish; other languages represented include Portuguese, KiSwahili, Hindi and Arabic. The teacher is a White, francophone Québecoise with several years of teaching experience. Although the main focus and purpose of the class is on learning French, the students do devote some time to other curriculum subjects, including mathematics.

The existence of the *classe d'accueil* is a result of the centripetal language forces in the Québec education system. The fact that students are expected to learn French and are provided with a special class, the express purpose of which is to teach students French arises from a unitary ideology that positions French as the main language of education and of society. The origins of this policy are in concerns that French in Québec was or is being eroded by its proximity to the surrounding English-dominant provinces of Canada and the United States and a political project to reinforce the role of French in Québec society. While the defence of French is understandable, it is important to note that there is little role in the *classe d'accueil* for students' expertise in other languages. As the teacher says:

> Le but c'est qu'ils soient capable de communiquer en français assez pour être capable d'apprendre dans une classe régulière une année après [...] Nos deux grands axes qu'on doit développer dans notre programme c'est communiquer en français et s'intégrer a son milieu (.) d'accueil. Donc oui on doit faire des mathématiques mais axé sur le français. C'est sur le vocabulaire.

> The goal is that after a year they'll be able to communicate in French enough to be able to be in a mainstream class […] The two main thrusts of the program are communication in French and integration into their new environment. So yes we have to do mathematics but focused on French. On vocabulary.

This general unitary approach to the use of French has an impact on the learning and teaching of mathematics. For example, less time is allocated to mathematics than is the case in mainstream classes for this age group: language policy has a direct impact on the learning of mathematics. Furthermore, much of the emphasis during mathematics lessons tends to be on vocabulary development. The teacher also reported that at the start of the year, when the students could not speak French, the class was very quiet. Hence, during mathematics lessons, there could be little discussion. This changed as the students became more proficient and by the end of the year there was considerable lively discussion.

The heteroglossia in the class is readily apparent. Even when students were only speaking French, they did so with a variety of non-standard accents, reflecting the languages they spoke at home. For example, during a lesson on convex and non-convex shapes, one exchange involved multiple pronunciations of the (French) word angle. The students are sorting a set of regular and irregular shapes:

Teacher: pourquoi tu les mets comme ça [why are you putting them like that?]
S6: c'est plus parce que (…) attends attends (…) [it's because … wait, wait …]
Teacher: s'il vous plait dans deux groupes tu vas me les laisser tantôt [in two groups please, I'll look at them in a minute]
S6: c'est que tout ça ont des angel (…) angle (.) toute ça (.) il y a des angles et [it's that all that have an'gel … angle (.) all that (.) there are angles and]
S7: ^awngles^
S6: ungles et ici (…) deux angles ici [ungles and here … two angles here]

Similarly the use of non-standard forms, including vocabulary, led to frequent repair sequences, in which the participants came to a common understanding. In the following example, a student is justifying his claim that the letter L is a non-polygon:

Teacher: bon (.) L (.) S40 (…) L tu le mets ou? [good, L, S40, where would you put L?]
S40: polygone
Teacher: dans polygone? [in 'polygon']
S40: non non-polygone
Teacher: dans non-polygone (.) pourquoi? [in 'non-polygon' (.) why?]
S40: parce que le carré des (cous) des [because the square…]
Teacher: des cous?
S40: non (…) il y a un (cou) [no (…) there's a (cou)]
Teacher: un trou ok la ligne n'est pas fermée (.) très bien bravo [a hole, ok, the line isn't closed, good, well done]

The student appears to say "cou" (neck) or possibly "coups" (hits or blows) which has the same pronunciation, which the teacher struggles to interpret, eventually revoicing it as "trou" (hole) and praising the student for a correct choice.

On some occasions, students used their home languages, particularly in the case of Spanish-dominant students. The teacher, however, explicitly enforces a norm of students not using their home languages, as arose during an exchange about the meaning of convex, in the same lesson:

S46: euh si on peut espagnole [er, if I can Spanish]
Teacher: non non
S46: el el le convexe les cotés ont une forme des V [el el the convex the sides are in a V]
Teacher: comme un V? [like a V?]
S46: ouais un V mais très [yes a V but very]
Teacher: ok si moi je dessine ça comme ça? [ok if I draw it like that?]
S46: ah
Teacher: il n'y a pas de V [there isn't a V]
S46: oui erm
Teacher: ah si je tourne? [ah if I turn it]
S46: oui

Here, the student asks to use Spanish but is not given permission. His next utterance begins in Spanish, then moves more into French, though with apparent effort and with many accompanying gestures which the teacher interprets verbally. The teacher's role therefore consists of refusing the use of Spanish and of interpreting and voicing the student's efforts.

In the *classe d'accueil*, then, the teacher mediates between a French-only policy and the varieties of French and other languages that the students bring. In the second example, above, the trouble arises because the student is confined to using French, despite struggling to find an appropriate word to justify a mathematical decision (his categorisation of the letter L). The teacher's mediation involves maintaining French as the language for their exchange, as well as seeking to interpret what the student is saying, offering a possible word and hearing his words and gestures as mathematically correct.

10.7 Case 4: Immigrant Students and Teacher in a Mathematics Class in South Africa

Since 1994, South African cities have become primary destinations for migrants from around the continent. As migrants cross borders into South Africa, they bring diverse languages into an already linguistically diverse context. Some of the languages spoken by African migrants are not spoken in mathematics classrooms in South Africa. For example, some of the migrants are teachers and children from

French-speaking countries. These children join mainstream classes where the language of instruction is English. Research has shown that some mathematics teachers in these classrooms use the learners' home languages during the teaching of mathematics to support learners who are in the process of learning English (Adler, 2001; Setati, 2005).

This case concerns a multilingual Grade 11 mathematics class with immigrant learners from the Democratic Republic of Congo (DRC), in a school located in an inner city residential neighbourhood of Johannesburg. The students were learning linear programming. The class had 26 students: 3 from the DRC, 1 each from Malawi and Zimbabwe and the rest from South Africa. Therefore, languages represented that were not official languages in South Africa included Shona, Lingala, and Chichewa. The teacher was from the DRC and has many years of teaching experience, including more than 10 years teaching in South Africa. The teacher shared English with all learners in the class. He also shared French (the language of instruction in the DRC) and Lingala, with the immigrant learners that are a focus in this case.

During an interview, the unitary approach to language was emphasised by the teacher when he stated that everything is taught in English and, therefore, immigrant learners had to learn it. However, in the course of the interview (which was conducted in English) centripetal forces were evident as he recognised that he sometimes switches to French when teaching even though the language of instruction is English as shown below:

Teacher: Everything is in English, they have to try and learn English. We are teaching in English but for mathematics sometimes I switch to French for those who understands little English.

Interviewer: Are there French or Portuguese lessons for these learners?

Teacher: There is none. There are no French lessons but for me I try and explain in French but because but for those from Portuguese-speaking countries they have to learn English. Maths sometimes you can play with language, they understand but if they do not understand I try and explain to them in French. But they have to learn English. Generally they have to speak English.

Interviewer: Do you allow them to present their mathematical knowledge in French?

Teacher: For me I try, but they have to learn English everything is in English.

The unitary approach means that English should be used to teach linear programming to immigrant students and immigrant students have to learn English in order to participate. The use of another language like French was clearly not desirable during the teaching and learning of linear programming.

However, centrifugal forces were evident, in the teacher's use of French during the teaching of linear programming to sensitise immigrant learners who understood French. For example, the teacher introduced the French version of the word aquarium to aid the learners with understanding this word. Heteroglossia is noticeable in his pronunciation of the word in French and the way that he positioned himself as someone who can read English but cannot pronounce it properly in English.

Teacher: A school wants to take learners on an outing, a school wants to take learners on an outing to an aquar-aquarium, eh! How do you pronounce it I don't know in English aquarium or aquarium? Because in French we pronounce it as aquarium [teacher introduces French pronunciation] in an outing to an aquarium

The role of centrifugal forces was clear during lesson observation where learners were constructing inequalities from given statements in English. The teacher reformulated his explanation of some words into French, so that the linear programming content would be meaningful to immigrant learners. This practice led to the development of immigrant learners' knowledge of the mathematical skills needed when solving a linear programming task. The translation of selected words was based on immigrant learners' prior knowledge of French regardless of the fact that it was not the language of instruction in South Africa:

… a workshop is available for 20 days each month … at least 2 units of Ralto must be produced each month …

Teacher: As I cannot go beyond this less or equal to and here they say availability. Aaah! Availability … this is a constrain when they say availability … Availability implies less or equal to et cela signifie moins de (*and that means less than)* therefore will have the constrain here…

Teacher: At least two unit of Ralto must be produce each month, y a au moins deux at least two units of Ralto. Alors … sens … hein plus grand que deux Ralto … This is the first constrain … they say at least two units, at least two units means what?

Learners: Greater than

Teacher: Greater than or equal, at least two units of Ralto … It means x must be at least 2. I cannot produce less than two units for Ralto. Je ne peux produire moins pour que deux Ralto …

The teacher switches between English and French in order to enable immigrant learners to understand the mathematical meaning of 'available' and 'at least' in relation to the linear programming task. This approach may support immigrant learners to learn to communicate in mathematical English. In this multilingual mathematics classroom, the teacher mediates between English and the French the immigrant learners brought to the class. This form of mediation is a result of the centripetal and centrifugal forces present in this specific context.

10.8 Discussion and Conclusions

In each of the four cases described in this chapter, tensions are apparent between prevailing assumptions that reflect a unitary language perspective and the heteroglossia of mathematics classrooms in the context of language diversity. Centripetal forces are felt through language policies, parental pressure, and teachers' or students' beliefs and preferences. In Malaysia, the national language policy to use English as the language of mathematics instruction; in Québec, a provincial language policy requires immigrants to learn in French; in South Africa, school language policies follow parental preferences for English. There are some interesting contrasts however: in Malaysia, parental pressure led to officially sanctioned diversification of the languages used for mathematics instruction, while in South Africa, the diversity permitted in national policy is reduced by a widespread preference for English. Hence, while language policy tends to act as a centripetal force in mathematics classrooms, the uptake of such policies is shaped by local responses, notably from parents.

Heteroglossia is apparent in all four cases, and includes the use of multiple languages, multiple discourses and variations in accent, word choice and the use of gesture and a range of representation systems. In the Malaysian and South African examples, code-switching was, despite teachers' stated preferences, regularly deployed to enhance students' access to mathematical meaning or the meaning of mathematical problems. In the Québec case, by contrast, there was little evidence of code-switching, despite some of the students sharing a common language. Instead, the students in the Québec case used accented, highly idiosyncratic French, and required substantial mediation on the part of the teacher to interpret their mathematical thinking and guide their expression of their ideas.

It is important to note, however, that heteroglossia is not only about code-switching; for example, several of the cases shown here exemplify the use of a range of levels of formality in mathematical expression. In the first case, the teacher encourages students to consider an everyday context for their problem, and in the third case, the students' efforts to describe convex shapes draw on a variety of informal forms of expression, including gestures and the use of nonspecific forms of reference (e.g., this, those). Again, the teacher plays a key role in mediating between these different levels of formality.

The centripetal and centrifugal forces are clearly in tension in our four cases. The teacher in case 1, for example, says that students should learn in their main languages so that they understand mathematics, but also that they should learn in English, because it is an international language. In case 2, a single language policy is negotiated into a dual language policy, but in practice a mixture of languages is used in mathematics classrooms. In case 3, a student expresses a desire to use Spanish to explain his mathematical thinking but the teacher, the program, and the province all insist on French. And in case 4, English is the medium of instruction, but the teacher mixes it with French. These tensions, moreover, are more complex than they are described here. In three of the cases, not all students share all languages in their mathematics classes and in cases 3 and 4, the only language shared by everyone including the teacher is English.

Teachers' mediation of this tension takes several forms. In some cases, it involves translating, code-switching or using language mixtures to provide additional forms of mathematical meaning making for students. In all cases, it involves policing the language policy, sometimes strictly (e.g., case 3), sometimes not (e.g., case 1). It can also involve the use of multiple discourses (e.g., everyday vs. mathematical, case 1) and systems of meaning (e.g. gesture, case 3). Finally, the teacher plays a central mediational role in all cases in interpreting and making mathematical meaning with students' utterances, including gestures, such as by re-voicing them using different words, a different language or mathematical symbols.

Acknowledgments RB's data collection was funded by the Social Science and Humanities Research Council of Canada. RB is grateful to Adil Dsousa and Élysée Cadet for their assistance. All four authors are grateful to the participants in their research projects.

References

Adler, J. (2001). *Teaching mathematics in multilingual classrooms*. Dordrecht, The Netherlands: Kluwer Academic.

Bakhtin, M. M. (1981). *The dialogic imagination: Four essays* (M. Holquist (Eds.); C. Emerson & M. Holquist, Trans). Austin, TX: University of Texas Press.

Barwell, R. (2009). Summing up: Where next for multilingual mathematics classrooms. In R. Barwell (Ed.), *Multilingualism in mathematics classrooms: Global perspectives* (pp. 161–168). Bristol, England: Multilingual Matters.

Barwell, R. (2012). Heteroglossia in multilingual mathematics classrooms. In H. Forgasz & F. Rivera (Eds.), *Towards equity in mathematics education: Gender, culture and diversity* (pp. 315–332). Heidelberg, Germany: Springer.

Barwell, R. (2014). Centripetal and centrifugal language forces in one elementary school second language mathematics classroom. *ZDM Mathematics Education, 46*(6), 911–922.

Chitera, N. (2009). Code-switching in a college mathematics classroom. *International Journal of Multilingualism, 6*(4), 426–442.

Choong, K. F. (2004). *English for teaching of mathematics and science (ETeMS): From concept to implementation*. Retrieved from http://files.embedit.in/embeditin/files/sqcN7d3DYy/1/file.pdf

Clarkson, P. C. (2007). Australian Vietnamese students learning mathematics: High ability bilinguals and their use of their languages. *Educational Studies in Mathematics, 64*(2), 191–215.

Clarkson, P. C. (2009). Mathematics teaching in Australian multilingual classrooms: Developing an approach to the use of classroom languages. In R. Barwell (Ed.), *Multilingualism in mathematics classrooms: Global perspectives* (pp. 145–160). Bristol, England: Multilingual Matters.

Clarkson, P., & Dawe, L. (1997). NESB migrant students studying mathematics: Vietnamese students in Melbourne and Sydney. In E. Pehkonen (Ed.), *Proceedings of 21st Meeting of the International Group for the Psychology of Mathematics Education* (Vol. 2, pp. 153–160). Lahti, Finland: University of Helsinki.

Dlamini, C. (2008). Policies for enhancing success or failure? A glimpse into the language policy dilemma of one bilingual African state. *Pythagoras, 67*, 5–13.

Duranti, A. (1998). *Linguistic anthropology*. Cambridge, England: Cambridge University Press.

Farrugia, M. T. (2009). Reflections on a medium of instruction policy for mathematics in Malta. In R. Barwell (Ed.), *Multilingualism in mathematics classrooms: Global perspectives* (pp. 97–112). Bristol, England: Multilingual Matters.

Gee, J. P. (1999). *An introduction to discourse analysis: Theory and method*. London: Routledge.

Halai, A. (2009). Politics and practice of learning mathematics in multilingual classrooms: Lessons from Pakistan. In R. Barwell (Ed.), *Multilingualism in mathematics classrooms: Global perspectives* (pp. 47–62). Bristol, England: Multilingual Matters.

Khisty, L. L. (1995). Making inequality: Issues of language and meaning in mathematics teaching with Hispanic students. In W. Secada, E. Fennema, & L. B. Adajian (Eds.), *New directions for equity in mathematics education* (pp. 279–297). Cambridge, England: Cambridge University Press.

Lim, C. S., & Ellerton, N. (2009). Malaysian experiences of teaching mathematics in English: Political dilemma versus reality. In M. Tzekaki, M. Kaldrimidou, & H. Sakonidis (Eds.), *Proceedings of the 33rd Conference of the International Group for the Psychology of Mathematics Education* (Vol 4, pp. 9–16). Thessaloniki, Greece: PME.

Lim, C. S., & Presmeg, N. (2010). Teaching mathematics in two languages: A teaching dilemma of Malaysian Chinese primary schools. *International Journal of Science and Mathematics Education, 9*(1), 137–161.

Moschkovich, J. N. (2008). "I went by twos, he went by one": Multiple interpretations of inscriptions as resources for mathematical discussions. *The Journal of the Learning Sciences, 17*(4), 551–587.

Parvanehnezhad, Z., & Clarkson, P. C. (2008). Iranian bilingual students' reported use of language switching when doing mathematics. *Mathematics Education Research Journal, 20*(1), 52–81.

Planas, N., & Civil, M. (2013). Language-as-resource and language-as-political: Tensions in the bilingual mathematics classroom. *Mathematics Education Research Journal, 25*(3), 361–378.

Planas, N., & Setati, M. (2009). Bilingual students using their languages in the learning of mathematics. *Mathematics Education Research Journal, 21*(3), 36–59.

Planas, N., & Setati-Phakeng, M. (2014). On the process of gaining language as a resource in mathematics education. *Zentralblatt für Didaktik der Mathematik, 46*(6), 883–893.

Setati, M. (2005). Teaching mathematics in a primary multilingual classroom. *Journal for Research in Mathematics Education, 36*(5), 447–466.

Setati, M. (2008). Access to mathematics versus access to the language of power: The struggle in multilingual mathematics classrooms. *South African Journal of Education, 28*, 103–116.

Setati, M., & Adler, J. (2000). Between languages and discourses: Language practices in primary multilingual mathematics classrooms in South Africa. *Educational Studies in Mathematics, 43*(3), 243–269.

Setati, M., Adler, J., Reed, Y., & Bapoo, A. (2002). Incomplete journeys: Code-switching and other language practices in mathematics, science and English language classrooms in South Africa. *Language and Education, 16*(2), 128–149.

Setati, M., & Barwell, R. (2006). Discursive practices in two multilingual mathematics classrooms: An international comparison. *African Journal of Research in Mathematics, Science and Technology Education, 10*(2), 27–38.

Chapter 11
Purposefully Relating Multilingual Registers: Building Theory and Teaching Strategies for Bilingual Learners Based on an Integration of Three Traditions

Susanne Prediger, Philip Clarkson, and Arindam Bose

11.1 Introduction

The purpose of this chapter is twofold: to build theory as well as suggesting teaching strategies that could be used with multilingual mathematics learners. The teaching strategies are based on the approach of purposefully relating multilingual registers by which we integrate three research traditions and approaches.

In the mathematics education research literature there are three different strong ideas related to different language registers and discourses that have not been closely linked. Indeed they are often treated as quite distinct entities. These are code-switching between first and second languages, transitions between informal and academic (mathematical) forms of language within a given language, and transitions between different mathematical representations. Exploring the overlap between these three ideas, and in particular by articulating their interconnections, new insights and implications are gained. For this, we draw upon the sociolinguistic construct of registers as functional varieties of language use being associated with different contexts. Rather than three apparently discrete sets of ideas, an integrated set of ideas for teachers and researchers is presented that has the potential to drive theory, curriculum, and teaching developments.

S. Prediger (✉)
Institute for Development and Research in Mathematics Education, TU Dortmund University, Vogelpothsweg 87, Dortmund 44227, Germany
e-mail: prediger@math.uni-dortmund.de

P. Clarkson
Australian Catholic University, Fitzroy, VIC, Australia
e-mail: Philip.Clarkson@acu.edu.au

A. Bose
Homi Bhabha Centre for Science Education, TIFR, Mumbai, India
e-mail: bosearin@gmail.com

R. Barwell et al. (eds.), *Mathematics Education and Language Diversity*, New ICMI Study Series, https://doi.org/10.1007/978-3-319-14511-2_11

Mathematical ideas are not always obvious to the learner; indeed often they are quite opaque and nonintuitive. Once the move away from traditional teaching that promoted rote learning gathered pace, teachers turned to strategies which would bridge the gulf from what starts as being nonintuitive, but in the end becomes obvious to students. One of the enduring strategies over the last five decades is the use of manipulatives. These can act as excellent scaffolding devices that reflect the targeted mathematical relationships, and may well bring students to a place where they are predisposed or begin to engage in particular learning. But for most students, the learning experience needs to be focused by using more scaffolding than that provided by just manipulatives. This additional scaffolding is normally a purposeful entanglement with manipulatives of language and pictorial representations of the ideas.

The intertwining of such components for good teachers just seems to happen, but young teachers need to learn what emphasis to give each component at appropriate times. We suggest that the teacher's role is not to encourage students from moving from sloppy language to a language of precision, but rather to recognize that students will often need to use everyday language as they grasp for the underlying mathematical relationship that they are trying to understand. Hence the teacher will privilege the dynamic of students moving between everyday, school and technical registers, as they start to understand more deeply the central core of a mathematical idea, the limits of the idea, and its relationship to other mathematical and nonmathematical ideas that already exist within their ideas network. To play this role, the teacher needs a thorough knowledge of the mathematics under study and the appropriate language forms so that their support is encouraging, generative, and productive as the students engage in sustained thinking.

Guidance for promoting the use of the technical mathematical register has been developed in the literature over the last decade or so. One main thrust of the Cognitive Guided Instruction (CGI) project was to encourage teachers to use language that guided students to using an appropriate mathematical register to discuss mathematics, with which they were engaged (Carpenter, Fennema, Franke, Levi, & Empson, 1999). Watson and Mason's (1998) different categories of questions, and the use of open-ended questions (Sullivan & Lilburn, 2004), prompt the use of technical language. However there seems little in the literature that guides teachers in transitioning their language from the everyday to the school register, although this has received increasing importance in the discussion about linguistic difficulties (see below). The emphasis is to move directly to a technical language register without looking in depth at better ways of moving students through these transitions, let alone privileging the dynamic between them with an expectation that this dynamic will be an ongoing feature of the classroom, with students free to move between registers as they feel the need to do so.

We return to these deliberations later, but at this point it is useful to consider one of two specific teaching snapshots that will allow the later theoretical argument to be grounded. The two empirical snapshots from very different language contexts (India and Germany) suggest that the teaching strategy of purposefully linking appropriate registers seems to have a high potential to initiate substantial student activities that lead to deeper understanding. It is interesting to reconstruct how teachers and students gradually developed their strategies to clarify the mathematical relations by using different representational registers. In particular, the graphical register supported the processes of linguistic and conceptual clarification. Although only a very short insight into the rich data can be given here, these snapshots show that the teaching strategy adopted from integrating language and mathematics education experiences seems to be promising as it focuses directly on the conceptual core of linguistic and mathematical challenges. We choose to present one snapshot here to ground our discussion, follow this by our theoretical argument, and then return to a second snapshot to position the argument fully in the context of a classroom.

11.2 Relating Registers for Common Multiples: Snapshot from India

The context for the following snapshot was a camp held at the Homi Bhabha Centre for Science Education, Mumbai (HBCSE, Bombay). The students were low-achieving sixth graders in the age-group of 10–11 years from a neighborhood English-medium school. The camp classes were held over a period of 2 months every Thursday for around one and a half hours. Altogether 21 students (12 boys and 9 girls) participated in the camp. They were selected by their class teachers on the basis of their performance in a school mathematics test with less than 40 % marks. All the students knew Hindi and English while some also spoke Marathi and Tamil. Marathi is the official language of the state of Maharashtra of which Mumbai is the capital city. The teaching for the entire camp was undertaken by an HBCSE member-researcher who speaks English, Hindi, and Marathi fluently. In urban India, the phenomenon of code-mixing is very common and is a regular feature in daily-life conversations (similar to urban Pakistan as described in Halai, 2009). However, most English-medium schools (including the one these students came from) promote a monolingual English practice, and the use of other languages in the classrooms is actively discouraged. The medium of instruction in the camp was English, but teacher and students also used *Bambaiya Hindi* (the local dialect of Hindi used in and around Mumbai), which we consider here as the students' first language.

The following two Episodes, A and B, are from two different camp lessons that depict interesting student–teacher exchanges involving transitions between verbal registers with significant amounts of code-switching and mixing. The lessons were videotaped, and selected clips were transcribed for the analysis. For the two transcripts, an English translation has been provided on the right of each utterance, and the numeral before each utterance indicates the line number in the respective transcript. In the right-hand translation, the underlined script is the translation, and the non-underlined script is English words used in the original verbal exchange. "T" and "C" stand for teacher and student respectively.

11.2.1 Tasks and Situation in the Camp Lessons

In the classroom Episodes A and B, the teachers wrote the following word problems on the blackboard and then started explanations and discussions with the students:

- *Problem A.* A cat take [sic] a jump of 4 and a rat takes a jump of 2. Now, (the) cat is at number 8 and (the) rat is at number 20, will (the) cat (be able to) catch the rat?
- *Problem B.* In the year 1997, price of a rubber ball is [sic] Rs. 7. If the price increases every year by Rs. 0.25, then in which year the price of the ball will be Rs. 9?

The analysis focuses on how the teachers and students interactively made sense of the word problems.

11.2.2 Code-Switching Between Registers for Better Explanation

The teacher began by writing the "warm-up" Problem A as the problem of the day on the board in English. Then she started a verbal exchange with the students to ensure they have understood the task well:

a38	T	samajh mein aaya kya likha hai? (*indicates the warm-up problem for the day on the black board*)	T	understood what I wrote? (*indicates the warm-up problem for the day on the black board*)
a39	T	hum Hindi mein bolenge.	T	We'll speak in Hindi
a40	T	cat kitna jump leti hai ek saath?	T	how much jump does cat take at one go?
a41	C	Four	C	Four
a42	T	four ka? aur rat kitna jump leta hai?	T	in fours? And how much jump rat takes?
a43	C	Do	C	Two
a44	T	do ka/	T	in twos/
a45	T	to samjho aisa yeh ek number-line hai/ haan, aisa ek tree hai number wala/	T	then understand this way a number-line is taken/ yes, there is a tree of numbers
a46	T	one, two, aisa wala/	T	one, two, likewise/
a47	T	to abhi cat kidhar hai?	T	so where is the cat now?
a48	C	four, four pe/	C	four, at four/
a49	T	nahin cat is at number eight/	T	No cat is at number eight/
a50	C	eight/	C	eight/

The teacher used a code-switch from English to Hindi to ensure that the students clearly understood the task, and that they were able to work on their own. But while doing so, she used technical terms (suitable for the problem-tasks) in English such as, "numbers," "number-line," and some number names. Interestingly she also used English for "cat" and "rat" which although not mathematical-technical terms, are crucial for understanding the context of the original English written problem. Hence her code-mixing addresses the technical register as well as relevant parts of the everyday register. By code-mixing (using the mathematical-technical terms but in English, embedded in Hindi sentences) the teacher tried to ensure that the students can understand and appropriately use these key terms in English.

11.2.3 Use of Gestures and Nonverbal Cues

The presentation of the problem-task was accompanied by gestures and drawings on the board for everyone to see. In Problem A, the teacher used her hand-gestures indicating a wave-like flow of "jumps" (read: numbers) that the rats and cats take; these gestures can be assigned to the base level of concrete representations (Fig. 11.1). She also tried helping students visualize the jumps by drawing a similar "wave" number-line on the board, and marked the initial positions of the cat and the rat. In the graphical representation, the nonverbal cues supported the process of understanding by visualizing the subsequent jumps and the point where the cat meets the rat.

At one early point in the lesson, the following interaction occurred:

a55	T	idhar samjho cat hai? hmm.. aur twenty main nikalti nahin/ wahan pe rat hai/	T	<u>Consider</u> cat <u>to be here? Hmm.. and I haven't marked</u> twenty/ Rat's <u>there/</u>

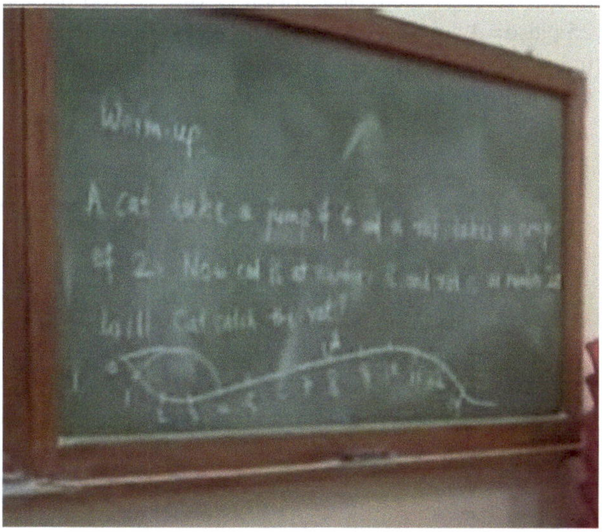

Fig. 11.1 Text and drawing on the board (screenshot)

a56	T	to kya cat, aise woh kud raha hai aur/ hamesha rat kitna jump le raha hai? do/	T	will cat then, she's jumping like this/ and how much jump does rat take always? Two/
a57	T	aur cat kitna le rahi hai? Four/	T	and cat takes how much? Four/
a58	T	to kya cat rat ko pakad sakti hai? Karke dekho/	T	then can cat catch the rat? Show by doing/
a59	C	yes teacher/ yes teacher/ pakad sakti hai/	C	yes teacher/ yes teacher/ can catch/

Importantly the style of verbal language changed immediately when a drawing was introduced into the teaching context. The introduction of a graphical representation allowed the language to change to a more deictic form ("she's there," line a56). The comment by the teacher "Show by doing" (line a58) allowed the linguistic load to be lightened for the students, and allowed them to develop first their solution ideas which they later expressed in a more elaborate verbal language. In this way adequate language in the school and/or technical registers were built successively. However, the existence of and the relation between these registers all took place without the teacher's explicit reference to them.

11.2.4 Use of Technical and Everyday Registers

In Episode A, the teacher connected the technical and the everyday register by repeatedly using terminologies likes, "number-line" and "jump," "catch," etc. In contrast, in Episode B, she encouraged using everyday registers that involved the use of nonformal units like "anna," which are old currency-units no longer in use. However, such nonformal currency-units are still part of the everyday register and people refer to them quite often (1 rupee = 100 paise = 16 anna; 50 paise = 8 anna; 25 paise = 4 anna). This transcript shows how the use of these informal "everyday" units helped the students to make sense of the abstract value 0.25 (that is to be added to 7 until reaching 9):

b67	C	teacher, maine na, inn charon ko aath anna, char anna, sabko plus kiya tho equal to one rupees aaya/	C	teacher, I took all these four, eight anna, four anna, I did plus and it came equal to one rupees/
b68	T	kaise?	T	how?
b69	C	yeh char anna - char anna se aath anna aaya/ isko bhi char anna - char anna se aath anna aaya/ one rupees/	C	this four anna-four anna make eight anna/ this too gave four anna-four anna eight anna/ one rupees/
b70	T	means?	T	means?
b71	T	nahin, yeh dono milake kitna hua?	T	No, how much these two taken together?
b72	C	aath anna/	C	eight anna/
b73	T	hmm, aur yeh aage ke do?	T	hmm, and this further two?
b74	C	aath anna/	C	eight anna/

b75	T	tho kitne aath aane milane hain tumhe? Do/	T	then how many eight anna do you have to put together? Two/
b76	C	do aath aane/ tho ek rupaya banta hai/	C	two eight anna/ then that makes one rupee/
b77	T	hmm, ek rupaya banta hai/	T	hmm, makes one rupee/

By using transitions between the mathematical-technical and everyday registers, and by transitions between their languages (code-switching and mixing), students were able to arrive at the solutions. Such transitions appeared to help in reducing the cognitive load for the students. Hence this seems to have encouraged all the students to take part in the classroom activity and contributed to the solving process, correct or not. This also helped in building confidence among the students.

With this analysis of the Indian snapshot in mind, we now return to the theoretical issues that have been used and discuss them in more depth.

11.3 Revisiting Three Traditions of Reflecting on Linguistic Transitions

11.3.1 Transitions Between First and Second Languages: Code-Switching

The role of language in thinking and learning is well accepted. In Post-Piagetian constructivism, it is argued that a child's mind is not a *tabula rasa*. Children come to school with prior knowledge drawn from their environment and everyday experiences, including knowledge of languages, both home language and/or local language, which in many multilingual societies are different languages. In many countries like India, Pakistan, South Africa, etc. it is not unusual to find children entering elementary grades with knowledge of more than two languages (Bose & Choudhury, 2010; Halai, 2009; Setati, 2005). A natural "movement" between languages occurs in the forms of speech and thought during the conversation flow between conversant bilinguals (or multilinguals). This phenomenon of movement between languages or language-switching is known as "code-switching": the practice of switching between two or more languages in a conversation or an utterance (Farrugia, 2009). Code-switching is generally accompanied by "code-mixing" where the latter refers to the case when only "one or few words" are borrowed from one language and embedded into a sentence predominantly in the alternate language (Farrugia, 2009). Code-switching is widely recognized as a social phenomenon that can facilitate effective mathematical learning since both mathematics and language are socially constructed and hence the linguistic and social nature of mathematics allows it to develop together with language (Barton, 2009; Choudhury & Bose, 2011).

Different contexts of mathematics classrooms where learners and teachers are engaged in classroom activities, and therefore also engaged in language negotiations, have been carefully described in terms of code-switching (see Barwell, 2005;

Bose & Choudhury, 2010; Clarkson, 2007; Farrugia, 2009; Halai, 2009; Setati, 2005, 2008). Such instances occur while engaging with difficulties in comprehension of the problem-tasks or keywords and phrases therein; search for alternate strategies; presentation of explanations; achievement with correct solutions accompanied by exuberant exclamations with a few words/phrases drawn from a different language; and while using mathematical terminologies. In addition, teachers code-switch while providing scaffolding to make the subject-matter comprehensible or to develop certain mathematical abilities among the students; to reduce students' cognitive load; and while enforcing authority and discipline in the classroom. Code-switching is also used to facilitate the connection of verbal languages with visual representations (see Sect. 11.2.3). Furthermore, the use of code-switching by either teacher or students seems to allow simultaneous learning of both languages, as well as mathematics (Barwell, 2005; Bose & Choudhury, 2010; and many others).

Building upon this empirical work, it is time not only to argue that code-switching should be *allowed* and descriptively analyzed concerning its conditions and effects, but it is also the time to develop and promote teaching strategies that make use of code-switching and other links between first and second languages more purposefully. In particular, teachers should be enabled to guide students to make conscious choices to use transitions between their languages as a possible solution strategy for co-learning mathematics. An aspect of such a model will be to privilege students' competences in all their languages.

11.3.2 Transitions Between Everyday and Technical Language

When turning to the second key idea of this chapter, that of students moving between informal and formal registers within their dominant language, we extend this idea to multilingual learners. Within the mathematics education research literature, Pimm (1987), Freudenthal (1991) and many others have advocated a careful transition from everyday language to the technical language of mathematics as an important teaching strategy that enhances conceptual understanding of mathematical concepts and ideas. Empirical studies (e.g., Van den Heuvel-Panhuizen, 2003) show the power of this teaching strategy.

However, some students seem to experience serious difficulties in making such a transition and it appears that the source of the problem is connected to an intermediate register between the informal everyday register and the formal technical language. For a long time, this intermediate register has been underestimated. For a theoretical explanation of this issue it is useful to use Cummins' distinction between Basic Interpersonal Communicative Skills (BICS) and Cognitive Academic Language Proficiency (CALP). Cummins (2000) has suggested that there is a distinction between surface fluency in an everyday language register and the language skills needed in a context of high cognitive academic demands. He developed the construct BICS to describe the situation when there are contextual supports for language. Face-to-face conversations, for example, provide actions with hands and eyes, instant feedback, and other cues to support meaning. Such situations are said to be "context embedded" (Koch & Oesterreicher, 1985) and conceptually oral, and surface language skills are sufficient.

On the other hand, where higher-order thinking skills such as analysis and evaluation are required (for example in problem solving contexts), language becomes "disembedded" from a meaningful supportive context and becomes more abstract. Often, it appears conceptually written even if used orally. Such a situation can be thought of as "context reduced" (Koch & Oesterreicher, 1985; Schleppegrell, 2004) and needs more explicit linguistic means to be mastered. The skills required to become fluent in this style of language falls in the domain of CALP. The corresponding language register is here termed the "school register." By "school register," we refer to the term language of schooling as explored by Schleppegrell (2004) and discussed in many political contexts (as in the European Council; see Thürmann, Vollmer, & Pieper, 2010).

Although many overlaps exist, we can, for analytical reasons, distinguish the school register from the everyday register (in which Pimm's informal language and Cummins' BICS is located), as well as the technical register (which comprises mathematical technical language of school mathematics), hence giving a three-tiered model as shown in Fig. 11.2.

Most teachers are aware that the technical register needs to be acquired in school, whereas the school register (to which students of privileged socioeconomic background are often already acquainted) is sometimes treated as *a learning condition*, instead of *a learning goal*. This distinction suggests quite a different set of teaching strategies. One crucial implication for learning is that students with weaker language background, either because they come from a lower socioeconomic background, or they are from a migrant community not speaking the language of schooling, or both, experience difficulties.

Hence for language acquisition (of the everyday register, of the school register as well as of the technical register), it turns out to be important not only to transit once from the everyday register using the school register to the technical register, but to move flexibly forward and backward between all the three, as emphasized by Freudenthal (1991) and elaborated by Clarkson (2009). While extending the model for multilingual learners,

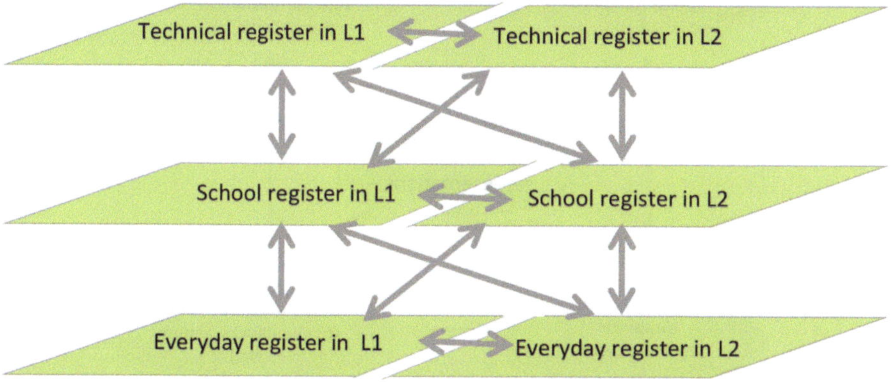

Fig. 11.2 Three-tiered model of registers (adapted from Clarkson, 2009)

Clarkson (2009) adds the important perspective that the three registers might exist in more than one language for multilingual learners (see Fig. 11.2).

He suggested that as well as multilingual students switching codes, they should also be encouraged by teachers to recognize and move between everyday register(s), school register(s), and technical register(s). It is not enough, he has argued, to concentrate only on the mathematical-technical register, as has so often happened in the past. Students bring to their mathematical learning their own lived experiences outside the classroom, which will be encoded mainly in their everyday language, which in turn may well be in their first language (L1). This is why a dynamic transitioning between registers and in switching codes needs to be encouraged by teaching strategies.

Figure 11.2 clearly outlines many possible connecting paths. Whether all pathways are useful for some students is an open question. For example, if a student's L1 does not support some aspects of school mathematics, then the top left hand cell may be empty. Some pathways that students use have been identified in the literature (Halai, 2004). But which of these pathways should be encouraged by the teacher is still a question for research and is probably quite bound to the specific language context.

11.3.3 Transitions Between Different Mathematical Representations: Relating Mathematical Registers

Bruner (1967), Dienes (1969), Lesh (1979), Duval (2006) and many others have pointed out that relating different mathematical representations is an important activity for developing students' conceptual understanding (Fig. 11.3). Many examples from the research literature indicate how this is employed as a fruitful teaching strategy, for example in task design (Swan, 2005).

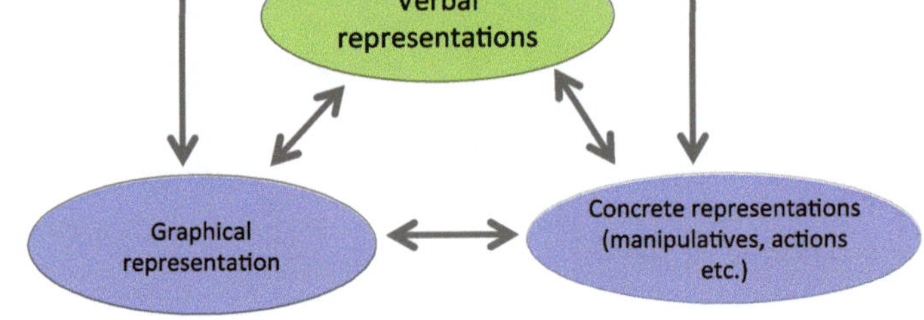

Fig. 11.3 Transitions between different representations (Bruner, 1967; Duval, 2006; Lesh, 1979)

Historically, the crucial notion of using manipulatives, an influence of Dienes (1969) in particular, in teaching mathematics, was a huge step forward in the 1960s that changed the very nature of teaching mathematics. Coinciding with this was a growing emphasis on visualizing, and making or using pictorial representations of mathematical ideas in teaching, an influence of Bruner's (1967) enactive, iconic and symbolic modes of representation.

These new ideas began to offer an alternative to the traditionally accepted teaching style when the teacher delivered small portions of mathematics that needed to be learnt for a particular lesson followed by students completing individually symbolic-based exercises. With the introduction of manipulatives and the growing emphasis on visual representations, learning by discovery and with understanding emerged as an alternate teaching practice, and has now become the dominant paradigm in the discourse on teaching mathematics.

More recently, Duval (2006) in particular has given a semiotically grounded theoretical foundation for why transitions between different modes of mathematical representations are crucial for the acquisition of conceptual understanding. He has emphasized that the abstract nature of mathematical concepts is one reason for these necessities.

Although from the early days in the 1960s, language was seen to be an integral part of mathematical semiotic registers, often the manner in which teacher and students interacted with language, both verbally and in written modes, was left unexamined. In particular the "linguistic" registers were not differentiated for L1 versus L2 and hence notions such as code-switching that might occur could not be recognized. Nor were the differences between everyday registers, school registers, and technical registers included in their models to any extent. These differences need to be acknowledged as yet another variation within the communications between classroom participants.

11.3.4 Integrating Three Transitions Between Languages, Registers, and Representations

Following Leisen (2005), Prediger and Wessel (2011) formulated an integration of the above three different perspectives on language transitions as the "relating register approach" (see Fig. 11.4). In Fig. 11.4, the different registers and representations are ordered hierarchically according to their abstractness (as proposed by both Leisen, 2005 and von Kügelgen 1994). The type face used in both the upper and the base levels are printed in grey in order to sketch that they are not always used: primary and lower secondary classrooms do not refer to the symbolic-algebraic register, upper secondary classrooms usually do not refer to concrete artefacts and manipulatives.

Recently this integrated model has proved to be very useful as a heuristic tool that was used to guide the practical design and support of learning processes for multilingual learners (Leisen, 2005; Prediger & Wessel, 2011) as well as its empirical investigation (Prediger & Wessel, 2011). However, the model requires further theoretical and empirical exploration.

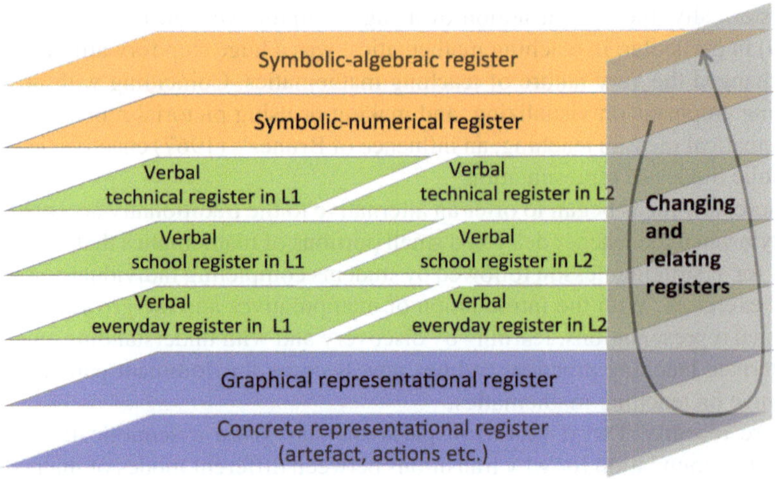

Fig. 11.4 An integrated model which relates the three transitions (Prediger & Wessel, 2011)

11.4 Theoretical Exploration: Representations, Registers, or Languages?

11.4.1 Different Conceptualizations

In the first integrated model (Fig. 11.4), the different registers and representations are ordered hierarchically according to their increasing degree of abstractness (as proposed by Leisen, 2005; von Kügelgen 1994). However, it has become apparent in Prediger and Wessel (2011) that the degree of abstractness also depends on the mathematical topic under study and the concrete representations used. As well it would appear that the different levels are not of the same quality. For building an adequate theoretical conceptualization, different authors have suggested different theoretical constructs including the following:

- The linguist von Kügelgen (1994, p. 34) offered the construct "concept levels" with a strong emphasis on the hierarchy which has not been proved to be useful for all cases and which has too narrow a focus on words without grasping the complexity of language.
- The psychologist Bruner (1967) suggested the classical conceptualization of different "modes of representation" (enactive, iconic, symbolic), which has proved to be useful for designing learning sequences. However the construct "representation" lacks important dimensions such as references to contexts, functions, and social embeddedness. Hence the construct "mode of representation" might be understood in the way that there are one-to-one translations between all modes of representations without shifts in meaning and function. This is definitely not the case.

- The language education researcher Hallet (2012) focuses on "symbolic languages" and emphasizes their different semiotic functions to describe and explain the world in specific ways. This has proved useful for formulating this latest model.
- Some mathematics education research focuses on different semiotic functions. In his construct of "semiotic registers," Duval (2006) emphasizes that the meaning ("content") of a mathematical object can change with a shift of representation: "The content of a representation depends more on the register of the representation than on the object represented. That is the reason why passing from one register to another changes not only the means of treatment, but also the properties that can be made explicit" (p. 111). With this observation, Duval points to an aspect that is equally important for the sociolinguistic construct of "register." This distinction has proved to be useful to us.
- The sociolinguist, Halliday (1978), defines register as "set of meanings, the configuration of semantic patterns, that are typically drawn upon under the specific conditions, along with the words and structures that are used in the realization of these meanings" (p. 23). Halliday distinguishes registers from dialects by defining dialects as ways of saying the same things differently (which in a certain way also applies to representations), whereas he describes registers as "ways of saying different things" (p. 35). As a consequence, a change of register also implies a shift in meanings. Additionally, Halliday emphasizes the social embeddedness of the communication situation for characterizing registers: "A register can be defined as the configuration of semantic resources that a member of a culture typically associates with the situation type. It is the meaning potential that is accessible in a given social context" (p. 111). Hence, registers are characterized by the types of communication situations, their field of language use, and the discourse styles and modes of discourse. This has also been useful in the present model building exercise.

11.4.2 Registers with Different Representations

The everyday register, the school register, and the technical register can thus be characterized as registers in Halliday's sense that are used in different communication situations (with of course some overlapping, especially between the technical register and the school register). Graphical and symbolic representations can also be conceptualized as registers, but only in Duval's sense. The decision not to mix the different conceptualizations of registers led us to give up the symmetry of all levels in Fig. 11.4, and to develop a more complex 3×4 matrix structure (Table 11.1) that shows how the different representations are used in the different registers.

The everyday register has been characterized by its contextual embeddedness and lacking mathematical explicitness in face-to-face-communication. It encompasses verbal and concrete representations, but rarely graphical and symbolic representations. An example is given in Table 11.1 (adapted from Meyer & Prediger, 2012).

Table 11.1 Linguistic registers and their representations—illustrated for percentages in a shopping situation (from Meyer & Prediger, 2012)

<table>
<tr><th></th><th>Everyday register</th><th>School register</th><th>Technical register</th></tr>
<tr><td>Representation in words</td><td>Yesterday I was at a sale, in my favourite shop. The sale meant I received a 10 € discount for the trousers. Since I paid in cash, the sales clerk gave me another discount of 3 %. In all I only paid 77.50 €. How much was the original price?</td><td>In a sale, a pair of trousers was reduced by 10 €, and further 3 % discount was offered if you paid in cash. Hence, the final price was 77.50 €. What was the original price of the trousers?</td><td>If the original value is reduced by 10 € and then by 3 %, the new value is 77.50 €. What was the original value?</td></tr>
<tr><td>Graphical representation</td><td>
</td><td>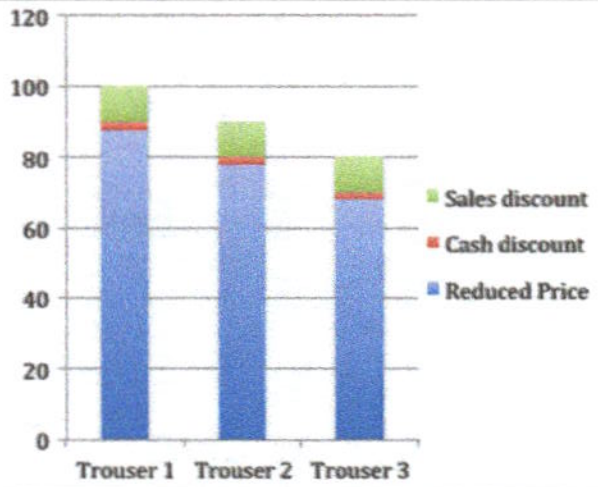
</td><td>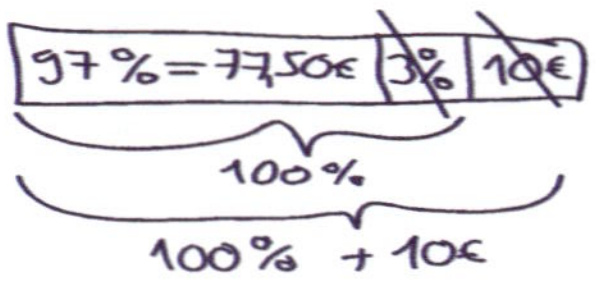
</td></tr>
<tr><td>Symbolic-numerical representation</td><td>–</td><td><table>
<tr><th></th><th>Original price (€)</th><th>Price with sales discount (€)</th><th>Price with 2nd discount (€)</th></tr>
<tr><td>Trouser 1</td><td>100</td><td>90</td><td>87.30</td></tr>
<tr><td>Trouser 2</td><td>90</td><td>81</td><td>78.57</td></tr>
<tr><td>Trouser 3</td><td>80</td><td>72</td><td>69.84</td></tr>
</table></td><td><table>
<tr><th></th><th>Base value</th><th>First reduction</th><th>Second reduction</th></tr>
<tr><td>Trial 1</td><td>100</td><td>90</td><td>87.3</td></tr>
<tr><td>Trial 2</td><td>90</td><td>82</td><td>78.57</td></tr>
<tr><td>Trial 3</td><td>80</td><td>72</td><td>69.84</td></tr>
</table></td></tr>
<tr><td>Symbolic- algebraic representation</td><td>–</td><td>–</td><td>$(G - 10\,€) \times 0.97 = 77.50\,€$</td></tr>
</table>

The school register has been characterized by its context-disembeddedness, higher explicitness, less personal references, and higher complexity of its grammatical structures (Schleppegrell, 2004). It appears in newspapers, books, and in mathematics classrooms especially in textbooks. Although predominantly written, this register clearly is found in the verbal speech of the teachers. It uses not only verbal and graphical, but also numerical representations. In contrast, symbolic-algebraic representations are rarely used in the school register.

The technical register can be characterized by similar communication situations and similar modes of discourse as the school register, but it is optimized for even higher economy and unambiguousness for very restricted purposes; in the case of mathematics, for structural and quantifiable relations. In mathematics classrooms, the technical register is mostly used for intra-mathematical contexts, or for grasping the mathematical structures in real-world situations. The mathematical-technical register encompasses all registers, including the symbolic-algebraic register, which is exclusively used in the mathematical-technical register. Only the symbolic-numerical and symbolic-algebraic registers allow the symbolic treatments (e.g., in formal algorithms) from which mathematics gains its calculation and logical power.

The example, Table 11.1 shows that by using Halliday's construct "register," we can emphasize different situations of language use: whereas the situation in the everyday register is very concrete and vivid, it is more depersonalized in the school register, and further decontextualized in the technical register. For example, talking about lucky sales would not be appropriate in the school or technical register, but in contrast using the symbolic-algebraic representation of this problem found in the technical register in the shop window is highly unlikely to increase sales of the trousers. Thus the example sketches how acquiring language proficiency is indeed connected to situation-adequate use of registers and representations.

11.4.3 Some Teaching Implications

We suggest that developing quite specific examples to show moves from one register to another can be a useful exercise both for researchers and teacher educators. But giving such fine-grained specific advice to teachers, particularly young teachers, is probably not very helpful in changing their practice for the better. Expecting teachers to employ such micromanaging will make the task of teaching unmanageable. It may be better to suggest particular holistic ways of using language in the mathematics classroom and then, over an extended period of time, work with teachers on examples from their own classrooms to examine such dynamic language use in a context of which they have ownership. But this point needs further exploration by intense development and detailed research.

Asking students to swap between their languages is not simple. The model (Fig. 11.4) suggests not only a vertical and horizontal movement in its middle language component, but implies diagonal movement is also possible, with switching between language registers happening rapidly. No teaching strategy we know of seems to have attempted to guide teachers through this complex maze. For making practical use of these general theoretical considerations in mathematics classrooms, extensive further development and research is needed to learn how purposefully relating registers can be elicited in classrooms. We make some suggestions in the following.

A first step involves the specification of concrete activities by which students are encouraged to relate registers. By using activities that students find easy to relate to, it will be easy to ask students to translate from one register into another, like the example given in Table 11.1. Hence:

- Verbal everyday register L1 → graphical register: Here is a shopping situation (Table 11.1). Draw a picture that shows you the relevant quantities.
- Verbal technical register L2 → verbal everyday register L1/L2: The situation is explained very succinctly (Table 11.1). Retell the story, using your preferred language, giving more details.
- Symbolic → picture: Find or draw a picture for the equation $(x - 10\,€) \times 0.97 = 77.50\,€$.

As Swan (2005) has shown, more guided activities can also be fruitful; for example finding the same mathematical content in different registers or examining if different representations fit. For example:

- Does this drawing fit with this symbolic expression for the calculation? Explain why (or why not).

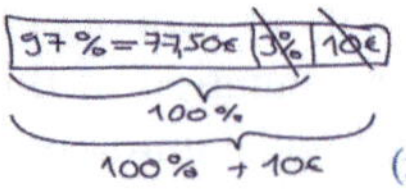

$(x + 10\,€) \times 0.97 = 77{,}50\,€$.

- On these 15 file cards, you find symbolic expressions, texts with situations, and drawings for percentage tasks. Group those that belong together. Devise and then add any missing cards.

Further examples of concrete activities, which have been investigated by Prediger and Wessel (2013), cover:

- Translation from one register into another (freely chosen or determined)
- Finding and fitting registers
- Examining and then deciding whether correct registers fit given situations
- Explaining how to find a mathematical relation or structure in a certain register
- Collecting and reflecting on different means of expression within one register

These activities can be used by teachers in the preparation of lessons, and also spontaneously for initiating students' moves between registers and representations during teaching. If the teacher has a model such as in Fig. 11.4 available as a core referent in their suite of stable teaching strategies, then they will be able to make further spontaneous interventions that allow students to bring all their language abilities to bear in the moment of engagement.

We now turn to the second snapshot from the classroom that will give more insight into the theoretical position we have outlined.

11.5 Transitions Between Registers for Developing Conceptual Understanding of Fractions: Snapshot from Germany

11.5.1 Research Context

The following empirical snapshot from the MuM-Project[1] has been documented by Prediger and Wessel (2011). In the German MuM-Project, the teaching strategy "relating registers" is developed and researched within the paradigm of design research by means of design experiments (Gravemeijer & Cobb, 2006). The empirical research focus is on the situational potential to initiate substantial mathematical and linguistic activities of students.

Twenty-five design experiments in interview settings were conducted with pairs of students in grade 6. Although these students grew up in Germany, their parents were immigrants. The students had Turkish as their first language and good German BICS. The teacher only spoke German. The interviews were video-recorded, transcribed and analyzed qualitatively. The processes of relating registers were coded and then analyzed with respect to the opportunities and challenges for conceptual understanding.

The learning situation started with a text in the second language school register that encodes complex information of a UN-report about rates of analphabets, i.e., people who cannot read (see Fig. 11.5). The main conceptual challenge of this text was to refer 2/3 to the adequate referent whole, which is neither all adults nor all women, but the group of all analphabets (being a quarter of all adults).

To give students the opportunity to construct the necessary conceptual relations, five activities of relating registers were initiated:

Step 1: Translate the given difficult text (in school register L2) into own words (everyday register L1/L2).

Step 2: Check if the text matches to another utterance of a fictitious student, Tobias, "Wow, two-thirds of all women cannot read? Is that possible?"

Step 3: Translate the texts in school register and in everyday register into own drawing.

Step 4: Check if the given drawing (Fig. 11.5, right side) matches their own text and picture.

Step 5: Assign simpler texts, pictures, and fractions (this step is not discussed here).

[1] Within the long-term project MuM ("mathematics learning under conditions of multilingualism"), the study "Understanding fractions for multilingual learners. Development and evaluation of a language- and mathematics-integrated teaching strategy by relating registers" (Prediger & Wessel, 2013) was funded by the ministry BMBF (Grant 01JG1067).

Analphabets in the world

According to a UN report, 1/4 of all adults in this world are analphabets, that means, they cannot read. Due to this, they cannot learn many professions, 2/3 of all analphabets are women.

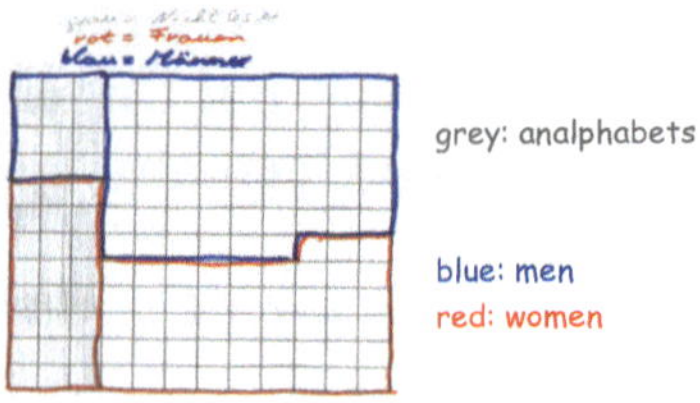

Fig. 11.5 Text and drawing by a fictitious student

11.5.2 The Case of Amir and Ekim: Challenges and Insights While Relating Registers

The case of Amir and Ekim, two 12-year-old boys in grade 6, illustrates some exemplary phenomena that were reconstructed in many cases. Having read the text (in Fig. 11.2) and contextualized it by talking about analphabets, the boys were asked to find a simpler formulation for it. Amir wrote down: "1/4 of all adults cannot read. That is why they cannot get many jobs." The first excerpt of the transcript (translated from German) starts when Ekim suggested the next reformulations for the phrase "2/3 of all analphabets are women":

59 Ekim: Oh wait, shortly. One quarter [*whispers something not understandable*]
60 Amir: Loud!
61 Ekim: Of one quarter are two-thirds
62 Amir: Women
63 Ekim: Who cannot read. Eh, we will write: thereof are two-thirds
64 Amir: Shall I comma? Mhm. [*negating his own question*]
65 Ekim: No, don't think so. Thereof are tw-two-third [*whispering*], [*louder:*] twothirds women, who cannot read.
66 Amir: [*writes down a slightly changed phrase: "Thereof are two-thirds* ***of*** *women who cannot read"*]

Ekim understood which whole the two-thirds must refer to in line 59, and formulated it explicitly in line 61: The two-thirds refer to the quarter. In his second suggestion in line 65, he correctly substituted "the quarter" by the undetermined adverb "thereof" and specified them linguistically by juxtaposing "women who cannot read."

When his partner Amir wrote the phrase, he added another preposition "of." By this little word, he referred two-thirds to two different wholes, so his sentence was linguistically and mathematically incoherent. The subtlety of these details becomes visible in (the nonprinted) line 70: When Ekim read the modified text aloud, he did not seem to realize the divergence embedded in his own suggestion.

In the second step, the boys were asked to evaluate Tobias's fictitious wrong interpretation ("Wow, 2/3 of all women cannot read? Is that possible?"):

89 I: And if—ehm—you think about, that he has the same data as you have, Tobias, he has the same data as you. That means, the same text or an easier text. Does the data match the text then? The numbers?
90 Amir: [*break 11 sec*] Yes, doesn't it? [to *Ekim*]
91 Ekim: Well, of one quarter are two-thirds of—ehm—of all women cannot read and then [*break 8 sec*]
92 I: Let us just draw it. [....]

Ekim correctly explained the relation between part and whole (line 91), but related the two-thirds to the group of all women as the whole by adopting Tobias's wording ("of all women" in line 91). He did not yet succeed in identifying Tobias's mistake. The episode shows the challenge of mentally constructing adequate relations between the part and its whole, and of finding an adequate verbalization for it.

In the third step, the boys were asked to draw the situation in a square (reconstructed in Fig. 11.6).

164 I: Yes, that is very good. And when we refer it back to the text now, can you explain to me what it means for the situation in the text? Again with the information—ehm—that the whole square are all humans in the world?
165 Ekim: Well this [*hints to the whole square*] are all adults and that [*hints to the red quarter*] are all adults who—eh
166 Amir: who cannot read
167 Ekim: exactly. who cannot read. And thereof, now two-thirds are women who cannot read.
168 I: Mhm. [*agreeing*] Do you draw that, too? Can you draw that into it?
169 Ekim: Two-thirds
170 Amir: Thirds. [*break 4 sec*] Yes.
171 Ekim: Shall we do that here? [*hints to the red quarter, but the interviewer does not react. Ekim answers himself without any break*] Yes, don't we? We must do that.

Ekim's verbalizations in lines 165 and 167 show how the transfer to the graphical register strengthened his understanding and he verbalized the structural relationship successfully. In line 165, he hinted at the graphical elements and used deictic means ("this" and "that") for expressing his ideas about the part and the whole precisely, although not yet explicitly. Using the drawing to clarify his thinking, he found a successful explicit verbalization of the structural relation between part and whole in line 167. The scene indicates how the struggle for explicit verbalization is associated closely with a stabilization of their mathematical insights. After that, they draw the picture as scanned in Fig. 11.7. They subdivided the quarter into three parts (intended to be of equal size) and colored two of them in blue. Their own drawing helped to transfer the situation to the context of an alphabet and to translate the mentally constructed relations into verbal formulations.

Fig. 11.6 First drawing

Fig. 11.7 Amir's and Ekim's drawing

In the end, the boys succeed in assigning all three representational registers and thus solved a cognitively demanding task cooperatively. When coming back to Tobias's fictitious utterance, Amir could now explain what Tobias did wrong: "Mmh, he has not the one quarter… drawn." (line 354, not printed here). Still having difficulties in expressing his thinking in the technical register, he emphasized that the quarter must not be neglected; hence he constructed the essential meaning through the long process of relating registers.

11.6 Final Remarks

In this chapter, we have outlined three language registers: the *everyday register*, *school register*, and *technical register*. In the mathematics classroom these registers are used in different communication microsituations, but are often used in combination across the context of a whole teaching session. It is important to stress that although the core of each register is clearly distinguishable, the boundaries between them are not hard and rigid, but permeable and at times quite fluid. This is due to the clear overlap between the contexts within which each register should be used, especially between the technical register and the school register. We also note that with the recent emphasis on authentic and/or real problems, which we applaud, teachers often deliberately seek to have students' transition between their everyday language register in describing and understanding such "real" problems before reconceptualizing it in the technical register. We encourage this, but clearly one of the key messages of this chapter is that such transitions by students do not always happen naturally. We need to look for teaching strategies that identify the language registers that the student may be using at any one time, and show teachers that there can be deliberate choices by students concerning what language register they wish to work in and what they could work in. Once a problem is solved mathematically, then a transition transporting the solution from the technical language back into everyday language is required so that students can appreciate the full power of their solution by linking it and embedding it into the network of their everyday experiences.

We have also commented on the movement between languages that multilingual students will naturally undertake (code-switching), whether school policy approves of such transitions or not. We have noted that such code-switching may occur when using everyday, school or the technical registers of language. Transitions between registers can be within the same language or between languages.

It is our contention that many mathematics teachers do not appreciate the complexity of the language, landscape their students' travel through mathematics lessons. Complex though this is, far from being a necessary impediment to learning, we contend that this rich landscape offers affordances that can be grasped to the advantage of deep student learning. We have noted that these affordances are often clearer when teachers plan to use manipulatives and visualization to contextualize the learning environment they attempt to create.

This first practical and empirical exploration shows the potential of bringing the three transitions together, theoretically as well as in terms of curriculum development and teacher strategies. Nevertheless this beginning brings with it a range of questions, both, theoretical and practical. We therefore conclude by asking a number of these questions which may prove to be crucial for an ongoing research agenda:

- Do students who are taught using this model develop better understanding of the subject-matter and/or perform better than other students? (Some empirical evidence has already been collected that pertains to this question: see Prediger & Wessel, 2013).
- What are the similarities and differences in language processes/strategies that students use while moving between L1 and L2 as compared to moving up the tiered model?
- Which teaching strategies can possibly guide/encourage students to use multiple languages and language forms?
- Are there central teaching principles at the core of different teaching strategies, with the strategies changing in response to the differing language contexts found in different countries and classrooms?
- Are there effective teaching strategies already in the literature that can be adapted to meet the conditions of this model? For example, can open questions, authentic questions, and an adapted use of think boards/place mats (Cunningham, 2002), be used as teaching strategies that are useful/effective in implementing this model?
- Are strategies aimed at making "language" more explicit in mathematics classrooms, such as, for example, using displays such as "word walls," helpful in focusing students' attention on language issues?
- What changes to mathematics curriculum documents, beyond including vocabulary lists and glossaries, will make issues such as language and the way it is taught and used for both monolingual and multilingual students one of the core components around which content and procedures coalesce, rather than the present structure that has content as the central component of such documents?

References

Barton, B. (2009). *The language of mathematics: Telling mathematical tales*. Dordrecht, The Netherlands: Springer.

Barwell, R. (2005). Integrating language and content: Issues from the mathematics classroom. *Linguistics and Education, 16*, 205–218.

Bose, A., & Choudhury, M. (2010). Language negotiation in a multilingual mathematics classroom: An analysis. In L. Sparrow, B. Kissane, & C. Hurst (Eds.), *Proceedings of the 33rd Conference of the Mathematics Education Research Group of Australasia* (pp. 93–100). Fremantle, Australia: MERGA.

Bruner, J. (1967). *Toward a theory of instruction*. Cambridge, MA: Harvard University Press.

Carpenter, T., Fennema, E., Franke, M., Levi, L., & Empson, B. (1999). *Children's mathematics: Cognitively guided instruction*. Reston, VA: NCTM.

Choudhury, M., & Bose, A. (2011). An investigation of the role of language-negotiations in a multilingual mathematics classroom. In M. Setati, T. Nkambule, & L. Goosen (Eds.), *Proceedings of the ICMI Study 21 Conference: Mathematics Education and Language Diversity*, São Paulo, Brazil (pp. 28–37).

Clarkson, P. C. (2007). Australian Vietnamese students learning mathematics: High ability bilinguals and their use of their languages. *Educational Studies in Mathematics, 64*, 191–215.

Clarkson, P. C. (2009). Mathematics teaching in Australian multilingual classrooms: Developing an approach to the use of classroom languages. In R. Barwell (Ed.), *Multilingualism in mathematics classrooms: Global perspectives* (pp. 145–160). Bristol, England: Multilingual Matters.

Cummins, J. (2000). *Language, power and pedagogy: Bilingual children in the crossfire*. Clevedon, England: Multilingual Matters.

Cunningham, S. (2002). A process for understanding mathematics. *Australian Primary Mathematics Classroom, 7*(2), 4–6.

Dienes, Z. P. (1969). *Building up mathematics* (revised ed.). London: Hutchinson Educational.

Duval, R. (2006). A cognitive analysis of problems of comprehension in a learning of mathematics. *Educational Studies in Mathematics, 61*, 103–131.

Farrugia, M. T. (2009). Reflections on a medium of instruction policy for mathematics in Malta. In R. Barwell (Ed.), *Multilingualism in mathematics classrooms: Global perspectives* (pp. 97–112). Bristol, England: Multilingual Matters.

Freudenthal, H. (1991). *Revisiting mathematics education*. Dordrecht, The Netherlands: Kluwer.

Gravemeijer, K., & Cobb, P. (2006). Design research from a learning design perspective. In J. van den Akker, K. Gravemeijer, S. McKenney, & N. Nieveen (Eds.), *Educational design research: The design, development and evaluation of programs, processes and products* (pp. 17–51). London: Routledge.

Halai, A. (2004). Teaching mathematics in multilingual classrooms. In M. Hoines & A. Fuglestad (Eds.), *Proceedings of the 28th Conference of the International Group for the Psychology of Mathematics Education* (Vol. 1, pp. 240–243). Bergen, Norway: PME.

Halai, A. (2009). Politics and practice of learning mathematics in multilingual classrooms: Lessons from Pakistan. In R. Barwell (Ed.), *Multilingualism in mathematics classrooms: Global perspectives*. Bristol, England: Multilingual Matters.

Hallet, W. (2012). Semiotic translation and literacy learning in CLIL. In D. Marsh & O. Meyer (Eds.), *Quality interfaces: Examining evidence and exploring solutions in CLIL*. Eichstätt, Germany: Eichstätt Academic Press.

Halliday, M. A. K. (1978). *Language as social semiotic: The social interpretation of language and meaning*. London: Edward Arnold.

Koch, P., & Oesterreicher, W. (1985). Sprache der Nähe—Sprache der Distanz. Mündlichkeit und Schriftlichkeit im Spannungsfeld von Sprachtheorie und Sprachgebrauch. *Romanistisches Jahrbuch, 36*(85), 15–43.

Leisen, J. (2005). Wechsel der Darstellungsformen. Ein Unterrichtsprinzip für alle Fächer. *Der Fremdsprachliche Unterricht Englisch, 78*, 9–11.

Lesh, R. (1979). Mathematical learning disabilities: Considerations for identification, diagnosis, and remediation. In R. Lesh, D. Mierkiewicz, & M. G. Kantowski (Eds.), *Applied mathematical problem solving* (pp. 111–180). Columbus, OH: ERIC/SMEAC.

Meyer, M., & Prediger, S. (2012). Sprachenvielfalt im Mathematikunterricht—Herausforderungen, Chancen und Förderansätze. *Praxis der Mathematik in der Schule, 54*(45), 2–9.

Pimm, D. (1987). *Speaking mathematically: Communication in mathematics classrooms*. London: Routledge.

Prediger, S., & Wessel, L. (2011). Relating registers for fractions: Multilingual learners on their way to conceptual understanding. In M. Setati, T. Nkambule & L. Goosen (Eds.), *Proceedings of the ICMI Study 21 Conference: Mathematics Education and Language Diversity* (pp. 324–333). São Paulo, Brazil.

Prediger, S., & Wessel, L. (2013). Fostering German language learners' constructions of meanings for fractions: Design and effects of a language- and mathematics-integrated intervention. *Mathematics Education Research Journal, 25*(3), 435–456.

Schleppegrell, M. J. (2004). *The language of schooling: A functional linguistics perspective*. Mahwah, NJ: Lawrence Erlbaum.

Setati, M. (2005). Learning and teaching mathematics in a primary multilingual classroom. *Journal for Research in Mathematics Education, 36*(5), 447–466.

Setati, M. (2008). Access to mathematics versus access to the language of power: The struggle in multilingual mathematics classrooms. *South African Journal of Education, 28*, 103–116.

Sullivan, P., & Lilburn, P. (2004). *Using "good" questions to enhance learning in mathematics*. Melbourne, Australia: Oxford University Press.

Swan, M. (2005). *Learning mathematics through reflection and discussion*. Unpublished doctoral dissertation, University of Nottingham, UK.

Thürmann, E., Vollmer, H., & Pieper, I. (2010). *Language(s) of schooling: Focusing on vulnerable learners. The linguistic and educational integration of children and adolescents from migrant backgrounds studies and resources*. Strasbourg, France: Council of Europe.

Van den Heuvel-Panhuizen, M. (2003). The didactical use of models in realistic mathematics education: An example from a longitudinal trajectory on percentage. *Educational Studies in Mathematics, 54*(1), 9–35.

von Kügelgen, R. (1994). *Diskurs Mathematik. Kommunikationsanalysen zum reflektierenden Lernen*. Frankfurt, Germany: Peter Lang.

Watson, A., & Mason, J. (1998). *Questions and prompts for mathematical thinking*. Derby, England: ATM.

[illegible]

Chapter 12
Using ICTs to Facilitate Multilingual Mathematics Teaching and Learning

Paul Libbrecht and Leila Goosen

12.1 Consideration of Language Diversity in the Literature on ICTs for Mathematics Teaching and Learning

The field of ICTs in mathematics learning is an important area for ongoing research. Mathematics education conferences, such as the Congress of European Research in Mathematics Education (CERME), often include *Working Groups* on the use of ICTs in mathematics teaching and learning. Research published in this area of scholarship includes the presentation of new learning tools, design methods, training or support concepts, the investigation of methods to enhance the quality of the use of the tools, and results in an impact on the use of computers in mathematics teaching. In this section, we review studies about mathematics education with technology that inform our main aim for this chapter.

Borba, Clarkson, and Gadanidis (2013) provide evidence that new practices are enabled by the introduction of information technologies in teaching. In particular, they sketch teaching practices that allow a much richer communication, either peer-to-peer or in the form of performances. From observations such as these, one sees that the introduction of ICTs into mathematics teaching brings different ways to express and perceive mathematical activities, concepts, and phenomena. One can thus expect the use of ICTs to offer multiple new opportunities for learners to employ the different languages of their environments. However, the literature about

P. Libbrecht (✉)
Informatics, Weingarten University of Education Kirchplatz 2,
Weingarten 88250, Germany
e-mail: paul@cermat.org

L. Goosen
University of South Africa, Johannesburg, South Africa
e-mail: GooseL@unisa.ac.za

R. Barwell et al. (eds.), *Mathematics Education and Language Diversity*,
New ICMI Study Series, https://doi.org/10.1007/978-3-319-14511-2_12

ICTs in mathematics learning does not contain, yet, many reports that show support for their use in multilingual environments.

ICMI Study 17 (Hoyles & Lagrange, 2010) focused on the use of technologies in mathematics education and provided a broad "state-of-the-art" survey of the field, which we shall employ in this chapter. Drijvers, Kieran, and Mariotti (2010), as part of the study, surveyed the theoretical frameworks that are applicable to learning with digital tools and summarized their contribution, noting that "some aspects remain underexposed, such as the role of language in instrumental genesis" (p. 121). Indeed a simple text search through the study book shows that the word *language* is used very often to denote other meanings than that used in everyday language, such as a symbolic or programming language (in about 40 % of the occurrences). Even the chapters on inclusion and equity barely address the inclusion issues that are typical in western classrooms, such as the language barriers of children of immigrants' families, who need to master the language of teaching. Hence even in this definitive study of ICT and mathematics education, issues of learning in a multilingual context rarely rated a mention. The remainder of this section will highlight some aspects of research addressing the transformed nature of communication between multiple persons induced by tools of ICTs, followed by research on the learning affordances that ICTs are known to offer.

12.1.1 ICTs-Mediated Human Communication for Mathematics Learning

Endrizzi (2012) published a broad literature review of the French and English research literatures concerning the use of computer-based tools in higher education. The review underlined the scarce literature reporting on the use of computers for learning. The report also underlined the emergence of newer forms of didactical organization, notably the *flipped classroom*, where plain dissemination-like courses are relayed using online videos, and all face-to-face meetings in a classroom or tutorial room are used for opportunities to exchange, ask questions, and collaborate. These new forms of didactical organization employ technology as a central means of communication, which in this context is at least as important as the textbook is in most mathematics classes in industrialized countries. Although there is little in the literature that delves into the affordances or disadvantages of such an approach, one feature is directly relevant to this chapter: the more individualized nature of the communication is likely to impact on the use of different languages. It allows, for example, groups of speakers of a language other than the language of teaching to communicate in their own language. This context is similar to the study of Setati, Molefe, and Langa (2008) where students were assigned to groups sharing a home language and were shown to transparently leverage both the language of teaching (English) and their home language in their learning of mathematics.

The report of Endrizzi (2012) also provides hints on how recommendations of the use of online learning resources were exchanged and received. It noted that, while multiple learning resources were available for many of the topics to be learned, the choice to use them was rarely reflected and was stimulated by two factors: firstly, the necessity to work with a group of peers and, secondly, to take into account the recommendations of the teachers. This is important for teachers in a multilingual environment because it shows the importance for them to maintain a collection of quality resources, which are likely to help each of the learners take their differences into account, be they language, affinities, or expectations. The availability of such a collection of recommendations could be compared to the current practice of students that are more comfortable in a different language than the language of teaching and gather their own collections by simply crawling the web.

Beaty and Geiger (2010), in a contribution to ICMI Study 17, underlined the novel collaboration possibilities enabled by digital technologies and supported by multiple social learning theories. The new social dynamics of these collaboration methods is likely to exploit language diversity in a different way than a classroom where a single language is often the best choice to ensure a consistent presentation. Beaty and Geiger, citing Sfard, explained how "learning mathematics is an initiation into a certain well-defined discourse" which technological tools can carry. They described scenarios of computer supported collaborative learning (CSCL), but did not describe the potential effect in multilingual classrooms.

12.1.2 With and Beyond the Language in Learning Tools

Learning tools within ICTs employ a language, including words and mathematical formulae, to display the mathematical concepts they manipulate. However, that language is generally not as rich as that of a teacher or of classroom peers, and it often includes symbolic or graphical representations. This is noted by Hardman (2005) who stated that, "In the computer laboratory … there is less reliance on language as a tool to explain mathematical content, and more reliance on language as a tool to regulate behaviour" (Hardman, 2005, p. 10).

Banyard, Underwood, and Twiner (2006) studied the use of computers "at an inner city primary school … in central London" where 57 % of the learners had "English as an additional language" and 40 registered languages were spoken in the school. They noted "that the ICT rich learning environment removed the language hurdle that many children experience in their school work" (p. 482). In contrast Kozma (2005) indicated that there is a body of consistent evidence that indigenous minority language speaking learners "all experience growth in their sense of self-esteem and autonomy in their learning when given access to computers in the context of student-centred pedagogy" (p. 15).

Drijvers et al. (2010), having introduced several theoretical frameworks to describe learning processes with computers, citing Noss and Hoyles, gave the example of two girls Cleo and Musha, who employed a dynamic geometry system to solve the task of finding where the symmetry axis of two flags was when they were a reflection of each other (see Fig. 12.1). The girls solved the exercise with this sentence: "The mirror line

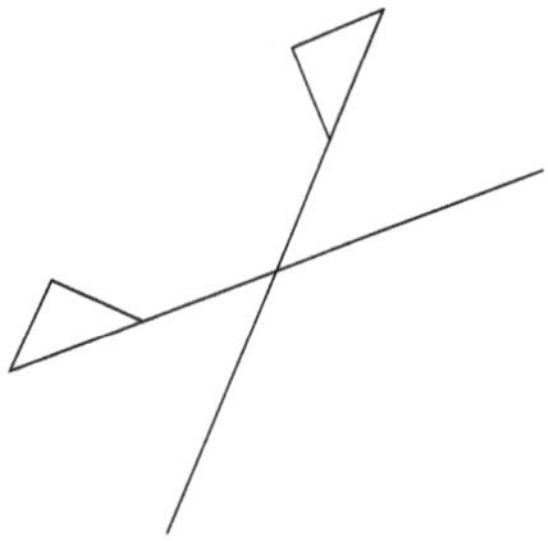

Fig. 12.1 Two symmetric flags (based on Noss & Hoyles, 1996, p. 115)

is what you see on the screen if you drag points and their reflection together" (p. 101). The learning tool gave them the opportunity to explore the geometry freely and formulate in their own words their understanding of a solution.

Olive et al. (2010) studied the representations of knowledge in technological tools and discussed how the representations that are *inside* the tools related to the conceptual representations of the learners. As an example, they showed that dynamic geometry software allows learners to attempt (almost) *all* triangles by simply dragging. This gives the learners an opportunity to illustrate multiple cases without having to use, and maybe be impeded by, language, which would otherwise be necessary to discuss the various triangles by naming them.

The studies cited here are a few examples of those that have studied the impact of ICTs in mathematics learning. These studies show that the diversity of the learners' interactions with computers, compared to classroom interactions, brings new opportunities to build understanding. But as indicated earlier, few of them highlighted opportunities for learning within a multilingual environment. We have not found any that have systematically considered the impact of ICTs on learning in multilingual environments. Having noted the scarcity of such literature, we now explore how multilingualism is considered in the ICTs literature in general, before returning to the opportunity to support mathematics learning in a multilingual environment in particular in the last section.

12.2 Multilingualism in the ICT Literature

The main focus of multilingualism in the ICT literature is on the ability to employ multiple languages for any given software. Most of the software economy is at a global scale and it is common to develop software packages so that interaction with them can occur in several languages. The process of refactoring and translating software to a new language is called *localization*.

The dimensions of localization have been studied by multinational corporations. Among the best-known approaches to what needs to be localized are those developed by Hosfstaede (1991). After interviewing IBM employees across the world, he developed a set of dimensions, which change according to the culture. Software designers used these extensively in the localization process. The dimensions Hofstaede developed qualify perceptions of social relationships and are summarized in his web site.[1] From these relationships, recommended user-interface adjustments can be formulated. A panorama of issues that may be involved in localization (including the dimensions of Hofstaede) is provided in a report by Marcus (2008):

- Typographical aspects (e.g., the rendering of the decimal or thousands separator, including differences in mathematical notations)
- Verbal aspects (the use of different words)
- Symbolic aspects (e.g., the immediate recognizability of a power outlet sign), which can go as far as employing the right color to stimulate a particular reaction
- Placement aspects (what is easily found where, on a screen)

While Marcus is not entirely backed by a systematic ethnographic analysis for each different dimension, he echoes a rich experience of practical software localization.

Several refinements of Hofstaede's dimensions are being attempted, including an exploration of how the *adaptability* of learning software is perceived (Stewart, 2012), the varying degrees of readiness to disclose personal information in different cultures (Blanchard, 2012). The Hofstaede dimensions provide a basic estimate, but they are not sufficient to create fully relevant software. For this, localization teams still need to evaluate and revise software language-by-language (that is communicative language, not software language), country-by-country, and culture-by-culture.

Software localization generally fits into the design of a model-view-controller architecture, which is one of the main software design patterns. For the most elementary localization, code for *views* is written, where each *message* (any piece of text) is transformed through its translation using a dictionary of localization messages. For this case, the translators' work is to write the translations of each message. These dictionaries include all the textual messages and also include the choices of colors, the patterns for numbers, or the choice of symbols to denote particular functions. However the dictionary approach is insufficient to tackle more complex cases; for example the display, input, and validation, of a postal address will need different algorithms if done in Canada where the zip code is a sequence of letters and numbers, compared to France where the zip code is made by five digits, or in Switzerland where the zip code is composed of four digits.

Localization is often not fully achieved. Kleiner (2012) indicated that while the ability to speak a local language is a requirement for any company wanting to reach more than a local customer base, it seems a challenge not yet commonly faced to translate e-commerce web sites across European countries: only 2 % of e-commerce web sites speak more than four languages (Europe has 27 main languages). The same can be said of education. One example is that of Dalvit, Thinyane, Muyingi, and Terzoli (2007) who noted that "many projects involving the implementation of ICTs in rural areas in Africa have failed because of the language barrier posed by the use of English" (p. 13). These authors experimented "with the use of both

[1] http://geert-hofstede.com/dimensions.html.

English and isiXhosa, the local African language". This work included "the development and use of teaching material in isiXhosa, to be used alongside the existing material in English" (p. 13). This initiative gave rise to a model of ICT deployment in rural areas, which appears to represent best practice.[2]

From the above, we see that the adaptability of software to different languages is a documented process but that it may be difficult to reach full generality. The next section explores the localization achievements for various software types relevant to mathematics education.

12.2.1 Mathematics Learning Tools with Some Multilingualism

Having described some aspects of recent research on localizing ICTs, we now describe various mathematics learning tools that support some contexts of multilingualism. Mathematics learning software that has attempted to speak multiple languages include:

- Dynamic geometry software and other "simple" manipulation software: for most of these, the names of operations are translated with simple dictionaries. Most of them also offer a small set of notational adjustments, which depend on the (mathematical) culture and purposes of the user.
- More specialized tools such as MoveIt-M (Fest, 2011) support a similar localization but specific refinement of language is needed for particular languages. For MoveIt-M, for example, the code which produces a verbal description of a transformation given by its fixed points set, needs adaptation for each new language to produce the appropriate word sequence. The different representations of the glide-reflection are shown in Fig. 12.2.
- Pocket calculators have started to display localized behavior from the simple distinction between the "," and "." as decimal separator, and through to the name of particular functions (e.g., *gcd* in English and *ggT* in German, *tg* in French and *tan* in German and English).
- The interactive exercises of the WebALT project have been designed to achieve multilingualism at scale. They are based on an abstract representation of the languages and of the interactive exercises. Caprotti and Seppälä (2006) indicated that:

> [these] exercises are produced by the authors using WebALT software that allows creating a language-independent representation of the kind of sentences used in the statements of typical mathematical problems... By allowing the student to view the exercises in a preferred language, multilingual tests in mathematics overcome language barriers in bilingual communities or in communities where there are large minorities speaking a language not supported at schools. (p. 3)

[2] More about this initiative can be read at http://siyakhulall.org.

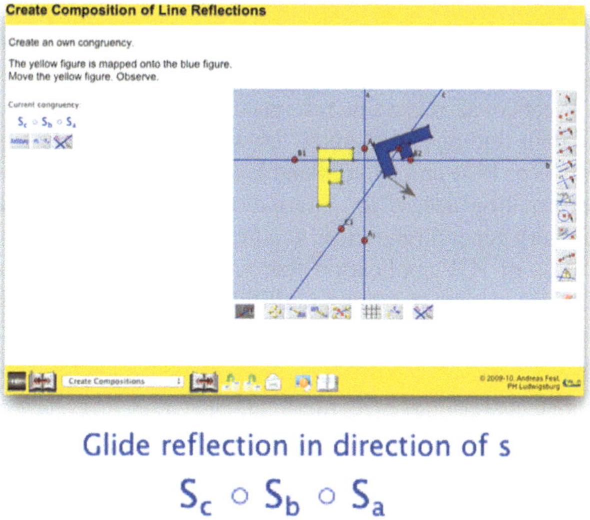

Fig. 12.2 Three representations of the glide-reflection in MoveIt-M: graphical, verbal, and symbolic. The transformation can be freely input by the learner

- The ActiveMath web-based environment is a learning environment (Melis, Goguadze, Libbrecht, & Ullrich, 2009) providing texts, tools, and interactive exercises for learners. ActiveMath's content is made of texts in multiple languages with mathematical formulae in an abstract language. It adapts the display of mathematical notations to each language, whether they come from a computation engine, the user input, or the content fragments.

These learning tools illustrate how far learning tools have been able to go concerning multilingualism, and how demanding a strong multilingualism could become. In Sect. 12.3, other ICT tools will be described for their potential contribution to multilingual learners. However before considering more ICT tools another challenge needs to be addressed: the multilingual abilities of search tools, which are an important family of content management tools.

12.2.2 Search Engines: Beyond a Dictionary of Translations

Although translating the user-interface messages from one language to another is an important first step, it is not sufficient to fully localize software. One further challenge is the development of search tools that work well in multiple languages. There seems to be little research into this issue: "Multilingual search, although sophisticated algorithmically, is not yet interesting from an interface perspective, but this may change with time" (Hearst, 2009, Sect. 12.5). As a rare work in this direction, Peters, Braschler, and Clough (2012) sketched out how current cross-lingual search engines can be built and evaluated. In most cases, such search engines are based on the availability of parallel corpora of documents in all considered languages, which are often created using automatic translators. From the perspective of this chapter, it is a pity that Peters et al. make few references to students. They do cite Wu et al. indicating that "their findings highlighted the wide use of multilingual resources by Chinese students" (p. 192) and Clough and Eleta showing "the potential benefit of multilingual information access for international students studying abroad", but clearly these are very general references to education, although it is clear that search tools are one class of ICT application that are crucial for multilingual students. Peters et al. cited no studies involving multilingual search tools that supported the learning or teaching of mathematics. We suspect the inability of current automatic translation tools to produce satisfactory learning texts in mathematics to be one of the reasons for this.[3]

The role of a search tool is to present a sequence of relevant matches based on a query consisting of a few words. Multilingual challenges that need to be considered in a search tool targeting mathematics in particular include:

- Different algorithms are used for different languages (for example, *sitting* would be equivalent to the word *sit* in an English search engine, but is a word in French with a separate meaning, close to *demonstration*, and hence is treated differently compared with how it is in English).
- Some words have a different semantic field, and thus a different importance, in different languages. For example the term *direction* may have a didactical meaning in English, but has none in French.
- Concepts expressed in multiple words are often used in mathematics. A good search engine should spot these concepts so that variants are found (e.g., *natural numbers* and the mathematical symbol N, or *the right-angled triangle* and *this triangle is right-angled*).

Differences in the ways search engines behave between languages are evident in widespread web search engines such as Google.com or Bing.com. For example, with Google selected for French, inputting *implication* (which has the same meaning in English) will automatically match the verb *impliquer* (imply) and will also suggest related searches (in logic, in law…). However, inputting the word *hlephula* (implication in Siswati) with a browser configured to prefer Siswati will only show exact matches. Thus the search engines have a different ease of use depending on the language. To resolve such differences, the language technology research community is progressively assembling *language resources* so as to encode the knowledge that supports such features in multiple languages, but there is still a long way to go.

[3] Although no formal research has been conducted on the validity of automatic translators for mathematical texts, we note that there are challenges. For example, the theorem of Thales was translated into French by all the automatic translators we could find as the théorème de Thalès, but these two theorems do not state the same fact (the first states that points on a circle span a right angle to the ends of the diameter, the second is the intercepting lines theorem, stating proportionality of measures). While the knowledge to perform such translation may emerge, we have not observed projects that aim at the completion of such a task.

This challenge of searching for mathematical information in multiple languages is well illustrated by the rich online service Wolfram|Alpha.[4] For it to speak another language, it should, for example, also index datasets present in other cultures (e.g., nutrition information in some cultures of that language, which is far from transformable faithfully to US food facts). As well it should present the formula codified appropriately in the target language, and it should also introduce parsing rules for that language. This might weaken the fine tolerance currently available. For example, the input *triangle rectangle* is understood in English as two figure names and thus displays information comparing these two figures side by side (as in the query, *triangle vs rectangle*). However if French was fully supported, it would be understood as the query for *right-angled triangle*.

The above brief discussion of different software aspects that are used to support multiple languages has mapped out some of the spectrum of software diversity and the inherent difficulties still present that learners and teachers in multilingual environments are likely to meet. But the scene is not completely devoid of good news. We now turn to the features of currently available software and describe how they can support, at least to some degree, the learning process in multilingual contexts.

12.3 Supporting Mathematics Teaching and Learning in Multilingual Environments

In this section, we describe a few available methods where ICTs can help mathematics learners and teachers in a multilingual environment. This description is complementary to the previous sections in that it focuses on the role of supporting the learners or teachers, instead of the computer-specific or learning-specific aspects. We highlight the present immaturity of common knowledge about the need of multilingual learning classrooms: for example only incomplete support is given for the diverging mathematical traditions supported within each tool. In most cases, learners need to be (self) informed of such differences, and adapt themselves to the language of the ICT tool.

12.3.1 Language Switch

For multilingual learners, the ability of a software application to speak multiple languages offers them a flexibility that may support them. For example a student with a mathematics background in one language, but now in a classroom where the teaching language is another, would be helped by performing the exercise first in his own language, to leverage previous knowledge, then moving to the language of the classroom

[4] The web site www.wolframalpha.com is a combination of a search engine and a computer algebra interpreter. Each query is translated into a parameterized Mathematica programme which is displayed to the user.

to perform as well as his peers.[5] This requires a language switch to be accessible. Such a switch constitutes a way for learners to employ any of their languages as a *transparent resource*, similarly to the bilingual tasks described in Setati et al. (2008). She describes how learners were given a statement both in their home language and in the language of teaching and how they solved the problem in part by discussing the solution process in a mix of the language of teaching and in their home language.

In rare cases in the environments, the users are able to activate a change of the language with a simple action. This keeps all possible states, translating any feedback and any user input that would support the student. Hence parts of an exercise where background knowledge is useful could be entered in the student's own language, and then the process continued in the different language of teaching. Thus far, this is only possible within the ActiveMath environment (Libbrecht, 2010a) and some dynamic geometry software applications (see below). Very little is known of the potential advantages of such language switches for multilingual learners in the ICT environment, although this is a potentially fruitful area of research. These actions are related to the practice of code-switching, which is addressed by other chapters in this volume. We will therefore not explore the notion of switching any further, but move directly to the context of computer use.

One should note that the existence of such a language switch imposes a parallelism between the languages, which is sometimes difficult to achieve. As expressed in Melis et al. (2009), such concepts as *instant slope* in the English language are not fully translatable to the French language: the only correct translation would be the *pente de la tangente* (the *slope of the tangent to the curve*) which carries a completely different set of prerequisites and thus would be connected to quite different concepts in an exploration, demonstration, or a navigation through knowledge.

12.3.2 Multilingualism in Dynamic Geometry Systems

Dynamic geometry software and other interactive learning tools are often able to speak several languages, and a language switch is commonly supported within the application's settings. However, such switching is often incomplete in that it does not perform all changes needed to become relevant to the other language. The *content* of such software is not multilingual, since it is considered to be an input by authors, and authors need to deliver different versions for different languages. To adapt a piece of content, authors are tasked to specifically translate texts and mathematical notations for each language. Typically, this adaptation is done in part with other adjustments that follow the different mathematical traditions. The adaptation of notation often goes as far as changing the letter of variables. For example, the letter *G* is often used in German to denote the summit of a curve (*G* stands for *Gipfel* which means summit), however an English text may use the letter *S* (for *summit*) so as to be as easy as possible to remember in such a sentence like *Let S be the summit of the mountain.*

[5] See a number of other chapters in this volume that address this issue extensively.

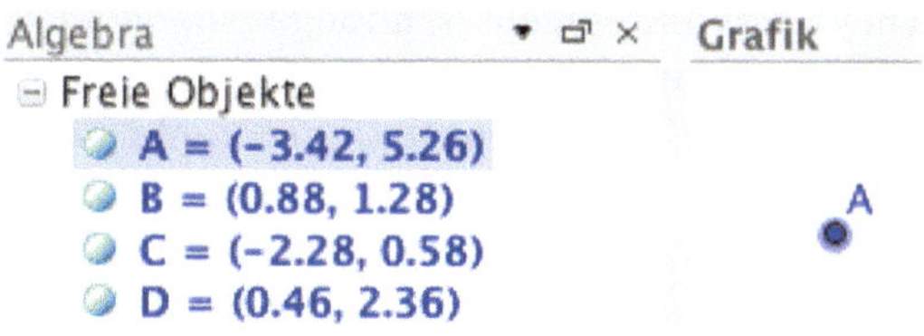

Fig. 12.3 Display of coordinates in GeoGebra in German (but with English notations)

Dynamic geometry systems are not as flexible at fully supporting the notational differences between languages. One such example is shown in Fig. 12.3, which shows the GeoGebra display of the coordinates of points in the construction. The default language used is English, hence it uses the English notation for numbers and coordinates (comma between coordinates, period as decimal separator) even though the environment is in German (where a vertical bar should be used to separate the coordinates and a comma should be the decimal separator).

12.3.3 Multilingualism in Computation Tools

Computers are often seen as tools to perform computations and are particularly useful in performing financial or engineering complex calculations. Computational tools are also commonly used in learning. Three important classes of computational tools used in school mathematics learning are calculators, algebra systems, and spreadsheets. The impact of multilingualism on these three is varied.

Calculators are widely used in schools and are mandatory parts of the curriculum in many industrialized countries. Calculators range from basic four-operation calculators to elaborate algebra systems. The calculator market, until recently, has ignored the language differences with, for example, the only option to press the "." to enter the decimal separator. However a recent trend has appeared in school-oriented calculators where some of the functions are expressed in a local language, including a variation in what symbol to use for decimal separation, depending on the language zone you live in.

Computer algebra systems (CAS) have a similar calculation role as calculators, but with a much richer set of functions and interface. Traditional CASs include commercial applications such as Mathematica or Maple, and multiple open-source projects such as Sage, Yacas, or Macsyma. The set of functions of these systems is so broad that a complete translation almost never exists. Teachers find it normal, to our knowledge, to explain the English names and notations to students. Attempts at delivering multilingual interfaces to computer algebra systems are, however, emerging. Saludes and Xambó (2012) proposed such an approach for Sage. The CATO system proposes a unified user-interface in German for multiple computer algebra systems.[6] We anticipate such moves will continue to grow.

In contrast, interestingly, spreadsheets are generally available only in local languages. They include MicroSoft Excel, OpenOffice Calc and Apple Numbers.

[6] More about the CATO System can be read from http://computeralgebra.biz/.

They allow the student to arrange data in a table-like fashion and input formula in cells that compute their values from other cells. The formula language does use words of the local language (for example the "SUM" function (in English), taking a range of cell-references, is written as "SOMME" in French or "SUMME" in German). One could speculate that the choice of translating all formulae matches the expectations that the mathematical knowledge of regular business users is mostly taught in a local language at school. However, probably due to the incomplete capacities of mathematical input on regular keyboards, the language used in such software is far from the normal mathematical notation of everyday-school mathematics. It seems that little is known about their language binding which can be a challenge for users who, for example, may search for the appropriate function representing the *average* function, but do not know its name or search by a synonym such as *mean*. In our experience, teaching the language of a spreadsheet is often accepted as a duty of mathematics teachers.

12.3.4 Learning Resources Repositories

Many tools are document-based in that they can open documents that authors have created; all the categories discussed above are examples. This paradigm, as well as the paradigm of describing a learning scenario by describing the use of learning tools, has given rise to learning resource repositories. These repositories are collections of educational resources, which support teaching and learning by providing access to lesson plans, documents, software, and packages. It is expected that these learning resources can easily be opened and distributed to learners for them to use. A particular family of educational resources are the open educational resources (OERs), which are exchanged together with a license that allows a free redistribution and, often, redistribution in modified form. This family of resources is important, as it allows regular teachers to become part of an economy of exchange of resources, each being adapted to the different needs of the teaching situations.

Together with the software to open and manipulate the resources, repositories support teachers in the differentiated assignment of tasks by endowing them with a broad diversity of resources. Based on various search criteria, they may select learning resources in alternate languages, which they may reuse directly, reuse after a modification, or use as inspiration. This also supports such use scenarios as the parent trying to work with his or her child in a domain being currently studied at school.

Learning object repositories can also support teachers who are trying to find resources that help their teaching in languages other than the language of teaching. Indeed, some learning resources are almost instantaneously translatable to another language, even sometimes without understanding the original language. For example the learning resource in Fig. 12.4 from the i2geo.net repository can be used to demonstrate corresponding angles in multiple languages with, at most, the renaming of points (e.g., the use of different characters) or the change of colors.

Fig. 12.4 A demonstration of the corresponding angles' concept for the teachers

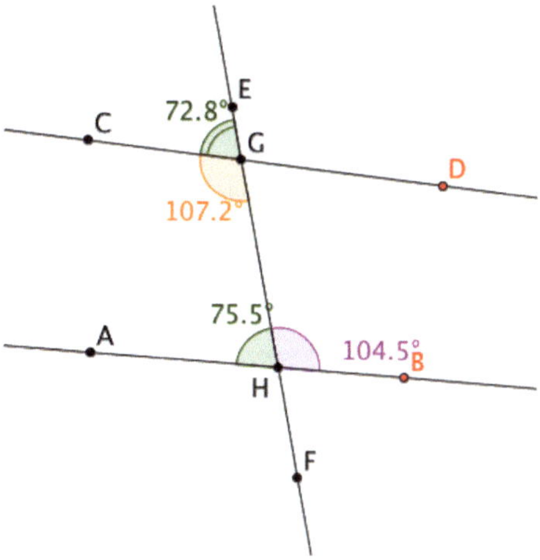

National cultures are often not completely defined simply by a common language. Hence adaptations are often needed even when teachers and students might be using the same language. For example, the first author has observed that secondary school teachers in the German state of Sachsen-Anhalt require their students to write the names of the vertices of a triangle ordered counter-clockwise, while teachers in the state of Bavaria write them clockwise. The teachers report that they warn their pupils when using resources from other states since this convention has implications in various reasoning descriptions.

The freedom to perform adaptations is an important technological consequence of using educational resources. It even allows curious teachers to observe the methods of teaching of other cultures and potentially adapt their resources. Moreover, obtaining and then adapting resources is important for teachers who are teaching in regions where resources in their home languages are rare. Of all editorial licensing models we have encountered, only open licenses seem to allow such an approach.

A huge challenge remains in the ability of teachers, learners, or parents who are seeking extra resources for learning mathematics, to formulate appropriate language queries in such a manner that the concepts they are interested in are identified and matched to appropriate learning resources within resource repositories, or more generally on the web. To this end, the approach of *controlled vocabularies* has been generally used for resource repositories: an editorial team creates a structure of the concepts or domains one expects in the learning resources. This structure enables users to choose from the concepts in this list when searching or contributing. This avoids such ambiguity of using the word *transformation* instead of using *congruence* and allows cross-language queries. Many learning repositories, which display the vocabulary as a hierarchy in each language, use this approach; an example in higher-education mathematics is the merlot.org repository. In the i2geo.net repository, however, one rather searches by entering fragments of the concept name and choosing the appropriate concept as in Fig. 12.5.

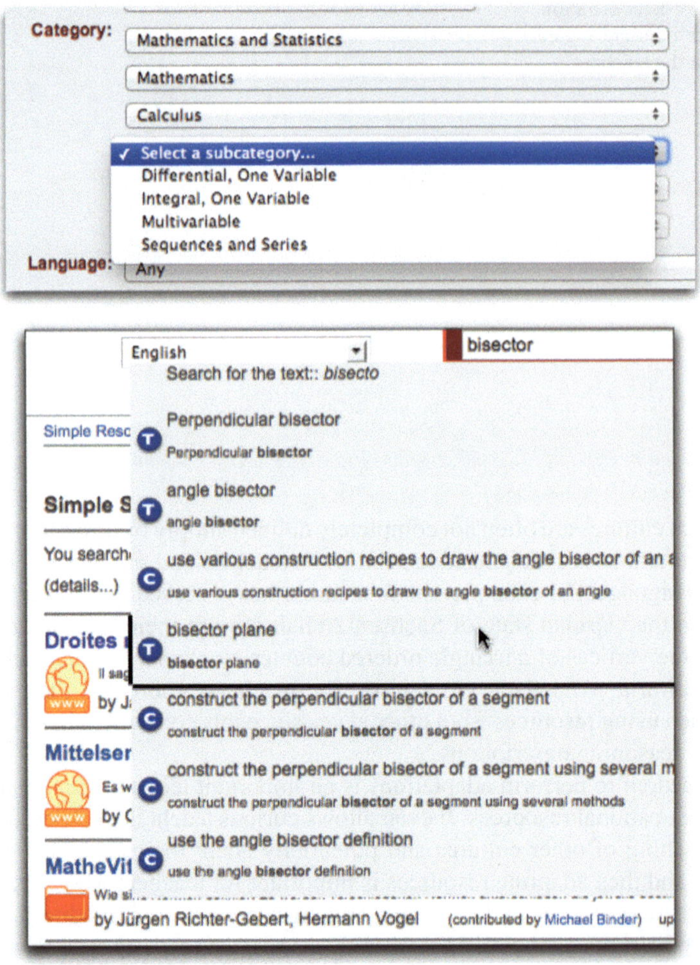

Fig. 12.5 Choosing a field in the merlot.org repository or a concept on the i2geo portal

Both approaches have their strengths and drawbacks. The hierarchy view is often criticized as being too shallow or too large to read (the example above, for example, requires knowing that *sequences and series* is part of *calculus*). On the other hand, with the search approach of i2geo, the users run the risk of not knowing the appropriate words and "miss the right topic". Nonetheless, both of these approaches succeed in making available search results from multiple languages (since the annotation vocabulary is multilingual). In particular, with i2geo.net, there are signs that some teachers dare to cross the language barriers.[7]

[7] An example evaluation from a French teacher on a learning resource in Spanish is at http://i2geo.net/xwiki/bin/view/QR/Coll_msadaall__TransformacionesDeFunciones_3.

12.3.5 Communication Tools

Empowered by worldwide networks, computers, and more recently mobile phones, are being used more and more as communication devices. Written communication in the form of emails for computers and text messages for mobiles are the most widespread. While such devices generally can use the English language, they can often be used to enter other languages as well. The glyph sets and input methods are often available for current languages even though it may be a challenge to acquire, for example, a French speaking operating system in Spain. Most input methods in these tools do not include mathematical formulæ; their input often uses quite separate components (e.g., WebEQ applet, Wiris Input Editor, or ASCIIMath widget[8]).

Even without inbuilt tools to facilitate detailed mathematics, these communication tools can be used for many aspects of mathematical work. For example, questions for assignments can be asked by email or short mobile phone messages. Similarly, web-forum-boards can be used to host discussions about mathematical problems. The multilingual nature of such communication may appear in several forms. However the constraints of entering the questions in a written form, using the limited text available, often requires students and teachers to reformulate the question. These constraints inevitably lead to the invention of a new language. This language adaptation is particularly well known in the use of basic mobile phones where the input takes considerably shortened forms (e.g., using the word *ur* instead of *your*, which can be mistaken for *you are*, unless the context is carefully considered, as in other language meaning making). Nonetheless, projects such as the one carried out by Waitayangkoon (cited in Kozma, 2005) and the blossoming of a school-helpers' industry shows that this can provide an important form of support. This space for more focused discussion allows individuals to use a language closer to their own and their peer's informally created language. We contend that this communication can also assume the important role of bringing the school-world closer to the homes of students where, often, a different language than the language of teaching is spoken. Indeed, in industrialized countries, the use of mobile phones has brought internet-based communication into the daily life of most of the current generation of teenagers.[9]

Even though forms of learning that employ mobile phones are still an object of active research, for example in the mLearn series of conferences, they are of importance for the widening of use of ICTs for learning, since the accessibility of these devices is likely to be orders of magnitudes higher than the accessibility of ICTs (Vosloo, 2012).

---

[8] For the Wiris Input Editor, see http://wiris.com/, for ASCIImath, see http://www1.chapman.edu/~jipsen/mathml/asciimath.html.

[9] At time of writing, two reports are worth mentioning to show the important penetration of mobile phones in the hands of the young generation: the JIM study in Germany (http://www.mpfs.de/?id=613) which indicates 72 % of the 12–19-years-old have a smartphone, a number which has doubled in 3 years. In the United States, the Mobile Mindset Study (https://www.lookout.com/resources/reports/mobile-mindset) provides similar numbers.

12.3.6 Reference Tools

An important and widespread use of ICTs is the use of online reference tools in the form of encyclopaedias or dictionaries. Compared to the reference works available in book form, online reference tools make it possible for both students and teachers to obtain knowledge virtually anytime and anywhere. Although the availability of such resources is sometimes criticized, as it seems to remove an invitation to learn, online reference tools have made their way into most branches of school teaching. Teachers commonly invite students, if possible, to obtain documentation about a topic. The Wikipedia encyclopaedia (wikipedia.org) is the foremost example of this knowledge source for its richness and coverage. However, as most community-based contribution resources (it is open for anyone to edit), Wikipedia frequently suffers from inhomogeneity and shallow editorial control, thus it is not rare to find contradictory articles, or articles which only represent the knowledge or practice of a small population. Other dictionaries valuable for school mathematics include the Maths Thesaurus (thesaurus.maths.org), the Online Encyclopedia of Integer Sequences (oeis.org), and the MathWorld dictionary (mathworld.com).

Among these reference tools, only Maths Thesaurus and Wikipedia are multilingual, and this multilingualism is supported by links between the translations. These links are important bridges as they allow acquisition or at least access to another language. Thus when the content given in one language proves to be insufficient, a user first accessing the encyclopaedia in her own language may well then read, for example, the English version, whose coverage could be far more extensive, or indeed may switch to the version in her own language from English for a more thorough and contextual understanding.

Finally, an online reference tool particularly aimed at teachers (and possibly learners) aiming to understand language differences in mathematical practices is the Notation Census (Libbrecht, 2010b).[10] This reference tool is a collection of observations from textbooks in multiple languages aimed at collecting mathematical notations from around the world. Persons wishing to know how mathematical concepts are written in different cultures can make use of this tool to discover the differences. This tool is particularly relevant to combat, in a small way, the general perception in many cultures that mathematical knowledge is universal and culture-free. An example of the content of the Notation Census is a comparison of the multiple use of brackets, including the style of bracket, for the half-open interval in the English, French and German, and Dutch languages (Fig. 12.6).

These notations are not strictly equivalent. This is because they are scans extracted from textbooks. The textbooks are employed as witnesses of mathematical notations in traditional practice, but like most extractions from original sources, the process of extraction means some diminution of meaning. We anticipate that teachers welcoming students from multiple origins in their classrooms may make use of the Notations Census to help them better understand and probably better

[10] The notation census is available at http://wiki.math-bridge.org/display/ntns/.

$$[a, b) \qquad [0, \tfrac{\pi}{2}[\qquad [0, \infty[$$

Fig. 12.6 Collected extracts from book-sources in the notation census (redrawn here for the sake of readability) from the census entry about half-open-interval (http://wiki.math-bridge.org/display/ntns/interval_co)

explain mathematical concepts. The teachers' explanations may well describe how the symbolic expressions being discussed are used to express, in the students' original languages, concepts, such as, in this instance, the half-open interval. Moreover, we anticipate that learning scenarios may be performed by learners who master well a mathematical topic, so as to see the concept decoupled from graphical notations, first visualizing the different practices (and maybe their history), discussing their individual advantages, and then inventing their own.

12.4 Outlook

This chapter has opened a broad set of potential avenues for future research. Within the maturing practices of ICT uses for mathematics learning, multilingual mathematics learners and teachers have a special set of tools that are growing to help them: from localized computational engines to cross-linguistic references, from language-specific interactive tools to the discovery processes empowered by the worldwide-web.

There is little in the mathematics education literature as yet to indicate the effectiveness of these tools for learners in a multilingual environment. We have presented the evidence we have found available, but clearly more research is needed to answer questions such as the following:

- What is needed so that the ICTs used by teachers on a day-to-day basis can provide language-specific support to individual learners in the modern classroom? In multilingual classrooms, can ICT tools support all languages?
- Should learners speak multiple mathematical languages so that they are ready to travel or should the tools be localized to their languages?
- At which age and how is it safe to require a student to use, say, an English input syntax to solve an equation or manipulate computation tools? (Diverging answers to this question were obtained depending on the expert we asked: generally, academic mathematicians consider it an important capacity to practice several languages, whereas practicing teachers very quickly answer that it is not feasible to teach their students another language.)
- What makes a mathematical learning tool easy to use in a given culture? Should the learning tool use a lexical and notational vocabulary that is consistent with the course? The encoding specificity principle described by Clark and Mayer (2002) seems to indicate so. What is the *cost* of not doing so?

To conclude, we provide an example of a difficult localization issue for which no general theory has been devised. Within the LeActiveMath EU project, an introduction to the concepts of derivative and difference quotient has been written using modelling tasks based on a hike through the mountains where the altitude function was considered. The chapter, including interactive exercises, was written in German and English by a German teacher, with a revision by a team of Scottish teachers. The Scottish teachers judged the content as appropriate and used the material with many of their students. What, however, would be needed in order to make such a collection of interactive resources applicable to learners in the United States? Clearly, a conversion of units would be useful since altitude is measured in feet in the United States rather than in meters. But the necessary changes are considerably broader than this. A veteran teacher in the state of Indiana, in the United States, indicated that the whole introductory section, including most of the interactive exercises would need to be replaced by something else since mountains are, to students in his state, a concept far away from their vision of the real world.

Is the message from this example that content always needs to be reviewed for each new culture? This is probably too strong. Is the conclusion that learning software should be disconnected from learning content so as to ease localization? This is more and more impossible in the worldwide-web, where content is bound to interactive tools so that the best consistency of expression is achieved and so that the most interactive learning experience is offered. We also know that learning is much more powerful if conceptualized within a student's context. Hence, if the software is not bound to any content, who provides the context? Is it really possible to have learning software devoid of content? Clearly, much work is still needed.

Acknowledgements This research has been partially funded by the first author's individual affiliations and the European projects Math-Bridge and Open Discovery Space. The opinions represented herein are, however, the authors'. We thank the reviewers of this chapter, among others Philip Clarkson, for their constructive critiques.

References

Banyard, P., Underwood, J., & Twiner, A. (2006). Do enhanced communication technologies inhibit or facilitate self-regulated learning? *European Journal of Education, 41*, 473–489.

Beaty, R., & Geiger, V. (2010). Technology, communication, and collaboration: Re-thinking communities of inquiry, learning, and practice. In C. Hoyles & J.-B. Lagrange (Eds.), *Mathematics education and technology: Rethinking the terrain* (pp. 251–284). Berlin, Germany: Springer.

Blanchard, E. G. (2012). *Is it adequate to model the socio-cultural dimension of e-learners by informing a fixed set of personal criteria*? Paper presented at the 12th International Conference on Advanced Learning Technologies (ICALT 2012). Rome: IEEE.

Borba, M., Clarkson, P. C., & Gadanidis, G. (2013). Learning with the use of the Internet. In M. A. Clements, A. Bishop, C. Keitel, J. Kilpatrick, & F. Leung (Eds.), *The third international handbook of mathematics education* (pp. 691–720). Dordrecht, The Netherlands: Springer.

Caprotti, O., & Seppälä, M. (2006). Multilingual delivery of online tests in mathematics. In *Proceedings of Online Educa*, Berlin, Germany. Retrieved from http://citeseerx.ist.psu.edu/viewdoc/download?

Clark, R., & Mayer, R. (2002). *e-Learning and the science of instruction*. San Francisco: Pfeiffer.
Dalvit, L., Thinyane, M., Muyingi, H., & Terzoli, A. (2007). The deployment of an e-commerce platform and related projects in a rural area in South Africa. *International Journal of Computing and ICT Research, 1*(1), 9–18.
Drijvers, P., Kieran, C., & Mariotti, M. A. (2010). Integrating technology into mathematics education: Theoretical perspectives. In C. Hoyles & J.-B. Lagrange (Eds.), *Mathematics education and technology: Rethinking the terrain* (pp. 89–132). Berlin, Germany: Springer.
Endrizzi, L. (2012). Les technologies numériques dans l'enseignement supérieur, entre défis et opportunités. *Dossier d'actualité Veille et Analyses*, n°78. Retrieved from http://ife.ens-lyon.fr/vst/DA/detailsDossier.php?dossier=78&lang=fr
Fest, A. (2011). Adding intelligent assessment: A Java framework for integrating dynamic mathematical software components into interactive learning activities. *Zentralblatt für Didaktik der Mathematik, 43*(3), 413–423.
Hardman, J. (2005). An exploratory case study of computer use in a primary school mathematics classroom: New technology, new pedagogy? *Perspectives in Education, 23*(4), 1–13.
Hearst, M. (2009). *Search user interfaces*. Cambridge, England: Cambridge University Press.
Hosfstaede, G. (1991). *Cultures and organizations: Software of the mind*. London: McGraw Hill.
Hoyles, C., & Lagrange, J.-B. (2010). *Mathematics education and technology: Rethinking the terrain*. Berlin, Germany: Springer.
Kleiner, T. (2012). Technological challenges of the multilingual European society (Opening address at the META-FORUM 2012). Retrieved from http://www.meta-net.eu/events/meta-forum-2012/report#kleiner_presentation
Kozma, R. B. (2005). Monitoring and evaluation of ICT for education impact: A review. In D. A. Wagner, B. Day, T. James, R. B. Kozma, J. Miller, & T. Unwin (Eds.), *Monitoring and evaluation of ICT in education projects: A handbook for developing countries* (pp. 11–18). Washington, DC: InfoDev/World Bank.
Libbrecht, P. (2010a). Customized and culturally enhanced rendering and search (Deliverable 4.1 of the Math-Bridge project). Retrieved from http://project.math-bridge.org/outcomes_deliverables.php
Libbrecht, P. (2010b). Notations around the world: Census and exploitation. In S. Autexier, J. Calmet, D. Delahaye, P. D. F. Ion, L. Rideau, R. Rioboo & A. P. Sexton (Eds.), *Proceedings of the Conference on Intelligent Computer Mathematics*: *Vol. 6167. Lecture notes in computer science* (pp. 398–410). Berlin, Germany: Springer.
Marcus, A. (2008). Global/Intercultural user-interface design. In A. Sears & J. A. Jacko (Eds.), *Human computer interaction handbook*. Hillsdale, NJ: Lawrence Erlbaum.
Melis, E., Goguadze, G., Libbrecht, P., & Ullrich, C. (2009). Culturally aware mathematics education technology. In E. Blanchard & D. Allard (Eds.), *Handbook of research on culturally-aware information technology: Perspectives and models* (pp. 543–557). Hershey, PA: IGI Global.
Noss, R., & Hoyles, C. (1996). *Windows on mathematical meanings: Learning cultures and computers*. Dordrecht, The Netherlands: Kluwer.
Olive, J., Makar, K., Hoyos, V., Kee Kor, L., Kosheleva, O., & Sträßer, R. (2010). Mathematical knowledge and practices resulting from access to digital technologies. In C. Hoyles & J.-B. Lagrange (Eds.), *Mathematics education and technology: Rethinking the terrain* (pp. 89–132). Berlin, Germany: Springer.
Peters, C., Braschler, M., & Clough, P. (2012). *Multilingual information retrieval: From research to practice*. Heidelberg, Germany: Springer.
Saludes, J., & Xambó, S. (2012). The GF mathematics library. In P. Quaresma & R.-J. Back (Eds.), *Proceedings of the THedu'11 Workshop. EPTCS 79* (pp. 102–110).
Setati, M., Molefe, T., & Langa, M. (2008). Using language as a transparent resource in the teaching and learning of mathematics in a grade 11 multilingual classroom. *Pythagoras, 67*, 14–25.
Stewart, C. (2012). *A cultural education model: Design and implementation of adaptive multimedia interfaces in eLearning*. Unpublished doctoral dissertation, University of Nottingham, UK.
Vosloo, S. (2012). *Draft policy guidelines for mobile learning (version 2.1)*. Bangkok, Thailand: UNESCO. Retrieved November, 2012, from http://www.unescobkk.org/id/news/article/inviting-public-input-on-the-unesco-policy-guidelines-on-mobile-learning/

Chapter 13
Language Diversity and New Media: Issues of Multimodality and Performance

George Gadanidis, Marcelo Borba, and Ricardo Scucuglia

13.1 Language Diversity and New Media

The discussion document of ICMI Study 21 on Mathematics Education and Language Diversity proposed the following central questions:

> Across the world, teaching and learning of mathematics occurs in contexts of linguistic and cultural diversity. How do we work with, and work within, this diversity to enhance the learning and teaching of mathematics? In particular, how can the range and complexity of learners' language backgrounds be most effectively used to promote their mathematical learning? (ICMI Study 21, p. 3—see Appendix)

We want to expand the discussion of linguistic diversity to include the role of new media in shaping how mathematics is, and how it might be, communicated. Students' out-of-school world is increasingly populated by digital and multimedia texts. Buckingham (2010) notes that,

> The term 'media' includes the whole range of modern communications media … Media *texts* are the programmes, films, images, web sites (and so on) that are carried by these different forms of communication … Media texts often combine several 'languages' or forms of communication—visual images (still or moving), audio (sound, music or speech) and written language. (pp. 3–4)

We want to consider these media and media texts as part of students' natural language environment and to explore the implications for mathematics teaching and learning. As we emphasize in this chapter, our focus is on the pedagogic production

---

G. Gadanidis (✉)
Faculty of Education, University of Western Ontario,
1137 Western Road, London, ON, Canada N6G 1G7
e-mail: ggadanid@uwo.ca

M. Borba • R. Scucuglia
Universidade Estadual Paulista (UNESP), Sao Jose Rio Preto, SP, Brazil
e-mail: mborba@rc.unesp.br; ricardos@ibilce.unesp.br

R. Barwell et al. (eds.), *Mathematics Education and Language Diversity*,
New ICMI Study Series, https://doi.org/10.1007/978-3-319-14511-2_13

of digital and multimodal texts for the communication and representation of mathematical ideas through the use of the performance arts.

We want to bring to light the contrast between students' out-of-school experiences with new media (Cummins, Brown, & Sayers, 2007), with multimodal forms of communication aimed at a wide audience, and the continued predominance in most mathematics education of textual and symbolic communication aimed at a narrow classroom audience. We focus on multimodality and performance for a wider audience as two important affordances of new media and their role in shaping language diversity in mathematics education. We show these affordances in action, through two case studies from our work in mathematics classrooms, in Canada and in Brazil.

13.1.1 Humans-with-Media

Borba and Villarreal (2005) suggest that humans-with-media form a collective where new media serves to disrupt and reorganize human thinking. Likewise, Levy (1993) sees technology not simply as a tool used by humans, but rather as an integral component of a *cognitive ecology* of the humans-with-technology *thinking collectives*. Levy (1998) claims "as humans we never think alone or without tools. Institutions, languages, sign systems, technologies of communication, representation, and recording all form our cognitive activities in a profound manner" (p. 121). As Levy (1993) suggests, technologies *condition* thinking.

According to Tikhomirov (1981), computers do not replace, substitute, or merely complement humans in their intellectual activities. Processes mediated by computers *reorganize* thinking. Tikhomirov argues that computers play a mediating role in thinking as language does in Vygotsky's theory. Regarding the nature of human–computer interaction in terms of feedback, the dimensions involving computational mediation provide new insights in terms of learning, development, and knowledge production. Tikhomirov claims that:

> With regard to the problem of regulation we can say that not only is the computer a new means of mediation of human activity but the very reorganization of this activity is different from that found under conditions in which the means described by Vygotsky are used. (p. 273)

Borba and Villarreal (2005) use Tikhomirov's theory to argue how the notion of mediation by computers is qualitatively different to the mediation involving paper and pencil, for example. Through digital mediation, information technologies reorganize mathematical thinking. Media shape meaning and knowledge production and transform mathematics. The idea that media tools we use condition our thinking is not new and it should not be surprising. McLuhan (1964) made this link several decades ago with his often quoted assertion that "the medium is the message."

Scucuglia (2012) argues that students-with-media engage in thinking and feeling collectives when they produce multimodal texts to communicate their ideas using drama and/or songs, as arts-based communication forms engage both cognitively and emotionally. The pedagogic synergy involving the use of new media and the arts offer an interesting cognitive and affective scenario for students' mathematical communication and digital representation at the elementary school level.

13.1.2 Multimodality

Pahl and Rowsell (2005) posit that the word *multimodal* "describes the way we communicate using a number of different modes to make meaning" (p. 27) in effect providing a context of language diversity. Rowsell and Walsh (2011) state that "multimodality is the field that takes account of how individuals make meaning with different kinds of modes" (pp. 55–56). According to Walsh (2011), *multimodality* is "a study of the communicative process, particularly how meaning is communicated through different semiotic or meaning-making resources and in different social contexts" (p. 105):

> Multimodality as in comprehension and competence with language through a variety of modes such as image, sound, touch, multi-dimensions, is the principle upon which digital environments work. This principle of multimodality needs to be understood for educators to apply and assess new modes of learning as a part of everyday classroom practice. (Rowsell & Walsh, 2011, p. 54)

In contrast to the increasingly multimodal nature of the web, many school experiences, especially in mathematics, continue to rely on discourses that are monomodal or bimodal (in cases where diagrams or graphs are employed). Kress and Van Leeuwen (2001) suggest that in a digital environment, "meaning is made in many different ways, always, in the many different modes and media which are copresent in a communicational ensemble" (p. 111). The shift from text-based communication to multimodal communication is not simply a quantitative change. It is not just a case of having more communication modes. It is a qualitative shift, analogous to the change that occurred when we moved from an oral to a print culture (Gadanidis, Hoogland, & Hughes, 2008). Print culture, for instance, supported the creation of fixed media and records.

The New London Group (1996) proposes a model formed by six designs. These are: Linguistic Design, Visual Design, Audio Design, Gestural Design, Spatial Design, and Multimodal Design. The "Multimodal Design is of a different order to the other five modes of meaning; it represents the patterns of interconnection among the other modes" (p. 78). Some elements of Linguistic Design are:

Delivery:	Features of intonation, stress, rhythm, accent, etc.
Vocabulary and Metaphor:	Includes colocation, lexicalization, and word meaning.
Modality:	The nature of the producer's commitment to the message in a clause.
Transitivity:	The types of process and participants in the clause. Vocabulary and metaphor, word choice, positioning, and meaning.

Nominalization of processes:	Turning actions, qualities, assessments, or logical connection into nouns or states of being [...]
Information structures:	How information is presented in clauses and sentences.
Local coherence relations:	Cohesion between clauses, and logical relations between clauses (e.g., embedding, subordination).
Global coherence relations:	The overall organizational properties of texts (e.g., genres) (p. 80).

The Visual Meanings refer to images, page layouts, and screen formats. The Audio Meanings refer to music and sound effects. The Gestural Meanings refer to body language, embodiment, and facial expressions. The Spatial Meanings refer to the meanings of environmental spaces/architectural spaces. However, the Multimodal Design "is the most significant, as it relates all the other modes in quite remarkably dynamic relationships" (The New London Group, 1996, p. 80). It is this component of the model that draws the other five together to form a whole.

More recently, Walsh (2011) theorizes how classrooms can become multimodal learning environments when students interact, collaborate, and produce multimodal texts in schools. Walsh actually emphasizes the role of intertextuality (combination of print-based and digital/multimodal texts) and dialogue in meaning production within educational and social purposes.

13.1.3 Performance for a Wider Audience

The multimodal texts that increasingly populate our digital environment are also conditioned by the pronounced sense of audience provided by new media. Using the Internet, and especially such tools as Youtube and Facebook, we can easily share our multimodal texts with the wider world and potentially reach a large audience. In a print culture, before the pervasiveness of the Internet, individuals had very limited access to a wide audience. Texts were produced by publishers to be consumed by us, the audience, without a reciprocal power to publish. With digital media, the audience now has access to production itself, through the availability of multimodal authoring tools (like blogs and Youtube) and access to a wide audience for self-published or collaborative work. The sense of audience that is associated with new media creates what Hughes (2008) has labeled as a "performative pull." We do not want to simply share our multimodal texts. We want to share them in performative ways. Hughes suggests that the new media that is infusing the web draws us into performative relationships with and representations of our "content." To use new media is to, in part, adopt a performative paradigm. The popularity of Youtube, whose slogan is "broadcast yourself," is a good example of the performative pull of new media.

Kress and Van Leeuwen (2001) also note that the multimodal nature of new media offers performative affordances. This is evident in the multimedia authoring tools used to create online content, which often use performance metaphors in their programming environment. For example, you program on what is referred to as the "stage," you use "scenes" to organize "actors" or "objects" and their relationships, and you control the performance using "scripts." Even more generally, Denzin (2003) notes that "We inhabit a performance-based, dramaturgical culture." Thus new media is helping amplify our cultural inclination for performance.

In the next section, we present two international cases of mathematics teaching and learning that illustrate the use and nature of multimodality and performance. Hill (1998) sees a performance-based pedagogy as a method of collaboratively making sense of experience. A performance-based pedagogy, coupled with the audience affordances of new media, position students as community storytellers of personal learning and growth and offers opportunities for them to experience "narrative reconstruction" as they reflect on their lives, their learning, their choices, their past experiences, and their goals for the future (Hull, 2003, p. 232). As Hull points out, "The ability to render one's world as changeable and oneself as an agent able to direct that change is integrally linked to acts of self-representation through writing" (p. 232). When students are given opportunities to share their "identity texts" with peers, family, teachers, and the general public through media, they are likely to make gains in self-confidence, self-esteem, and a sense of community belonging through positive feedback (Cummins et al., 2007). Hull (2003) urges collaboration among educators, researchers, and community organizations to "find space and time to think expansively about the interface of literacy, youth culture, multi-media, and identity" (p. 233).

There is ample research on the role of narrative in the construction of personal agency and identity (see Ochs & Capps, 2001). Bruner's (1994) studies of narrative indicate that changes in conceptions and representations of self are typically associated with "turning points" in personal narratives. Bruner identifies turning points as "thickly agentive … whose construction results in increasing the realism and drama of the Self" (p. 50). There is a dialogical relationship between narrative and self: to shape our narrative is to shape ourselves, and vice versa. There is also a dialogical relationship between narrative/identity and community. Narratives are social artifacts and "the narrated self is constructed with and responsive to other people" (Miller & Goodnow, 1995, p. 172). Stories change depending on the audience, and a personal knowledge story aimed for a school-based audience can change when the audience is the wider community. When the audience is the community, the narrative becomes more of a public performance. Hull and Katz (2006) note "the power of public performance in generating especially intense moments of self-enactment" (p. 47). Digital (unlike oral or solely print-based) stories, and hence more often than not using a language diversity context in the sense outlined in the first section of this chapter, potentially enhance the power of narrative to transform as they can be easily broadcast, creating a stronger sense of audience and performance (Hughes & Gadanidis, 2010).

13.2 Case Studies

In the case of mathematics education, Gerofsky (2006) notes:

> It is unusual (and energizing) to link mathematics and math education with performance, in no small part because many of the things that make a performance distinctive and interesting go squarely against many of the long-held traditions of mathematics. [It is important to] explore the human necessity of performance in mathematics. (p. 2)

In the two case studies that follow, one from Canada and one from Brazil, we explore the role that the new media affordances of *multimodality* and *performance for a wider audience* may play in shaping language diversity in mathematics classrooms. New media entered the classrooms in these case studies through the teachers' commitment to share a performance of student learning at the online Math and Science Performance Festival (available at www.mathfest.ca), which increased the sense of audience for both students and teachers. In addition, there was a focus in these classrooms on relating a good mathematics story. From our experience, when we ask parents how children respond when asked "What did you do in math today?" the common replies shared are "Nothing" and "I don't know" (Gadanidis, 2012). As we will discuss later in the chapter, the sense of audience for mathematics learning and the desire to relate a "good" mathematics story go hand-in-hand.

13.2.1 *Case A: Grades 2 and 4 Students in Canada Explore Optimization*

In Grades 2 and 4, students learn about perimeter and area at different levels of abstraction. Typically, these concepts are taught as definitions and formulas, with some concrete and pictorial representations. In the case below, Grades 2 and 4 students explore area and perimeter in the context of optimization. For example, "What is the largest possible rectangular area for a rectangular pen made with 12 m of fence?" and "What is the least amount of fence needed to enclose a rectangular pen whose area is 14 m^2?" Although such pedagogical directions are supported by reform curriculum documents, they are often not implemented. In a professional development session for Grades 1–6 Ontario teachers offered by the first author, all of the approximately 50 teachers present agreed with the statement that "rectangles with the same area must have the same perimeter," indicating that they most likely do not explore such relationships in their classrooms.

Grades 2 and 4 students explored optimization in the context of area and perimeter relationships: (1) "If you put 16 tables in a rectangular array, which arrangement fits the fewest chairs all around?" and (2) "What are the dimensions of the biggest rectangular pen that can be created with 20 m of fencing?"

In Grade 2 students first investigated rectangular arrays for numbers, using the story *The Doorbell Rang* (Hutchins, 1986). In this story, 2, then 3, 4, 6, and 12, children investigate how to share 12 cookies. Students used 12 linking cubes to

Fig. 13.1 Grade 2 song

We made 12 in a story

We made 12 in a Story
2 by 6 and 6 by 2
We made 12 in a Story
Have a look at the pictures we drew.

12 Kids on a bus
Have a look from the birds eye view
See my caterpillar
It has 12 polka dots too.

We made 12 in a story
3 by 4 and 4 by 3
We made 12 in a story
Look close at the pictures you see.

A cell phone with 12 buttons
12 dogs on a water slide
12 wheels on a bus
Boy can those dogs glide.

model their own solutions. They also used their imagination to illustrate the various arrays as everyday objects. For example, the 1×12 array might be a snake or a caterpillar (see Fig. 13.1), and the 3×4 array might be a calculator or a cell phone. They then used this knowledge to construct rectangular arrays for 16 patio stones (area), modeled using 16 linking cubes, and determine which arrangement would need the most or least fencing (perimeter), modeled using a piece of string. They explored different ways that 16 tables could be arranged to form a rectangle, and the number of chairs that would fit all around, and recorded their work on chart paper (see Fig. 13.3). Students also read the story *Wolf gets Hurt* (Gadanidis & Gadanidis, 2010), in which the Three Little Pigs capture the Big Bad Wolf and build a sturdy

Fig. 13.2 Grade 4 song

Need to fence a pen?

need a 20 metre fence
to make a sturdy pen
keep the dog in the yard
or catch bunnies off guard?

call us now, call us first
we're the professionals
we use math to optimize
no one beats our price

1 by 9, 2 by 8, 3 by 7, 4 by 6
rectangular designs
with four straight lines

but if times are hard
and funds are short
then order the 5 by 5 square
save money don't despair

rectangles are pretty
rectangles are nice
but to save you money
squares is what we advise
squares is what we advise

rectangular pen to hold him captive until the authorities arrive. The Three Little Pigs want to use the shortest fence possible for the pen with the 16 patio stones that they have available to build its base (so Wolf can't dig his way out). One of the songs written based on student thinking is shown in Fig. 13.2.

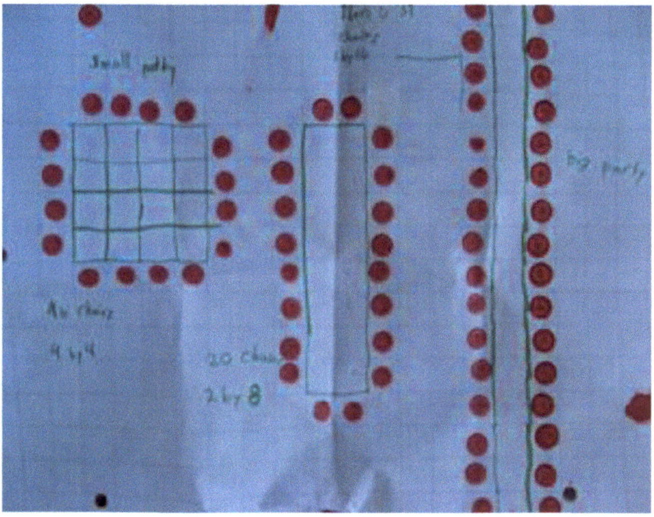

Fig. 13.3 Tables-and-chairs problem

In Grade 4, students also read the story *Scruffy's New Home* (Gadanidis, 2008), in which a young girl and her grandfather want to create the largest rectangular pen (in terms of area) using the 18 m of fence that they have available. Students graphed area vs. length and noticed that their solution (largest area) was the highest point on the graph. This was in contrast to their graph for the Wolf pen, where they graphed perimeter vs. length and noticed that their solution (shortest perimeter) was the lowest point on the graph. Students used the knowledge they learned to create an advertisement for a fencing company, and their ideas were compiled to create the class song shown in Fig. 13.2. Students also worked in small groups to author and practice performing dialogues (as skits) that they might have at home when someone asks: "What did you do in math today?"

A documentary of these Grades 2 and 4 activities is available at www.researchideas.ca/pen.html, with videos of teacher interviews and classroom action, music videos of the class songs, lesson plans, and an artistic representation of the activity (see Fig. 13.4). This documentary may be seen as a mathematics education research performance and as a mirroring of the classroom focus on mathematical performance.

The activities described above incorporated a variety of multimodal communication forms:

- Drawings
  - Illustrated children's literature used as a context for the activities.
  - Drawings of real-file objects based on arrays.
  - Illustrated advertisements for fencing companies.

Fig. 13.4 Artistic rendering of optimization

- Diagrams
  - Diagrams of number arrays.
  - Diagrams of tables-and-chairs arrangements.
- Tables
  - Tables that organized data from area–perimeter problems.
- Graphs
  - Bar graphs of area vs. length and perimeter vs. area.
- Songs
  - Lyrics based on student writing. (In this case, the lyrics were compiled by the teacher, and the students offered edits. In some other cases, students worked in small groups to summarize their learning in the form of stanzas, which were then compiled to form a song.)
  - A song set to music by a music teacher and performed by the students. (In some cases, students use common melodies like *Row, Row your Boat* or older students from the school set the songs to music for younger classes.)
- Videos
  - A music video of the song was shared publicly at the Math Performance Festival.

The activities also incorporated a variety of performative communication forms:

- Story
  - Illustrated children's literature as a context for the activities. (Story is a common way to organize or structure a performance.)

- Skits
  - Students created skits in preparation for sharing their knowledge with family and friends.
  - Math-based student advertisements based on their knowledge.
- Musical performance
  - Student performance of a song that summarized their learning.
- Video performance
  - Student musical performance shared with the wider world through the Math and Science Performance Festival.

What is the role of new media in connection to these multimodal and performative communication forms? We do not suggest that new media's affordances of multimodality and performance *caused* the students and teachers to use multimodal and performative forms of mathematical communication. Rather, we offer two important observations.

First, some of the communication forms listed above are already common in some classrooms and supported by curriculum documents, while others are atypical. For example, in a Canadian context, although it is not uncommon for a Grade 2 teacher to use a story as a starting point for learning, it is less common that it is used for learning of mathematics and it is generally less common as we move up the grades. Also, the use of performative forms of communication, such as skits and songs, is much less common. That is, there is a linguistic diversity in the case study classrooms, in terms of modes of communication, which is not common in mathematics classrooms. Most of these examples of multimodality or diversity can be done with other "old" media rather than with "new" digital media. However, old media does not lend itself as easily to such forms. To draw an analogy, it was possible for people relying on horse and buggy for transportation to live say 50 km away from their work, but this was a rare occurrence until the widespread adoption of the automobile.

Second, although Grades 2 and 4 teachers share student work with parents without the use of new media (by sending tests home or assigning take-home projects that involve parents), they typically do not do so with a wider audience, as did the teachers in these classrooms by engaging their students to create mathematical performances to be shared on a publicly available web site (www.mathfest.ca). In this context, it is also safe to assume that teachers want to share their students' best work. But what is "best work" changes with the audience. Narrow audiences, such as classrooms, tend to create work that is based on well-defined problems, fixed meanings, and symbolic representations. If the audience is widened beyond the mathematics classroom, to include "just plain folks," then the focus shifts to "emergent problems," "negotiated meaning," "causal stories," and generally narrative forms of communication (Brown, Collins, & Duguid, 1989). So, when the audience becomes the "just plain folks," what is "best work" is judged less in terms of right or wrong and more in terms of being a good story or performance.

Boorstin (1990), writing about movies, notes that good stories have the following characteristics (paraphrased to suit our mathematics context): they provide new mathematical perspectives; they offer mathematical surprise; they engage emotionally; and they help us experience visceral mathematical sensations, such as mathematical fit, pattern, and beauty. Without claiming causality, it is interesting to note the parallels between the goal of a good math performance and the multimodal and performative forms used in the activities. For example, by introducing constraints and focusing on optimization, the teacher created opportunities for students to experience the new and surprising idea (for young students and even many parents) that the area can remain constant while the perimeter varies (and vice versa) or that optimal solutions involve squares (assuming we are dealing with rectangular arrangements). Watson and Mason (2007, p. 4) "see mathematics as an endless source of surprise, which excites us and motivates us [...] The challenge is to create conditions for learners so that they too will experience a surprise."

Gadanidis and Borba (2008) and Scucuglia (2012) suggest that surprises are fundamental components for the production of conceptual mathematical performances. Mathematical surprises (at least surprises that are deeply and not superficially mathematical) require complex mathematical ideas and deep conceptual relationships, which is what the case study classroom focus on optimization provides. In contrast, a typical worksheet that asks students to calculate area or perimeter for rectangular shapes offers no mathematical surprise and little mathematical pleasure. By using stories and skits, students can experience emotional mathematical moments through the mathematical "adventures" of the characters involved. And, visceral sensations are introduced by engaging students with physical, diagrammatic, and tabular pattern and fit, offering opportunities to see beauty in mathematical ideas and representations, as well as musical performances. Boorstin (1990) adds that movie soundtracks offer visceral sensations to the audience.

13.2.2 Case B: Brazilian Students Explore Sequences and Series of Numbers

The overall conclusion that emerges from research is that the teaching of algebra is typically instrumental rather than relational, with a dominance of symbolic algebra over other representations (Borba & Confrey, 1996; Kieran, 1992; Kieran & Guzmán, 2009; Kieran & Sfard, 1999). Teachers seem to "hold a symbol precedence view of student mathematical development" and they seem to "overestimate the accessibility of symbol-based representations and procedures for students' learning introductory algebra" (Nathan & Koedinger, 2000, p. 209). Consequently, though they learn to manipulate algebraic expressions, students do not seem to be able to use them as tools for meaningful mathematical communication (Kieran & Sfard, 1999). The majority of students do not acquire any real sense of algebra and, early on in their learning of algebra, give up trying to understand algebra and resort to memorizing rules and procedures (Kieran, 1992). Such learning of algebra

Fig. 13.5 Odd numbers

creates a weak foundation for relating good math stories in response to the question. "What did you do in math today?"

In the case described below, from a Grade 7 classroom in Rio Claro, Brazil, students started exploring the problem of finding the sum of the first *N* odd numbers with concrete materials and noticed that they fitted together to form a square, as shown for the first four odd numbers in Fig. 13.5. Then they explored the sums of even numbers as well as combinations of the two. In doing so they did two things differently from the typical classrooms described above. First, they explored mathematics problems that are complex. Finding a formula for the sum of the first *N* odd or even numbers is typically a senior secondary school topic. Second, they used concrete and visual representations (as well as others, as we will soon see) to explore the problems, and develop and communicate their understanding.

Grade 7 students investigated odd numbers as L patterns (see Fig. 13.6) using several materials, such as songs, lyrics, video clips, manipulative blocks, and online applets. The L patterns approach to the study of sequences and series of odd and even numbers has the potential to offer surprise in generalizing patterns. Working with manipulatives and applets, students can construct sequences of numbers and they can connect blocks (or simulate the connection) to have a geometrical representation for the series which works as a visual proof. Playing with L patterns becomes a surprising and visceral experience when students connect Ls forming squares (odd numbers) and $N\times(N+1)$ rectangles (even numbers).

Noss et al. (2009) argue that the difficulties that mathematical generalization and algebraic expression pose for students have been thoroughly studied. For these authors, the difficulties students face when dealing with generalization activities are in some measure due to the way in which they are presented and the constraints of the teaching approaches used. Usually, teachers tend to teach the techniques isolated from all context to help their students find the rule. There is a need to introduce students to different approaches involving generalization of patterns (Noss et al., 2009). We do see the exploration of the L patterns in a digital-artistic environment as one more pedagogical approach to address this need.

From the humans-with-media perspective (Borba & Villarreal, 2005), experimentation-with-technologies and visualization play fundamental roles in mathematical thinking. Students–teacher-media were involved in figuring out a

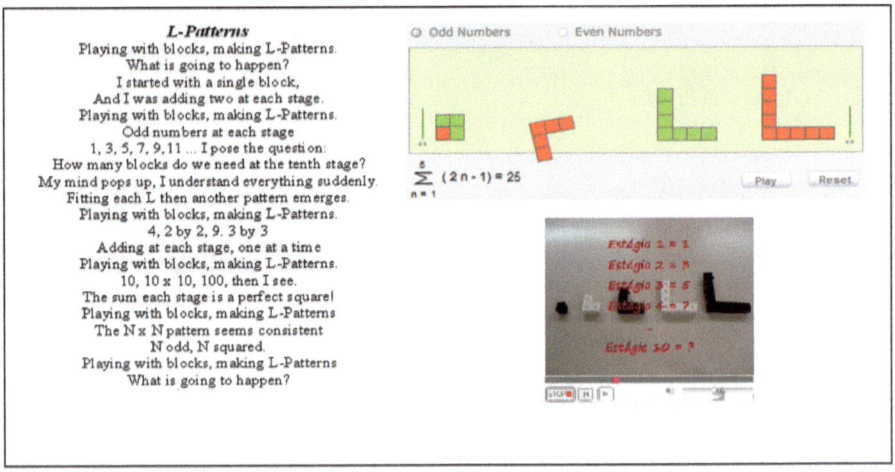

Fig. 13.6 L patterns

generalization for $S_n = 1 + 3 + 5 + \cdots + (2n - 1)$. By investigating the algebraic sum at each stage (writing, singing, and watching a video) and articulating it to the connections of Ls forming a square (constructing with blocks and simulating with an applet), the students–teachers-media identified a new pattern together, as thinking collectives. Then, one student came up with an interesting conjecture asking: "But, if we think about it, there are even numbers in rectangles, are there not?" The teacher said: "Perfect. What is the shape related to the series of even numbers?" The student then argued: "A rectangle, right? I can see it with the applet."

We suggest the student did an important articulation involving several representations and media in investigating the L patterns. Thinking-with-Applet-song-lyric-video-blocks-and-other-media, *the student* was able to conjecture a visual representation for both the sequences and series of odd and even numbers. The group of students had not developed an investigation about even numbers until that moment, but the student was able to visualize and manipulate the applet and the blocks, relate it to the lyrics, images, and sounds, recognize an approach relating to the sum of even numbers, connect it to the investigation of odd numbers, and communicate that the series of even numbers can be geometrically represented by rectangles. This moment revealed a significant role for technology and multimodal representation in shaping students' thinking and learning.

Based on the investigation of the L patterns, another pair of students decided to create a new sequence of numbers. These students created a first F using six blocks. They discussed the growth of the sequence and they decided to add three blocks at each stage, forming the sequence (6, 9, 12, 15,…) (see Fig. 13.7). The teacher brought up this discussion of F pattern to all students. He praised students' conjecture and imagination in creating a new pattern. He proposed that they develop a similar investigation on the F pattern as they had made to the L pattern. "What could be a generalization for the sequence? What would be the series? Would it be possible

Fig. 13.7 F patterns

to connect the blocks and create a visual proof to the series? Could we create a performance about F pattern?" One of the students said they were trying to figure out the series, but they did not find a regular shape (as a rectangle) to express an algebraic formula for the series. Figure 13.7 shows students' tentative start of creating a visual proof for the *F pattern* series. The teacher asked them if they had already found a formula to the sequence. After 15 min of dialogue between the pair of students, where they used blocks, and paper and pencil, one of them said: "I found it! It is n times 3 plus 3." The teacher said the generalization $3n+3$ seemed to be a good candidate. The teacher proposed to all students in the class to test it for several stages, and they confirmed the students' conjecture. The pair of students celebrated the confirmation of their conjectures greeting each other. This moment revealed students' pleasure in thinking mathematically. It reveals mathematics activity as an emotional, visceral, and surprising experience in learning (Gadanidis, Hughes, & Borba, 2008).

The pair of students came up with the idea of creating a new performance based on the theme of TV News. The idea was to present an interview with a mathematician who had discovered the F pattern. This was accepted by the group and students made suggestions such as: provide humor, use specific songs and sound effects, the kind of images they should portray about reporters and mathematicians, and how to explain the series.

The teacher emphasized students could explore a formula for the series of the F patterns, but students had difficulty in developing it. The teacher described what he called the "Gauss method" to figure out an algebraic formula for the series, which involves adding the first and last terms, then the second and second last terms, and so on, to see if a pattern emerges. After that, students started to write the script of the roles they would play in the skit. They collaborated and negotiated the meanings, reorganizing or rewriting their scripts. This negotiation and rewriting also happened during the performance. Students also negotiated their use of the manipulative blocks, the whiteboard, what should be written on the board, angles of video recordings and close-ups, moments of surprises and emotions, etc.

Students–teacher-with-media recorded the scenes in the same order they wanted to present it. The process of playing and recording involved students' speech repetition and improvisation through many takes of scenes. These actions revealed a significant aspect of improving communication and imagination in learning mathematics through creating a digital skit performance. After the recordings, based on other students' suggestions, the teacher edited the video called *The F Pattern*

Fig. 13.8 Making the F patterns video

News (see Fig. 13.8) with the participation of the students. The teacher worked on the Portuguese/English translations to create subtitles and some students were interested in learning how to say in English the words they said in the video (another aspect of language diversity). Students agreed that the mathematical idea was communicated clearly in the digital text. Then, the thinking collective created a final draft at the school's Lab and one of the administrators of the school submitted the video to the Math Performance Festival.[1] Below we present a transcription of the video:

Reporter 1: We are interrupting the TV show to tell you terrific news! One just discovered that beyond the L patterns, there is an F pattern!

Reporter 2: F pattern? What are you talking about?

Reporter 1: Our reporter has more information about it.

Reporter 2: Hey, can you hear us?

Reporter 3: Yes. I am here with Hypotenuse. She knows everything about math.

Reporter 2: What is this story about F pattern?

Hypotenuse: The F pattern is here represented by blocks.

Reporter 1: Could you please ask her what would be the sequence?

Reporter 3: Sure. What would be the sequence for the F pattern?

Hypotenuse: As we can see, the sequence increases three blocks each stage.

Reporter 3: Look, at the first stage you used six blocks. At the second stage you used nine blocks. At the third, 12 blocks. And so on. What about the 100th stage? How many blocks do you need to construct the 100th stage?

Hypotenuse: [using the whiteboard and the blocks] We have a simple formula to figure it out. You can notice the result each stage is equal to the index of the stage times three plus three. That is, $3N+3$. At the 100th stage we have 100 times three plus three. It is equal to 303.

Reporter 2: [with a surprising sound on the background] I am getting the information that the series was discovered. What would it be?

[1] Available at www.edu.uwo.ca/mpc/mpf2010/mpf2010-134.html.

Reporter 1: In other words, what would be six plus nine plus twelve and so on?

Reporter 3: Hypotenuse, what would be the series?

Hypotenuse: [in fast motion and using the whiteboard] I am going to use Gauss's ideas. The series would be the sum of these numbers. Gauss did it twice. Once from the beginning to the end, that is, $S=6+9+12+\cdots+3n-3+3n+3n+3$. And once from the end to the beginning, that is, $S=3n+3+3n+3n-3+\cdots+12+9+6$. Then, we can add these two expressions. So, we have $2S=(3n+9)+(3n+9)+(3n+9)+\cdots+(3n+9)$. Therefore, we have $2S=n$ times $(3n+9)$. It gives us the following formula… [Showing it on the board]. With this formula we can calculate the total number of blocks we need until the stage we wish.

Reporter 1: Last time we talked about the L patterns. Today, we talked about the F pattern. What about you? Which one is your favourite pattern?

Reporter 2: Which one will be our next pattern?

Reporter 1: Your local show is coming up next.

Reporter 2: Good night … have a good day … good afternoon.

Later, when these students were in Grade 8, we invited them to share their digital performance with a class of Grade 7 students in their school. Groups of students were engaged in exploring and discussing both the L and F patterns, in a context we potentially designed as a multimodal learning environment. Some students of these groups were thus exploring the ideas for the second time (Grade 8), and some of them for the first time (Grade 7).

After watching the digital performance, one of the Grade 7 students was manipulating the blocks, playing with the blocks, and making Fs using paper and pencil. The student said the group could explore the sequence 3, 6, 9, 12, 15,… instead of 6, 9, 12, 15,… There were interesting reasons that the student justified: (1) the new sequence with general term $3n$ is very similar to the original sequence with general term $6n-3$, and (2) the Fs made with blocks to represent the new sequence can be reorganized, rearranged, or reshaped in order *to fit*, forming a trapezoid. We see this event as a *visceral* moment in students' mathematical thinking, in terms of the esthetic pleasure of mathematical fit (Sinclair, 2006), as thinking about the problem was reorganized using manipulatives and other media. Another student noticed that the original F pattern could be seen as a combination of two patterns: the natural numbers 3, 4, 5, 6,… and the odd numbers 3, 5, 7, 9,…

The scenario is interesting in terms of the role of new media and the arts in students' mathematical investigation, meaning, and knowledge production. After watching a digital performance produced by students, a solution could be redesigned based on the use of several media. Students-with-blocks-video-and-other-media, dialogically and visually, conjectured that the series $3+6+9+12+\cdots+3n$ could be geometrically represented as a trapezoid. The (visceral) idea was further explored since students were engaged in producing a new digital mathematical performance. The use of new media (e.g., video camera and software to edit videos) and the arts (drama and music) to produce a digital performance offered ways for students to enhance their meaning production, because they were seeking to refine and improve their communication as they were producing the performance

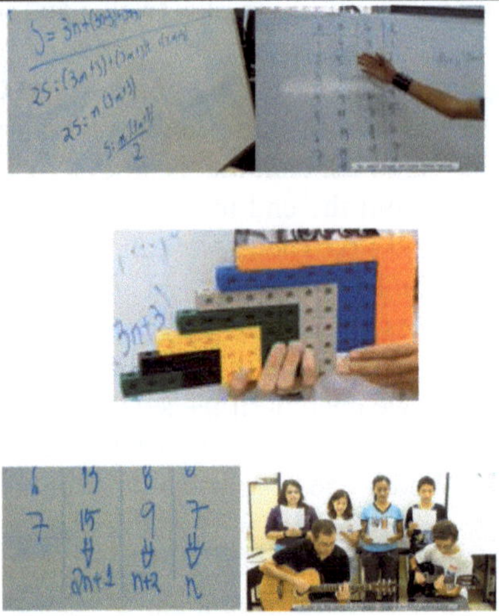

F Patterns

I will sing about what I've learned,

I will talk about what I've felt,

about Ls and Fs.

With Ls we have even and numbers,

many emotions.

With Fs we learned, more than once,

multiple of three.

Even numbers, 2n,

the series is a rectangle.

Fig. 13.9 A new performance: the series of F patterns represented as a trapezoid

(creating the story and the lyrics, practising and recording them, editing the video to share it online to the world). The process of expanding language diversity based on the use of new media and arts offered ways to better understand the conjecture and justifications, (re)organize thinking, connect representations, elaborate new conjectures, etc. It was a rich moment for mathematical meaning production, which is a significant aspect of learning.

The student performance explored the L and F patterns and introduced a geometric and algebraic investigation of the series related to the F pattern. Scucuglia and Gadanidis (2013) analyze aspects of this performance highlighting that the engagement with digital mathematical performance may open windows into mathematical thinking. The engagement in a multimodal learning environment offered ways for collective dialogues, new mathematical investigations and insights, and the production of a digital performance in which students communicated and represented aspects of their mathematical ideas learning. The new performance has the potential to open new windows for other groups of students, teachers, or classes. Figure 13.9 shows scenes of the performance and the lyrics.

The activities described above incorporated a variety of multimodal communication features:

- Concrete materials: blocks were used to construct odd, even, and F patterns.
- Drawings: students used drawings to record the concrete patterns they explored.
- Tables: students used tables to record sequences of numbers and investigate patterns.
- Interactive simulations: students explored a simulation of odd numbers fitting in a square.
- Songs: a song was used by the teacher to introduce the problem of finding the sum of the first *N* odd numbers.

- Drama: students authored and performed a skit about the F patterns problem.
- Videos: students explored a video about noticing patterns that might help to find the sum of odd numbers; students created a video of their skit about F patterns. They used gestures and movement to embody and communicate mathematical ideas.

The activities also incorporated a variety of performative communication forms, as elaborated above:

- The following performative modes of communication were used by the teacher:
  - Song
  - Video
  - Interactive simulation
- The following performative modes of communication were used by the students:
  - Drama
  - Video performance

It is interesting how easily the students took up alternative forms of mathematical communication, displaying great language diversity in an interconnected fashion. They used a variety of texts, materials, resources, and digital technologies: lyrics, paper, and pencil, manipulative blocks, computers, videos, Internet, and applets. It is also interesting that in this environment students were drawn to create and perform a dramatic skit to present their learning about F patterns. When students share their mathematical ideas by playing roles as actors they think about the audience. The production of a dramatic event (a skit) involving mathematics helps students on the (re)organization of their mathematical ideas to suit a wider audience. The development and creation of a mathematical narrative deals with sense-making in a way that may better appeal to "just plain folks," by using forms of communication that are commonly used in the media and narrative structures that draw attention to discrepancy and surprise. The process of recording a performance also involves speech repetition and helps students improve their mathematical oral/verbal communication skills and understanding. In addition, the production of drama-based narratives involves ways of imagining and reflecting about the "self" and its relation to doing and learning mathematics. When students perform mathematics they construct identities as performance mathematicians (Gadanidis & Borba, 2008).

13.3 Audience, Multimodal Communication, and Language Diversity

In this chapter, we have critiqued the issue of language diversity in terms of the traditional and persistently narrow mathematics education focus on textual/symbolic communication suggesting there are ways to expand the boundaries for this notion. We have contrasted this phenomenon with the multimodal and performative out-of-school environment that students experience due to the pervasiveness of digital media. We have also highlighted the new media affordances of multimodality and performance and illustrated what these might look like in the mathematics classroom through two case studies: one from Canada and one from Brazil. In this section, we discuss the potential interplay between performance, multimodality, and language diversity.

Reflecting on the two cases we have presented in this chapter, and keeping in mind our similar work in other classrooms in Canada and in Brazil, we note two important themes. First, students engage with multimodal forms of communication with little hesitation. Although we did not collect data to shed light on the reasons for this, we speculate that it might be the case because multimodal communication is part of students' digital environment, especially in out-of-school experiences, or because students find multimodal forms of communication interesting and engaging, and indeed for many students compelling. Second, what is different in these classrooms is more than simply the modes used to communicate information: there is also a relationship difference or shift with respect to mathematics learning, such as, from passive to engaged, from consuming to producing, and from thinking to thinking and feeling. Although our digital age is often labeled as an information revolution (as contrasted with the industrial revolution), Schrage (2001) suggests that this label misses the essence of the paradigm shift:

> *The so-called "information revolution" itself is actually, and more accurately, a "relationship revolution."* Anyone trying to get a handle on the dazzling technologies of today and the impact they'll have tomorrow, would be well advised to re-orient their worldview around relationships […] *When it comes to the impact of new media, the importance of information is subordinate to the importance of community.* The real value of a medium lies less in the information that it carries than in the communities it creates. (pp. 1–2, original emphasis)

Lankshear and Knobel (2006) suggest that the relatively recent "development and mass uptake of digital electronic technologies" represent changes on an "historical scale," which "have been accompanied by the emergence of different (new) ways of thinking about the world and responding to it" (pp. 29–30). These new ways of thinking can be characterized as "more "participatory," "collaborative," and "distributed" and less "published," "individuated," and "author-centric" … also less "expert-dominated" (Lankshear & Knobel, 2007, p. 9).

We suggest that the case studies we have shared offer glimpses into "what might be" when students are encouraged and supported in taking up the multimodal and performative affordances of new media. We further suggest that "what might be" is about a change in relationships between students and the world around them, such as taking up roles as producers of mathematical knowledge for audiences beyond their classroom. Change in how students communicate with a wider audience and how their use of language positions them in relation to the world around them involve change in their literacy identity. This change may be seen as increased language diversity, as students use new forms of communication and for new purposes (we mean "new" in relation to traditional mathematics teaching and learning contexts). Weber and Mitchell's (2008) notion of identity as "personal and social bricolage" views identity construction as "an evolving active construction that constantly sheds bits and adds bits, changing through dialectical interactions with the digital and non-digital world, involving physical, psychological, social, and cultural agents" (p. 43). New literacies, as defined by Lankshear and Knobel (2007), are not characterized solely by their digital or technical features. They also involve a new mindset or a new ethos that focuses on participation, collaboration, and distribution (Lankshear & Knobel, 2007). Traditional forms of literacy that focus solely on reading and writing in the dominant language typically fail to recognize the impact of digital media in students' lives or take into account students' multilingual and multicultural backgrounds and experiences.

We also suggest that once students focus their communication (at least in part) for an audience beyond the walls of their classroom and their school, the criteria by which "good" communication is determined changes, as the traditional "breach between learning and use" is bridged (Brown et al., 1989, p. 32). Students (and teachers) aim their communication not for their classroom, where the focus is traditionally on "precise, well-defined problems, formal definitions, and symbol manipulation" but for the "just plain folks" outside the classroom walls who reason with "causal stories" rather than "laws," act on "situations" rather than "symbols," resolve "emergent problems and dilemmas" rather than "well-defined problems," and produce "negotiated meaning and socially constructed understanding" rather than "fixed meaning and immutable concepts" (Brown et al., 1989, p. 35). Creating a math performance that outlines the common algorithm for adding two or three digit number or the rule of inverting and multiplying when dividing fractions does not make for a "good" story to share with the outside world.

Although not typically evident in mathematics classrooms, human cognition is story based and we naturally seek to experience and to relate good stories (Schank, 1990). We think in terms of stories, we understand the world in terms of stories that we have already understood, we learn by living and accommodating new stories, and we define ourselves through the stories we tell ourselves (Schank, 1990). Our lives make sense when shaped into narrative form (McIntyre, 1984). Story is a human symbol system used to comprehend events and entertain questions, and represent those events and questions in a sequence that offers a new perspective, surprise, emotional moments, and visceral sensations in a way that makes sense (Boorstin, 1990): that is, the story must provide its own justification, or prove itself.

Thus, as students and teachers use new media to share aspects of classroom learning with the wider world, they naturally seek to engage their audience through surprise, emotional moments, and visceral sensations, as did the students and teachers in our two cases. The simplest (and most common) way to do this is to keep the mathematics unchanged while couching it in a nonmathematical story context, a sugar-coating, that offers the story pleasures of surprise, emotional moments, and visceral sensations. But there is also the potential to create and relate truly mathematical stories that engage a wider audience. For this to take place, school mathematics needs to become more complex and less focused on procedural learning. As Gadanidis, Hughes, and Cordy (2011) note:

> We do not believe that artistic *mathematical* expression is possible in a mathematics program which focuses on procedural rather than conceptual knowledge. Students can add artwork to "decorate" procedural knowledge, thus adding a layer of "sugar-coating" to otherwise dry mathematical ideas, but mathematical art, like art in general, requires a deeper engagement and understanding. Thus, for us, challenging mathematics is a co-requisite for artistic mathematical expression. (pp. 423–424)

In the two cases we shared, the "challenging" mathematics took the form of a focus on optimization as a context for learning about area and perimeter and a focus on sequences and series of odd and even numbers as a context for learning about algebraic representation. These contexts offered Grades 2 and 4 students the mathematical surprise that squares maximize rectangular area and minimize rectangular perimeter and Grade 7 students the mathematical surprise that the series of odd numbers "hides" in squares (and in $N \times (N+1)$ rectangles for even numbers). The multimodal representations of these relationships also offer the visceral pleasure of mathematical "fit," as students slide the odd numbers to form a square, for example. The emotional mathematical moments experienced by the students are evident in their performances, through song and drama.

13.4 Concluding Remarks

When we expand the scope of what we mean by language diversity to include digital literacy we realize that new media not only offers students and teachers new ways of communicating about mathematics learning but it also offers the potential to shift literacy identities in ways that require new roles and new ways of thinking about what constitutes a "good" mathematics story or experience. Borba (2009) discusses some scenarios of the Internet being fully accepted in the classroom. Performance was one of the possibilities. It could be the case that mathematical performance could be a means that students who cannot speak, or students who do not speak the official language well, can express themselves. It could also be the case that students create performances that disrupt traditional school mathematics. In this manner, mathematical performance could gain a political dimension, which has not yet been explored in a theoretical manner nor in empirical research. For example, imagine exploring *Mathematics of the Oppressed* as a way of drawing a parallel between Boal's (1985) political work *Theatre of the Oppressed*, which introduces the term *spect-actor* to bring focus on the need for personal and collective agency, and the pervasiveness of mathematics education structures that minimize students' cognitive and affective agency. We believe such issues could be part of the research agenda for mathematics educators interested in political dimensions of mathematics education and mathematical performance.

The cases we shared in this chapter from Brazil and from Canada are early examples or first drafts of "what might be" when new media and its affordances are taken up by students and teachers. We believe that these examples, by exploring the new possibilities of digital media, combine mathematics and the arts, show the connection between media, emotions, and thinking, and show how language diversity may be expanded to include issues of multimodality and audience. We have argued that such changes reorganize thinking, generating changes in the knowledge produced by this collective of humans-with-media. We hope that the cases we have presented and discussed serve as objects for reflection and critique, so that better and more varied examples of "what might be" in mathematics teaching and learning may be developed.

References

Boal, A. (1985). *Theatre of the oppressed* (C. A. McBride & M.-O. Leal McBride, Trans.). New York: Theatre Communications Group.

Boorstin, J. (1990). *The Hollywood eye: What makes movies work*. New York: Harper Collins.

Borba, M. C. (2009). Potential scenarios for Internet use in the mathematics classroom. *Zentralblatt für Didaktik der Mathematik, 41*(4), 453–465.

Borba, M. C., & Confrey, J. (1996). A student's construction of transformations of functions in a multiple representational environment. *Educational Studies in Mathematics, 31*, 319–337.

Borba, M., & Villarreal, M. E. (2005). *Humans-with-media and the reorganization of mathematical thinking*. New York: Springer.

Brown, J. S., Collins, A., & Duguid, P. (1989). Situated cognition and the culture of learning. *Educational Researcher, 18*(1), 32–42.

Bruner, J. (1994). *Acts of meaning: Four lectures on mind and culture*. Cambridge, MA: Harvard University Press.

Buckingham, D. (2010). *Media education: Literacy, learning and contemporary culture*. Malden, MA: Polity Press.

Cummins, J., Brown, K., & Sayers, D. (2007). *Literacy, technology, and diversity*. New York: Pearson.

Denzin, N. K. (2003). *Performance ethnography. Critical pedagogy and the politics of culture*. Thousand Oaks, CA: Sage.

Gadanidis, G. (2008). *Scruffy's new home*. Whitby, ON: Brainy Day.

Gadanidis, G. (2012). *Trigonometry in grade 3? What works? Research into practice* (Research Monograph 42). Ottawa, ON: The Literacy and Numeracy Secretariat.

Gadanidis, G., & Borba, M. (2008). Our lives as performance mathematicians. *For the Learning of Mathematics, 28*(1), 42–49.

Gadanidis, G., & Gadanidis, M. (2010). *Wolf gets hurt*. Whitby, ON: Brainy Day.

Gadanidis, G., Hoogland, C., & Hughes, J. (2008). Teaching and learning in a Web 2.0 environment: Three case studies. *International Journal of Technology in Teaching and Learning, 4*(2), 117–133.

Gadanidis, G., Hughes, J., & Borba, M. (2008). Students as performance mathematicians. *Mathematics Teaching in the Middle School, 14*(3), 168–175.

Gadanidis, G., Hughes, J., & Cordy, M. (2011). Mathematics for gifted students in an arts- and technology-rich setting. *Journal for the Education of the Gifted, 34*(3), 397–433.

Gerofsky, S. (2006). Performance space & time. In G. Gadanidis & C. Hoogland (Eds.), *Digital Mathematical Performance: Proceedings of a Fields Institute Symposium*. London, ON: The University of Western Ontario.

Hill, R. T. G. (1998). Performance pedagogy across the curriculum. In S. J. Daily (Ed.), *The future of performance studies: Visions and revisions* (pp. 141–144). Washington, DC: National Communication Association.

Hughes, J. (2008). The performative pull of research with new media. *International Journal of Qualitative Methods, 7*(3), 16–34.

Hughes, J., & Gadanidis, G. (2010). Learning as community service: Thinking with new media. *Journal of Educational Multimedia and Hypermedia, 19*(3), 287–306.
Hull, G. A. (2003). At last: Youth culture and digital media: New literacies for new times. *Research in the Teaching of English, 38*(2), 229–233.
Hull, G. A., & Katz, M. (2006). Creating an agentive self: Case studies of digital storytelling. *Research in the Teaching of English, 41*(1), 43–81.
Hutchins, P. (1986). *The doorbell rang*. New York: Mulberry Books.
Kieran, C. (1992). The learning and teaching of school algebra. In D. A. Grouws (Ed.), *Handbook of research on mathematics teaching and learning* (pp. 390–419). New York: Macmillan.
Kieran, C., & Guzmán, J. (2009). Developing teacher awareness of the roles of technology and novel tasks: An example involving proofs and proving in high school algebra. In M. Tzekaki, M. Kaldrimidou, & H. Sakonidis (Eds.), *Proceedings of the 33rd Conference of the International Group for the Psychology of Mathematics Education (PME)* (pp. 321–328). Thessaloniki, Greece: PME.
Kieran, C., & Sfard, A. (1999). Seeing through symbols: The case of equivalent expressions. *Focus on Learning Problems in Mathematics, 21*(1), 1–17.
Kirkpatrick, A. (2007). *World Englishes: Implications for international communication and English language teaching*. Cambridge, England: Cambridge University Press.
Kress, G., & Van Leeuwen, T. (2001). *Multimodal discourse: The modes and media of contemporary communication*. New York: Oxford University Press.
Lankshear, C., & Knobel, M. (2006). *New literacies: Everyday practices and classroom learning* (2nd ed.). London: Open University Press.
Lankshear, C., & Knobel, M. (2007). Sampling "the new" in new literacies. In M. Knobel & C. Lankshear (Eds.), *A new literacies sampler* (pp. 1–24). New York: Peter Lang.
Levy, P. (1993). *Tecnologias da Inteligência: O futuro do pensamento na era da informática*. [Technologies of intelligence: The future of thinking in the informatics era]. Rio de Janeiro, Brazil: Editora 34.
Levy, P. (1998). *Becoming virtual: Reality in the digital age*. New York: Plenum.
McIntyre, A. (1984). *After virtue: A study in moral theory*. Notre Dame, IN: University of Notre Dame.
McLuhan, M. (1964). *Understanding media: The extensions of man*. New York: McGraw Hill.
Miller, P., & Goodnow, J. (1995). Cultural practices: Toward an integration of culture and development. *New Directions for Child Development, 67*, 5–16.
Nathan, M. J., & Koedinger, K. R. (2000). An investigation of teachers' beliefs of students' algebra development. *Cognition and Instruction, 18*(2), 209–237.
Noss, R., Hoyles, C., Geraniou, E., Mavrikis, M., Gutierrez, S., & Pearce, D. (2009). Broadening the sense of 'dynamic': An exploratory system to support students' mathematical generalisation. *Zentralblatt für Didaktik der Mathematik, 41*(4), 493–503.
Ochs, E., & Capps, L. (2001). *Living narrative: Creating lives in everyday storytelling*. Cambridge, MA: Harvard University Press.
Pahl, K., & Rowsell, J. (2005). *Literacy and education: Understanding the new literacy studies in the classroom*. Thousand Oaks, CA: Sage.
Phillipson, R. (2009). *Linguistic imperialism continued*. New York: Routledge.
Rowsell, J., & Walsh, M. (2011). Rethinking literacy education in new times: Multimodality, multiliteracies and new literacies. *Brock Education, 21*(1), 53–62.
Schank, R. (1990). *Tell me a story: A new look at real and artificial memory*. New York: Macmillan.
Schrage, M. (2001). The relationship revolution. Retrieved from http://web.archive.org/web/20030602025739/http://www.ml.com/woml/forum/relation.htm
Scucuglia, R. (2012). *On the nature of students' digital mathematical performances*. Unpublished doctoral dissertation, The University of Western Ontario, Canada.
Scucuglia, R., & Gadanidis, G. (2013). Performance matemática: Tecnologias digitais e artes na escola pública de ensino fundamental [Mathematical performance: Digital technology and the arts at the elementary public school]. In M. Borba & A. Chiari (Eds.), *Tecnologias digitais e educação matemática* [*Digital technology and mathematics education*]. São Paulo, Brazil: Editora Livraria da Física.

Sinclair, N. (2006). The aesthetic sensibilities of mathematicians. In N. Sinclair, D. Pimm, & W. Higginson (Eds.), *Mathematics and the aesthetic: Modern approaches to an ancient affinity* (pp. 87–104). New York: Springer.

The New London Group. (1996). A pedagogy of multiliteracies: Designing social futures. *Harvard Educational Review, 66*(1), 60–92.

Tikhomirov, O. K. (1981). The psychological consequences of computerization. In J. Wertsch (Ed.), *The concept of activity in Soviet psychology* (pp. 256–278). Armonk, NY: M.E. Sharpe.

Walsh, M. (2011). *Multimodal literacy: Researching classroom practice*. Sydney, Australia: Primary Teachers Association of Australia.

Watson, A., & Mason, J. (2007). Surprise and inspiration. *Mathematics Teaching, 200*, 4–7.

Weber, S., & Mitchell, C. (2008). Imaging, keyboarding, and posting identities: Young people and new media technologies. In D. Buckingham (Ed.), *Youth, identity and digital media* (pp. 25–47). Cambridge, MA: The MIT Press.

Chapter 14
Language Diversity in Research on Language Diversity in Mathematics Education

Faïza Chellougui, Hien Nguyen Thi Thu, and Carl Winsløw

14.1 Identifying the *Problématique*: Language(s) of Education, and of Research

In research on mathematics education, as in many other fields, "international" has become synonymous with "English." It is no surprise that the announcement (ICMI Study 21 discussion document, this volume, p. 307) for the present ICMI study conference stipulated that papers must be in *English*. This can be justified by the fact that, presumably, no other language is mastered (to some degree) by more researchers who are likely to be able to participate in a study of this kind. But it also raises a number of concerns, which this chapter aims to point out, together with a tentative approach to address them. We feel it is particularly appropriate to raise these concerns in connection with an international study whose subject—language diversity in mathematics education—seems, to us, is in need of a special sensitivity to this issue. Indeed, one of the sub-themes of the study being "Researching mathematics teaching and learning in multilingual contexts" (see Appendix), one could not help noticing that "contexts" in which English is not at all present (as a language of instruction) might indeed be somewhat underrepresented in a study where

F. Chellougui
Faculté des Sciences de Bizerte, Bizerte, Tunisia
e-mail: chellouguifaiza@yahoo.fr

H.N.T. Thu
Catholic University of Louvain, Louvain, Belgium
e-mail: nguyenhientt@gmail.com

C. Winsløw (✉)
Institut for Naturfagenes Didaktik, University of Copenhagen,
Øster Voldgade 3, Copenhagen 1350, Denmark
e-mail: winslow@ind.ku.dk

R. Barwell et al. (eds.), *Mathematics Education and Language Diversity*,
New ICMI Study Series, https://doi.org/10.1007/978-3-319-14511-2_14

English is expected on the part of researchers. For instance, a considerable body of research literature, including international journals, appears in other "old world" languages such as French, German, Russian, Spanish, and Portuguese, which, like English, became "dominant" in large regions beyond Europe, or have become common channels of publication, even where they are not dominant. One could also add to this list the vibrant body of mathematics education research literatures, which are formulated and published in non-English languages in India, Japan, and China.

The authors of this chapter share one commonly used research language, French, while our mother tongues are Arabic, Vietnamese, and Danish. In the heading of this first section, we have put the word *problématique* in italics to show that it has been appropriated into the English language.[1] We do so because there does not seem to be a genuine English word to designate what *problématique* means in French, an issue that occurs many times over in translations between languages. With some approximation we could explain the meaning of it as "a collection of related questions, phrased within a certain theoretical framework and, in particular, based on some fundamental assumptions related to this framework" (see also Wedege, 2006, for a thorough discussion of this point).

The reader will notice two things here: this supposed translation raises more questions than it answers, in the sense that we have not explained the more numerous terms used to explain the term *problématique*, in particular, what is (or could be) a "question" and a "theoretical framework." Moreover, our discussion of the subheading has taken us into a meta-discussion of terminology which, in itself, may seem to have rather little bearing on any question related to the teaching and learning of mathematics.

But we strongly argue that despite the clear reasons for communicating in "big" languages, one should not be naïve about theoretical frameworks—and the language which carry them—when formulating research questions. This, of course, is particularly true in a volume that bears on language diversity. Unlike the kinds of question we pose everyday, such as "How do you do?" or "What time is it?," we cannot allow research questions to grow on of what Chevallard (1999a, p. 7) calls "teashop English": certainly not when we address, and belong to, an international audience for which English of some sort may be the only available common ground, but in which "teashop English" remains the private business of a minority of so-called "native speakers."

The *problématique* we are trying to develop, bearing on research on mathematics teaching and learning, clearly requires a theoretical framework that allows us to talk *about research* and in particular *about* the language(s) used in such research (which we call the R-language(s)). It must be able to do so *in distinction* from language(s) appearing in the phenomena which the research bears on, i.e., the language(s) that are used by teachers and learners of mathematics in a particular context (these, we call the P-language(s)). The theoretical framework itself will employ a language, which we could call the M-language (with M for *meta*): in this paper, English.

[1] Indeed, the online service *Google translate*, http://translate.google.com, does recognize it to the level of providing reasonable translations of it into other languages.

We have so far mentioned, even if not entirely explained, some first terms of our M-language: besides *P-languages*, *R-languages* and *M-language*, also the term *problématique* and the terms used to explain it. We notice that except for the three abbreviations and one term (perhaps) imported from French, we have so far done with more or less standard terms from the English language, in accordance with the study call.

There is little chance, however, that a reasonable P-language can do with standard terms of any natural language, when the phenomena involve mathematics teaching and learning. While the R-language does in practice often remain naïve with respect to what is meant by "mathematics," mathematical practices in teaching and learning contexts will rarely fail to display crucial language-like items which are foreign to any natural language: number symbols, diagrams and, so on; the R-language may or may not contain systematic ways to refer to such items. The fact that these items are often similar in contexts where the natural languages appearing in the P-language differ greatly, may lure the researcher into the illusion that the P-languages are somehow equivalent or at least that the differences remain superficial. For instance, the solution of quadratic equations in symbolic form may be easy to recognize and follow on the blackboard of a classroom, even for an observer who understands nothing of what is said and written in natural language. In a very local sense, this impression of "following" may of course not be illusory at all. But the R-language can rarely, for a *problématique* bearing on teaching and learning, ignore the conditions and constraints which surround (often determine) these more or less transparent items. Thus, the R-language would need to be sensitive to the P-language in a wider sense, which almost invariably involves specific features of the P-language, such as (what we could, at this point, loosely term) *cultural connotations*; for instance, number words are often specific to and deeply rooted in cultural practices, and the differences may cause challenges for second language learners (cf. "Personal Case 2" below). We notice that "sensitivity" is, in general, not just a question of language, but also of the ways in which it is used, including a number of assumptions, specific rules and assertions. For example, in P-languages related to elementary mathematics, the English word *triangle* focuses on different defining aspects of a class of geometrical figures than the Danish word *trekant* (meaning, literally, "three-edge"). We shall term this "way of language use" a *theory* (so that we can speak of P-theory, R-theory, and M-theory).

The M-language, and in fact the M-*theory* which we are about to develop, must thus identify crucial features of the R-language and its capacity to deal with specifics of the P-language. Moreover, the M-language must enable us to ask relevant research questions about the R-theory itself. We note in passing the parallel between this preliminary discussion of M-language and a debate, more than a century old, on the foundations of mathematics as a science, in fact, as a theory. Here, logic—which, at least from a certain point of view, is part and parcel of mathematics—has been proposed as a framework for, at least in part, how to inquire into the basis of mathematics itself. While this approach is of course not free from problems, we notice that the particular problem of self-reference—taking a model from within a theory to model the whole theory—is not substantially more questionable than exhibiting an entirely foreign model, since the pains of infinite regress arise as soon as we ask how this "foreign" model would somehow equip us with safer grounds than the object we are about to model. We mention this because we will, in what follows, propose an M-theory for framing our problématique which is taken, almost wholesale, from a particular R-theory.

14.2 The *Problématique* and Its Meta-Language

Whether we talk of P-language, R-language, or indeed M-language, we refer to ways to talk about certain human *practices*. The R-theory on which this paper is based is the anthropological theory of didactics (ATD) takes as a basic assumption that P-languages serve to *talk about* certain practices related to *doing mathematics* or, more specifically, *solving tasks of a mathematical kind* using corresponding *techniques* (Chevallard, 1999b). The "mathematical kind" is, according to another fundamental assumption of ATD, determined by institutions.

As ATD is extensively developed in the literature (an excellent overview is given by Bosch and Gascón, 2006), we do not give a separate account here. We note, however, that the use of ATD as an M-theory is well established through a number of European projects aiming to compare and relate the variety of R-theories on mathematical education (Artigue & Bosch, 2014). This means that research is modeled as *research praxeologies* $[T/\tau/\theta/\Theta]$. Here T denotes a type of problems addressed in a research activity, τ the technique used, θ the technology (i.e., the discourse framing, explaining and justifying the techniques, based on what we have called the R-language), and Θ is a wider theoretical framework in which this discourse is defined (or at least makes sense), that is, the R-theory. We notice that the ATD model insists that R-practice $[T/\tau]$ and R-theory $[\theta/\Theta]$ (including scientific discourse) define each other mutually, as is the case for any praxeology: not only is theory and discourse created and adapted to explain and justify the challenges and methods used in research, but researchers' tasks and methods are themselves profoundly shaped by their scientific language and theories. This, in particular, situates the question of R-language(s) in a precise model of the research activity, at the level of technology: any R-technology is based on a R-language.

We can now formulate our *problématique* as precise research questions which one can undertake in order to design and assess a given research study with an emphasis on language sensitivity (particularly relevant, of course, for research on teaching in multilingual contexts!):

Q1. Are phenomena related to P-language(s) explicit in the types of problems addressed? If so, what techniques are used to tackle these phenomena? (even if P-language(s) are not explicit in the problems raised, we could ask the question about techniques, e.g., if the R-language and the P-language(s) are different).

Q2. At the level of technology, what means are used, if any, to justify the ways in which P-language phenomena are handled in the study?

Q3. Does the theoretical framework of the study provide means to situate and assess the study's assertions related to P-language, and how (if at all) are these means mobilized in the study?
Q4. Are institutional, societal or cultural specifics related to the P-language addressed by the technology? Could this be supported by the R-theory?
Q5. Does the technology consider the relation between P-language and R-language, especially as regards institutional, societal or cultural specifics? How?

We will address these questions for the research studies which were proposed and accepted for the ICMI Study conference, as well as in some further examples from our own research. But first, we illustrate the significance of our problématique by a case which spurred our initial interest in the questions above. We believe it demonstrates how the R-language issue, if ignored, may appear surprisingly (and inadequately) as results instead of questions.

14.3 Motivating Case: "Researching Mathematics Teacher Education"

The ICMI survey team on "researching mathematics teacher education," carried out by Adler, Ball, Krainer, Lin, and Novotna (2008) surveyed a certain corpus of research literature in order to provide a picture of "research that focuses on mathematics teacher education," and the team "saw as its responsibility to describe 'where we are' globally, in the field" (p. 127). The formulation of basic questions in the study (p. 128) does not explicitly mention P-language. However, the introductory remarks include the contention that "more and more learners should learn mathematics in English, a language that is not their main spoken language" (p. 124).

What is more surprising is the way in which R-language was handled:

> We selected from multiple outlets for this work, including peer reviewed journals, international handbooks and key conference proceedings. We looked across international journals as well as a handful of journals in Asia, Europe, i.e., published in languages other than English where it was possible to access these. In general, however, we did not have time and resources to investigate thoroughly journals written in e.g., French, German, Russian or Spanish. (p. 130)

So, what the team ultimately considered amounted to recent issues of three journals (ESM, JMTE, JMRE) and six congress proceedings (PME 1999–2003, ICME-9), where about 180 papers were considered relevant and "leading" relative to the theme.

As an outcome of scrutinizing these papers, the team then formulated four claims. One of them is on P-language: "Research in countries where English is the national language dominates the literature" (p. 135). This was demonstrated with a table showing, for instance, how papers from South and Central America make up only 3 % of the considered papers in PME, and are completely absent in JMTE. The authors admit that "these disparities are not surprising. The prevalence and increasing hegemony of English was referred to in the opening ceremony of the

Congress" (p. 135; the reference is to a Minister's opening address, where the importance of mastering English was underlined). As it appears, the "result" or "claim" is not in the least linked with the initial choice of considering only research produced in (select) English language journals. A lot of issues remain implicit here, like, apparently, the contention that "international" means "accessible to an audience who knows no other language than English." The possibility that a substantial literature on the theme could exist with other R-languages (in fact, it does!), no doubt with a higher share of contexts with other P-languages, has been ruled out from the beginning. It follows that almost all of our questions Q1–Q5 must be answered in the negative.

To be fair, we should mention the final "commentary" by Ball which shows some degree of alertness to this potential relation between P-language and R-language, at least when it comes to training new researchers in English-speaking countries: "It is important to develop a stance that avoids confusion between the local and the global. And so it is important to be able to work (read and speak) in more than one language" (p. 136).

14.4 Contributions to the ICMI 21 Study Conference: Overall Analysis

The proceedings of the study conference (Setati, Nkambule, & Goosen, 2011) contain a total of 54 papers, all written in English as required. In 52 of the papers, we can identify a "dominant" P-language in the context they study, in the sense that the papers study mathematical instruction, which occurs either solely or principally in this language. The dominant P-language turns out to be English in 34 of these papers, while a few others appear as well (Table 14.1; we notice that two of the 52 papers each treated two different contexts with different dominant P-languages, which is why the sum of the numbers in the table is 54, not 52). Relative to our questions, the ease with which the P-languages are identified implies that phenomena related to P-languages are indeed dealt with explicitly in virtually all papers, and most of them develop an explicit technology related to their treatment. The answers to Q1–Q3 are, therefore, positive for most of the contributions to this study.

The fact that 65 % of all papers described language diversity in contexts where English is the dominant P-language of instruction, is striking, because it certainly does not reflect an international situation. While it seems difficult to determine the percentage of the world's school children who are taught in English, less that 65 % of them are taught in English. In fact, only about 13 % of the world's population has English as their first or second language (according to Graddol, 1997, p. 10 the figure is about 750 million; here, having English as a second language is defined as the situation where English belongs "in a repertoire of languages where each is used in different contexts"). Adding to this a roughly similar number of people speaking English as a foreign language, we end up with roughly a quarter of the world popu-

Table 14.1 Dominant P-languages in ICMI Study 21 conference contributions and the corresponding average number of languages represented in references (as an estimate of R-language numbers)

Dominant P-language	English	Portuguese	Spanish	French	German	Catalan	Other[a]
# Papers	34	5	3	2	2	2	4
# R-languages	1.16	2.2	1.5	2	2	2	1.25

[a]Other dominant P-languages, each represented in the context of one paper, were: Swedish, Farsi, Chinese, and Russian

lation mastering English to some degree (and most certainly, much less than a quarter of the world's school children). While there are more first or second language speakers of Chinese and almost as many speakers of Hindi and Spanish (Graddol, 1997, p. 8), one might contend that English is of specific importance in multilingual settings, used as a language of instruction for children with another first language, particularly in some parts of Africa, North America, and Asia; indeed many of the papers address this situation. On the other hand, Spanish, French, and Portuguese have a similar role both in other parts of these same continents, and there is no doubt that these contexts are proportionally underrepresented in the contributions as listed in Table 14.1.

One might reasonably think that the choice of R-language for the study is linked to this bias. Naturally, in contexts where English is neither a first or second language, but where another language (like Spanish or Portuguese) is a dominant P-language, one is likely to find this other language also in higher education and, indeed, as an R-language. We think that indications of this could be seen in the references used, even in papers, which are, like the ICMI Study 21 conference papers, written in English. If R-language is, roughly, estimated according to the references cited by authors (as the basis for their presentation), it is interesting to note that 30 papers cite only English language references, while one paper cites only references in Spanish; the rest of the papers have references in more than one language (18 of these in two, and most of them with almost all references still being in English). Almost all (26) of the papers which have English as the only R-language (judging from references) deal at the same time, with contexts where English is the dominant P-language. On the other hand, English is the dominant P-language in only 5 of the 18 papers that have more than one R-language (in terms of referenced texts). In the last row of Table 14.1, the tendency also appears clearly: for papers dealing with contexts where English is the dominant P-language, it is also very often the only R-language, while in contexts where other languages are a dominant P-language, this other language typically appears in the references (except for the last case of "other languages") as evidence of its importance for the authors as an R-language.

It is, perhaps, surprising that while many papers discuss the institutional and cultural implications of a dominant P-language in contexts where this language is not the first language of students, virtually no papers specifically reflect on the significance of the R-language, particularly in the case where both are English (well over half of all contributions). This means that Q5 is hardly ever touched and that

Q4 may only be implicitly addressed (for instance, through discussions of more or less problematic translation issues). This is perhaps not surprising, given the dominance of English as an R-language in the countries of most of the participants. However, Q5 must be asked especially when the R-language is identical to the dominant P-language, to avoid the implicit naturalization of dominances of cultural and institutional norms in the latter by their dominance in the former.

At the same time, most of the ICMI Study, 21 conference contributions (43 of 54) study teaching contexts where two or more P-languages occur explicitly, typically because students use a second language in some way even when the dominant language is the only official language of instruction (as one might expect in a context of language diversity). We recall that in 34 papers, the dominant P-language was English, and that 30 of these 34 papers cite *only* literature in English.

When different P-languages occur, the use of English as an R-language presents some challenges, at least for exposing and analyzing exactly what children say, what their difficulties are etc.; however, this would be the case for any choice of R-language. More importantly, we contend that specific cultural and scientific perspectives tend to be imposed along with the R-language. For instance, a theoretical or methodological approach which is dominant in research contexts where English is the dominant (or only) R-language, tend to impose themselves with that language even where those perspectives are not otherwise dominant or even widely known.

To complement this picture, we now discuss, through close-up case studies, how these challenges appear in some of our personal work and experiences as researchers. We have found it most natural to use the first person in presenting these cases. We focus in particular on the more problematic questions Q4 and Q5.

14.5 Personal Case 1 (Faïza)

I am a Tunisian mathematics educator, with a Ph.D. in didactics of mathematics (joint degree between University of Tunis and University of Lyon). The roles of Arabic and French language in Tunisian mathematics teaching have changed over time but have remained relatively stable since 1988. The common daily language of Tunisians is a "dialectical" Arabic, while in school, only classical Arabic is taught. French is taught from third year in elementary school and on to the end of high school. For the first 6 years of elementary school, mathematics is also taught in Arabic. In the 3 years of lower secondary school, most oral parts of teaching continue to be in Arabic while the symbolic parts are written and read out in French (in particular, these parts are written from left to right, contrary to Arabic). This creates a certain divide in the mind of students. And from high school onwards, mathematics is only taught in French.

My research bears mainly on logical analysis of mathematical statements as they appear at different levels of teaching, and on the interaction between natural and symbolic (or more generally, "mathematical") language at the different stages mentioned above. In particular, for my thesis, I carried out an experiment with six pairs of students in their first year of university studies on mathematics and informatics (Chellougui, 2009). I want to discuss this case further as an illustration of the problems linked to the P-language (French) at this level, and then the challenges of treating the "implicit" P-language (Arabic) when the R-language is also French.

My experiment had two main phases: solving exercises (topic: basic features of lower upper bounds of subsets of R) and an interview about the solutions. For each pair, the students worked together to solve the exercises and to respond to the interview questions. The language used for both the solving phase and interviews was French. Among the specific and general phenomena identified, and what is most interesting in the context of this chapter, was the instability of the choice of vocabulary and symbols in the work of the students. As an example, consider the following dialogue from an interview with two students, J and T:

Interviewer: OK, we'll try to understand the proof together. A small remark, if you want to show that a statement is true and you say you want to use indirect proof, what do you mean?

T: It means that it is the inverse.

J: No, on the contrary.

(Silence)

Interviewer: OK, if you want, we go on to the next question [*one of the exercises, to show that* $B=\{x\in\mathbb{Q}_+ : x^2>2\}$ *is the set of upper bounds in* $\mathbb{Q}$ *of* $A=\{x\in\mathbb{Q}_+ : x^2>2\}$.

J: *A* and *B* is the negation. *A* is the negation of *B*.

T: *A* is the contrary of *B*.

J: They are different.

Int: Can you determine the intersection of the two sets *A* and *B*?

T: The empty set, *A* is the opposite of *B* or *B* is the opposite of *A*.

In this short excerpt, the vocabulary of oppositions (see Durand-Guerrier & Ben Kilani, 2004) appeared in disturbing ways; there was an inappropriate use of expressions like negation, contrary, different, and opposite. The terms "contrary" and "negation" apply normally to statements, but the students also used them when dealing with sets. Evidently, this and other mathematical distinctions in the P-language presented particular difficulties when it was a second language. In other points in the interviews it was noted that some students had a tendency to privilege (read aloud) symbolic language in their interchanges. This was possibly a way to circumvent language difficulties and gain precision. On the other hand, for developing a more familiar relationship to logical structures, it could easily be problematic for the students to be relying almost entirely on symbolic logic and algebra as a tool of conceptual clarification and support of reasoning. Without the support of informal "familiar" expressions, students in fact seem to have failed to deepen their own understanding of the mathematical processes in which they were involved.

The more general situation for Tunisia is not likely to change. There are no clear global solutions in sight. The constraints of our educational language policies are not likely or easy to change. As well the mathematical culture and literature in Tunisian universities has for many years been entirely based on French. For the didactician, therefore, the situation must be studied and possibly amended more locally.

At the level of the R-language, to address difficulties of the type mentioned, we need technology and theoretical tools that take into account the implicit interaction between the "official P-language" and the Arabic language in which the students live outside of mathematics classroom (their implicit P-language in this context). While the use of formal language (logical expressions, symbolisms, etc.) is a challenge also for students whose mother tongue is French, the research theories and technologies developed in French do not envisage the presence of an "implicit" mother tongue. Hence the spontaneous and supposed easy transition to the more formal language for French students in France cannot be interpreted in the same way for Tunisian students. In fact the asymmetry between French and Arabic in Tunisian mathematics education seems to reinforce the gap, noted in other contexts, between students' informal and informal reasoning in mathematics.

14.6 Personal Case 2 (Hien)

I am a Vietnamese Ph.D. student, studying in Belgium (French-speaking community). Even though the French language is not the dominant language used for scientific research and everyday life in Vietnam, the "French mathematical didactic" way of thinking is well diffused throughout Vietnamese society and plays an important role in the mathematical education field. The consequence is that references in the mathematical education field are mainly in the French language and from a French-speaking community. As I began my studies in Vietnam and then graduated for a Master degree in France, I used the French-speaking references more often for my research work. I am now studying in a Belgian French-speaking University where English dominates as an R-language. Consequently, English references have become dominant in my work environment. These new references are not just studies presented in another language, but they brought me new theories and more importantly, a new way of thinking, a new way of approaching other studies, quite different from the French ones.

My present work examines the influence of language and other factors such as the mathematical program or the learning context on the mathematical performances and competences of children. It is a comparison study between the Vietnamese and French languages. The research idea comes from the particular characteristics of Vietnamese language as a P-language for verbal number denomination. The verbal numbers in Vietnamese have a perfect correspondence with their decimal form, like in Chinese, Korean, or Japanese (e.g., 13 is said "ten-three"). The Vietnamese language also has peculiarities when the digit in the tens or hundreds position is a zero. This is not found in other Asian languages cited above. For example, the Arabic number 3 024 is named "three thousand zero hundred two ten four." There is also an exception when the zero is in the tens place where it is replaced by a word meaning "remainder." For example, the Arabic number 309 is named "three hundred remainder nine" in Vietnamese. Here, we can understand the word "remainder" as the remainder of the division of 309 by 100.

Our[2] research undertakes a comparison of Vietnamese and French P-languages. For my contribution to the ICMI Study 21 conference, we had to present in English, which in effect became the R-language. Even though French and English languages are in many respects similar, there are crucial differences such as in the verbal number structure. Producing a study in a third language cannot be reduced to a simple text translation. Obviously, we had to pay attention to explain the meaning of key terms so they could represent the exact meaning of the P-language; for example, terms such as "remainder" and "zero hundred" in Vietnamese (the first P-language) had to be clearly referenced in English. We also had to explain the original context in which those key terms were used. The study reported on also focused on the cultural impact of the use of large numbers. Therefore, precision was needed to describe the special context in which the P-language was used, in order for the reader to fully understand problems and situations related to language and culture that appeared in this context.

In our contribution to the ICMI Study 21 conference, even if the P-languages of the research task was Vietnamese and French, the R-theory is mainly based on literature using English as the R-language. I struggled to "transcode" between the P-languages and the R-language to produce a contribution. In the case where the P-language and the R-language are different, this "transcoding" process contains more challenges because it also must to keep the characteristics specifically related to P-languages, in order to comprehensively report the results of research.

Previous studies undertaken in Europe (Censabelle, 2000), have showed that the syntactic zero (see Granà, Lochy, Girellid, Seron, & Semenza, 2003) is a source of difficulty for children when they transcode a verbal number into Arabic code. This difficulty can be explained with reference to the masked character of the syntactic zero in its verbal form. In our study, we showed that the syntactic zero is, in contrast to what was found in European P-languages, an advantage for Vietnamese children during the same transcoding task. This contrast in the ways students come to understand just this small aspect of mathematics is hard to explain in English. One could consider taking the option of using Vietnamese as an R-language. But if research is conducted with a R-language that is not dominant, it will not be diffused and discussed on a wide scale. So both options limit the opportunities in different ways to examine and understand better some questions in a more general way.

14.7 Personal Case 3 (Carl)

My mother tongue is Danish and in most of my research, the P-language is Danish while the R-language is English or French. For several years, I have been somewhat split between the last two. Most of my colleagues and students do not read French and so I put a lot of effort into writing papers in English and finding good references

[2] "Our" is used here and in the following paragraphs to reference the research group I work in, not the authors of this chapter.

in English, even if the papers and many of these references are strongly linked to a research literature primarily written in French, in particular within the paradigms founded by scholars such as Artigue, Brousseau, and Chevallard. This is also evident in my contribution to the ICMI Study 21 conference. My experience during the conference in fact contributed to confirming a main point in my conference paper: that the use of an R-language is much more than the use of a medium, as it entails also a number of cultural and scientific biases, or at least implications. This impression was shared by a number of participants who, like me, regularly or mostly use other R-languages than English. Among the main biases are theoretical perspectives, and of course these will be found also in an international context where the shared R-language is, for instance French, German, or Spanish. In an international study on multilingual classrooms it is certainly natural to reflect on how these biases affect our work and what we can do to address them explicitly.

I now turn to a concrete example of how a delicate mixture of R-languages and P-languages can be handled explicitly (to take care of the questions Q4 and Q5 in an explicit, albeit not symmetric way). The case bears on certain practices related to Japanese mathematics teaching and hence the P-language is Japanese. I am lucky to understand this language reasonably well as I did my Ph.D. in Tokyo. But of course, this adds to the language complexity, even if I have never been able to familiarize myself thoroughly with—and much less draw on—the extensive research paradigms in mathematics education which are based on Japanese as a R-language.

In 2001, I met a Japanese colleague, T. Miyakawa, at the summer school of the French Association for Research in Didactics of Mathematics. Miyakawa was at the time writing his doctoral dissertation in Grenoble, and our common R-language is and remains French (even if we can communicate in English and Japanese, for other purposes). However, working with Japanese colleagues clearly entailed some influence and inspiration also from these sources, as will be illustrated by my case.

At the summer school in 2007, we were invited to present a Japanese format for teacher collaborative lesson planning, known in English-speaking countries as "Lesson Study." This led us to examine the parallels and differences of this format, and the "open-approach theory" by Nohda (1991), with the French research tradition in didactics, strongly linked to the theory of situations (Brousseau, 1997) and more specifically, a case of lesson study on proportions and the famous "puzzle situation" by Brousseau (1997, Chap. 4). Our exposition (Miyakawa & Winsløw, 2009) of this comparison in English was based on material in French and Japanese, and of course we had to explain very carefully how we had handled the corresponding methodic and methodological challenges related to the two P-languages. Indeed, we had to examine explicitly and critically the meanings of basic terms such as "research" and "question," as it could be rendered slightly differently, but potentially importantly, in the three R-languages involved, as well as the central terms from the P-languages. It appeared to be a considerable advantage to share our comparison of the two settings of "lesson research" in a third R-language, since no term, theoretical assumption, or indeed, cultural and societal specific, could be taken for granted (in the sense of being naturalized among readers). While my intention is not to claim that we achieved any kind of perfect neutral and transparent perspective, I do think the effort of taking the two contexts, and in particular the two R-languages, into a third one, was both rare and, as it transpired, rewarding for us as researchers as well as for our audience.

14.8 Conclusions

The main point of the ATD model of research praxeologies is to insist on the mutual, or co-constructive, relation between *practice blocks* (tasks, techniques) and *theory blocks* (technology, theory). Research tasks, or more generally problématiques, do not exist independently from technology and theory. Both are imbued with language: our research practice deals with contexts with particular P-languages, and our learned technologies and theories are formed in and by R-languages. Researchers focusing on the effects of language diversity in the mathematical praxeology of teachers and students should also be sensitive to the effects of their own choices and constraints when it comes to R-language. The five questions proposed in our framework can be used both to evaluate how a given study displays this kind of reflective sensitivity, and to shape future research in the area.

We have outlined some cases to demonstrate the multiple ways in which these questions may appear and be tackled, or in fact be ignored when we take our own technological and theoretical equipment to be somehow a natural standard. It is inevitable that we will always ignore *some* potential approach to research questions, simply because we do not know them. But in the context of studies of multilingual or cross-cultural contexts, and indeed in studies that claim to be "international," it is certainly not defensible to ignore the role of R-languages.

In fact, the absence of explicit reflection on the impact of an R-language seems to be particularly common, and unreasonable, in the case where the R-language coincides with a "dominant" P-language, as in the vast majority of studies surveyed in Table 14.1, and in Personal Case 1 of this chapter. The notion of asymmetry, introduced in that case, can be made more precise by pointing out the usual meaning of an R-language and more precisely an R-theory. If the R-task is to study some specific challenges raised by the multilingual contexts (even if only one dominant language is "allowed" or "practicable") then the R-theory can contribute to naturalize or even enforce the ignorance with respect to crucial challenges and opportunities as regards the use of P-language.

It is also evident that the study of phenomena linked to P-languages different from the R-language may cause problems in terms of communicating and transitioning between these languages, precisely because of these phenomena (see Personal Cases 2 and 3). What is perhaps less evident is that such difficulties may cause "international" studies to become somewhat regional, as illustrated by our motivating case and in fact to some extent the present study (see Table 14.1). This is because no language is a P-language to more than a small fraction of the world's population.

A researcher must master the P-language of the students and teachers in the context she investigates. A more subtle challenge, which is often hidden by formal restrictions to the use of just one R-language, is that no language is a functional R-language to more than a fraction of the world's mathematics educators. Any monolingual research community will therefore be faced with severe constraints in an attempt to address the role of P-languages in mathematics education at any level other than in her own language. This is especially so at an international level. Such limitations should be explicitly acknowledged in our research endeavors.

The role played by natural languages in this unjustified naturalization of scientific and cultural (including mathematical) perspectives is only the tip of an iceberg which is loaded with societal and political implications; but it is an important "tip" because it can be exhibited and analyzed in fairly obvious and objective forms when we are explicit about the *problématique* introduced and exemplified in this chapter.

References

Adler, J., Ball, D., Krainer, K., Lin, F.-L., & Novotna, J. (2008). Mirror images of an emerging field: Researching mathematics teacher education. In M. Niss (Ed.), *Proceedings of the 10th International Congress on Mathematical Education* (pp. 123–139). Roskilde, Denmark: IMFUFA.

Artigue, M., & Bosch, M. (2014). Reflection on networking through the praxeological lens. In A. Bikner-Ahsbahs & S. Prediger (Eds.), *Networking of theories as a research practice in mathematics education* (pp. 249–265). Cham, Switzerland: Springer.

Bosch, M., & Gascón, J. (2006). Twenty five years of the didactic transposition. *ICMI Bulletin, 58*, 51–65.

Brousseau, G. (1997). *Theory of didactical situations in mathematics*. Dordrecht, The Netherlands: Kluwer.

Censabelle, S. (2000). *L'écriture des nombres sous dictée chez l'enfant. Étude qualitative et implication de la mémoire de travail.* Unpublished licence dissertation, University of Liège, Belgium.

Chellougui, F. (2009). L'utilisation des quantificateurs universel et existentiel, entre l'explicite et l'implicite. *Recherches en Didactique des Mathématiques, 29*(2), 123–154.

Chevallard, Y. (1999a). *Didactique*? Is it a *plaisanterie*? You must be joking! A critical comment on terminology. *Instructional Science, 27*, 5–7.

Chevallard, Y. (1999b). L'analyse des pratiques enseignantes en théorie anthropologique du didactique. *Recherches en Didactique des Mathématiques, 19*(2), 221–266.

Durand-Guerrier, V., & Ben Kilani, I. (2004). Négation grammaticale versus négation logique dans l'apprentissage des mathématiques. Exemple dans l'enseignement secondaire Tunisien. *Les Cahiers du Français Contemporain, 9*, 29–55.

Graddol, D. (1997). *The future of English? A guide to forecasting the popularity of the English language in the 21st century*. London: The British Council.

Granà, A., Lochy, A., Girellid, L., Seron, X., & Semenza, C. (2003). Transcoding zeros within complex numerals. *Neuropsychologia, 41*, 1611–1718.

Miyakawa, T., & Winsløw, C. (2009). Didactical designs for students' proportional reasoning: An "open approach" lesson and a "fundamental situation". *Educational Studies in Mathematics, 72*(2), 199–218.

Nohda, N. (1991). Paradigm of the "open-approach" method in mathematics teaching: Focus on mathematical problem solving. *Zentralblatt für Didaktik der Mathematik, 23*(2), 32–37.

Setati, M., Nkambule, T., & Goosen, L. (Eds.) (2011). In *Proceedings of the ICMI Study 21 Conference: Mathematics Education and Language Diversity,* São Paulo, Brazil.

Wedege, T. (2006). Hvorfor stave problematik med Q? Hvad, hvordan og hvorfor i matematikkens didaktik. In O. Skovsmose & M. Blomhøj (Eds.), *Kunne det tænkes?—om matematiklæring* (pp. 312–331). København, Denmark: Forlag Malling Beck.

Chapter 15
Research Rationalities and the Construction of the Deficient Multilingual Mathematics Learner

Anjum Halai, Irfan Muzaffar, and Paola Valero

15.1 Introduction

Improving access to education and quality of achievement, especially in mathematics, has been a focus of educational research as well as of reform initiatives globally over the last two decades (e.g., EFA[1]). While advances have been made on access to education, the status of achievement, especially in mathematics, remains a major concern the world over. This is evident from the wide range of cross-national comparative initiatives and studies, such as SAQMEC,[2] TIMMS,[3] and PISA, that are undertaken to look at student achievement in mathematics. These studies regularly produce

[1] Education for All (EFA) is an international commitment first launched in Jomtien, Thailand in 1990 to bring the benefits of education to "every citizen in every society." EFA has six major goals (for details see http://www.unesco.org/education/efa/ed_for_all/).

[2] The Southern and Eastern Africa Consortium for Monitoring Educational Quality (SACMEQ) is an international nonprofit developmental organization of 15 Ministries of Education in Southern and Eastern Africa. A key purpose is to apply scientific methods to monitor and evaluate the conditions of schooling and the quality of education, with technical assistance from UNESCO International Institute for Educational Planning (http://www.sacmeq.org/about.htm).

[3] The Trends in International Mathematics and Science Study (TIMSS) is an international assessment of the mathematics and science knowledge of fourth-grade and eighth-grade students around the world. TIMSS was developed by the International Association for the Evaluation of Educational

A. Halai (✉)
Institute for Educational Development, East Africa, Aga Khan University,
Dar es Salaam, Tanzania
e-mail: anjum.halai@aku.edu

I. Muzaffar
Education and Social Research Collective, Pinner, UK
e-mail: imuzaffar@gmail.com

P. Valero
Aalborg University, Aalborg, Denmark
e-mail: paola@learning.aau.dk

R. Barwell et al. (eds.), *Mathematics Education and Language Diversity*,
New ICMI Study Series, https://doi.org/10.1007/978-3-319-14511-2_15

evidence that a proportion of the children studied are underachieving. Who are the underachieving students and what can be done to remediate their situation have been topics of research in general education as well as in mathematics education.

Nevertheless, the increasing importance given to large-scale international, comparative studies, due to their growth and expansion since the 1960s, has contributed to a conflation between the measurement of students' achievement and children's experience of mathematics education. The effects of this quantification have made it possible to think that there is in fact a correspondence between achievement and the mathematical experience of children, in particular for those who underachieve. A large proportion of the underachieving children are precisely those who belong to nondominant social groups in their respective countries (e.g., OECD, 2006). For example, "those whose parents have a low level of education, who have low socio-economic status, or come from an immigrant background, as well as boys, have a higher risk of low performance at age 15. Some 19 % of people at this age in OECD countries lack basic literacy skills, making it more likely that they will drop out of school with no qualifications" (OECD, 2012, p. 72). The quantification has also made it possible to see that it is the same students who fail year after year. Such observations are also connected to the construction of an image of cognitive, linguistic, and cultural deficit in those who fail, relative to those who succeed.

These studies do not necessarily provide a framework to understand and further investigate why and how current practices of teaching and learning in classrooms, especially when these processes are taking place in multilingual and multicultural classrooms, produce the systematic failure of many students belonging to nondominant cultural groups. This is a significant issue because of three main reasons. First, increasingly during the twentieth and twenty-first centuries there is a growing idea that success in school mathematics is a prerequisite for personal and national success. Second, in the globalized world today more children are multilingual than monolingual, and many classrooms bring together a diversity of children. When educational practices effect systematic exclusion and failure for some, the result is that a large proportion of children are in fact being excluded from participation in education. Third, language use in the classroom is not simply a matter of communication and cognition; language use as a political, social, and cultural tool is also important in understanding the teaching and learning processes.

For these three reasons and many others, mathematics education researchers have studied mathematics education practices in multilingual and multicultural classrooms. They have focused on the teaching and learning of mathematics of those students whose languages and culture (and also ethnicity) diverge from the language and cultural norms of dominant cultural groups in society. It is the hope of multilingual research in mathematics education to understand this situation in-depth, in order to devise better and more suitable educational practices that can help the children thus implicated in succeeding in school mathematics and in life. Research rationalities, with their theories and methodologies become an important

Achievement to allow participating nations to compare students' educational achievement across borders (http://timss.bc.edu/)

element in studying how researchers and their research can live up to the challenge of making changes possible.

In this chapter, we focus on the research rationalities made visible in published research papers studying mathematics teaching and learning in multilingual classrooms. We interrogate these rationalities from a point of view that highlights the implication of research practices in the construction of the very same problems it intends to study and address. We depart from the assumption that educational research, and within it, mathematics education research, is not a neutral player in this game. Increasingly during the twentieth century, educational research is regarded as the expert knowledge that is supposed to help people planning and building a better world. As research produces knowledge that is used in dealing with perceived problems of reality, it has a double effect of power. On the one hand, its concepts and ways of proceeding reify categories of exclusion by documenting their same existence. That is, research itself is implicated in the creation of the differentiation of multilingual/multicultural children. It constructs the idea of the "deficient multilingual child" inasmuch as it attempts to change the teaching and learning practices implicated in rendering it in the first place. The good intentions of researchers result in the construction of the very same category they want to dissolve. On the other hand, research provides the knowledge-based technologies for engineering how multilingual/multicultural children need to be helped, in order to diminish the gap between them and the dominant cultural groups. That is, the research sets up mechanisms of subjectification for children to become assimilated into the values, worldviews, and forms of being and knowing of those whose culture is the norm. The intention of helping diversity also has the effect of undermining its very existence.

This double effect of power in research allows us to evidence that the theoretical and methodological choices of research, educational or otherwise, are not value neutral. Rather they are built on assumptions about what is valued as knowledge and, therefore, what are valorized as approaches and frameworks for the generation of knowledge. No political examination of research and research methodology can leave this point unattended, particularly when the teaching and learning of mathematics for children of nondominant linguistic and cultural groups is highly political.

15.2 Examining Research Rationalities

Many researchers dealing with the study of mathematics education in multilingual settings declare that their research is political since it attends to the connection between language and power in the society. Gutstein (2007) summarizes this stance in these words:

> Thus, in a sense, politicizing the discussion around language usage is not a choice mathematics educators have the luxury of making—circumstances dictate that for us. Language is political in many ways [...] Language is about power, about who has the authority to designate the language of instruction and the "official" languages. It is also about students' identity and being, and to denigrate one's language is to disparage her culture, personhood, community, ancestors, and ways of making sense of the world. (pp. 244–245)

Even if not explicitly declared, many researchers adhere to the view expressed above. If the research on mathematics education in multilingual settings is political, the issue of the ways in which research addresses power becomes important to study. There could be different ways in which mathematics education research deals with the political. One way is by the implicit or explicit adoption of different views on power and how research argues that power enters mathematics education teaching and learning practices (Valero, 2008). Another possibility is to carry out "research on research": that is, analyzing how research itself is implicated in effecting power and constituting the political (Pais & Valero, 2012). Following this critical trend allows us to understand the ways in which scientific knowledge is more than the socially privileged form of getting reliable information about the state of affairs in the world, such as the percentage of bi-/multicultural and bi-/multilingual children who underperform in mathematics in the world. The knowledge of research is central in, on the one hand, maintaining a narrative of rational and planned progress in society on the grounds of objective and solid knowledge for devising better, evidence-based solutions and, on the other hand, constituting "social epistemologies" (Popkewitz & Brennan, 1997). Following Popkewitz and Brennan (1997), epistemology provides a context in which to consider the rules and standards by which knowledge about the world and self is formed:

> social epistemology locates the objects constituted by the knowledge of schooling as historical practices through which power relations can be understood. Statements and words are not signs or signifiers that refer to and fix things, but social practices that generate action and participation. (p. 293)

Mathematics education research, in general, is a field of study inserted in a particular logic that is defined by a growing tendency to limit the object of study to phenomena of learning and teaching. Biesta (2005) calls this tendency the "learnification" of educational research. One of the consequences of this tendency for mathematics education is the disavowal of the Political as a constitutive element in mathematics education practices and research (Pais & Valero, 2011). Researching research in the field of mathematics education in multilingual settings is an attempt to take a critical look at the research rationalities that contribute to the generation of ways of thinking about the people and practices of mathematics education in those settings, their problems, and the ways of addressing them. In the case of research in mathematics education in multilingual settings, critical research into existing research would invite us to examine the ways in which such research is not only embedded in "patterns of power relations" (Popkewitz & Brennan, 1997), but also and more importantly how it is implicated in generating the categories, distinctions, and forms of thinking about the practices of schooling and the people involved in multilingual mathematics classrooms. Critical research also constitutes an invitation to denaturalize the idea that the purpose of research is to directly propose better ways of dealing with teaching and learning practices and people in multilingual contexts. In other words, this type of critical research allows us to evidence how research contributes to the creation of ways of reasoning about the "problems" and the "solutions" to the very same problems it identifies and constructs.

In deploying our analytical strategy, we bring forward the term *research rationalities*. This is both a theoretical and methodological tool that allows us to link the intention of evidencing social epistemologies and the concrete examination of research texts in the area that we are studying. Research as a practice of knowledge is organized around the identification of problems that require knowledge to be addressed, as well as conceptualizations of the problems. The ways of proceeding of the researcher both in terms of the logic of the activity (the methodology) and the concrete forms of examination of evidence (the information collection methods) guide the process of constituting new knowledge. The research rationalities are the resulting forms of constructing problems, thinking about them, and devising solutions. As methodological tools, research rationalities allow us to examine the enunciations that repeatedly appear in papers and that constitute statements of truth about the problems that research addresses, the conceptualizations about the practices involved, and the solutions proposed.

More concretely, in examining research rationalities, we will deploy a Foucault-inspired discourse analysis (Arribas-Ayllon & Walkerdine, 2008) where we trace the enunciations that, repetitively in series of documents and texts, constitute forms of thinking about the objects that the researcher is focusing on. Such type of analysis, more than a procedure, is an analytical strategy that brings together the conceptual and theoretical tools along with a reading of the corpus under examination. We have looked mainly at mathematics education research on teaching and learning emanating over the last two decades, with a focus on multilingualism in mathematics education and related issues as noted above. We examined key journals published in English in the field of mathematics education and mathematics teacher education, and books and chapters in international collections. We have traced research rationality by identifying in the texts: (a) how the learner is portrayed; (b) how mathematics education (teaching and learning) is portrayed; and (c) the notion of language as formulated in the texts.

In the following, we address the elements outlined above in our analysis of the research rationalities in multilingual research in mathematics education. We distinguish between two types of studies: large-scale quantitative studies and small-scale qualitative studies. Even though we see these types to be connected in a broader discourse, they deploy different strategies to approach the study of multilingual settings, and as a consequence present slightly different, though not unrelated, rationalities.

15.3 The Logic of Achievement Gaps Research

Large quantitative studies have produced systematic evidence for certain truths about education: there exists differential achievement of students in school subjects. Since the mid-1960s, the time of the launch of the Coleman Report in the United States, discussion of the quality of education has focused on the relationship between the inputs in students' characteristics and students' outputs as measured through achievement in the dominant school subjects: that is, an education production function or an input–output model has been the dominant way of thinking about what makes a difference for school achievement. Quality of education and the differential gaining of different types of students had, from the 1920s until that time, primarily been measured in terms of the quality of resources put into schools (Coleman, 1972). One of these resources is "what students bring with them". The variable of students' socioeconomic status entered the scene of social and educational research in the early 1920s and since then has become a central parameter in these types of measurements (Sims, 1930). Coleman and his collaborators shifted the concept of equality of opportunity in education to rely on the connection between school inputs and outputs, by adding the connection between inputs to the outputs measured in terms of performance on standardized tests (Gamoran & Long, 2006). As part of a political trend at that time of starting to base policy on social research results, the Coleman Report opened a whole new way of understanding the effects of education in creating differentiation. The application of new statistical techniques and of social surveys, connected with the growth of both national and international large-scale banks of statistical measurements of educational effectiveness laid the ground for which facts such as "the achievement gap" emerged during the twentieth century. Realizing how social science and educational research (also mathematics education research) have constructed concepts and facts (Hacking, 1999) is important in understanding how certain research rationalities about the educational outputs of (multilingual) children in (multilingual) settings operate. Studies addressing how students, who operate with more than one language in situations where mathematics teaching and learning is carried out in one or more languages, are embedded in the logic and rationality of these types of studies.

Discussions of achievement gaps in mathematics in other countries have gained importance in a social context that not only increasingly emphasizes the connection between individual achievement in mathematics—but also science and mother tongue—with personal economic prosperity, and national economic welfare and competitiveness in global economies. Studies emphasizing the importance of examining the gaps (Lubienski, 2003), as well as those criticizing the effects of gap-gazing (Gutiérrez & Dixon-Román, 2011; Parker, 2000; Parks, 2009) have emerged in mathematics education. While probably in many national contexts, the issue of groups of students not performing as well as expected were taken as a matter of "the natural distribution of mathematical ability", the growing discourse of mathematics for all as part of political and economic agendas of progress and prosperity have posed clearly the achievement gap as a challenge for both equity and also productivity (Valero, 2013). Our point in mentioning these connections is that the growing association between mathematical achievement and economic and social prosperity is also part of the discursive network within which the quantitative research rationality for the study of mathematical achievement of multilingual students operates.

The discussion of the achievement gap in mathematics has been known to mathematics education research mainly through the publication of research in English-speaking countries, particularly the United States. Such research explains the statistical fact that white students as a group, independently of in-group variation, seem to systematically outperform students from other racial groups, particularly African Americans, Latino(a)s and Native Americans. In the context of the United States, the category "racial and ethnic groups" intersects with the category of "learners of English" or "second language learners". The latter may apply clearly to Asians and Latino(a)s, while it may not be clearly the case for African Americans or Native Americans. At the same time, the category of "language learner" quite often intersects in important ways with the variable of "poverty" (Brown, Cady, & Lubinski, 2011). When reviewing existing research that addresses the learning of mathematics of "English language learners", a majority of papers were found to concentrate on the experiences of Latino(a)s. Such prominence could be related to the fact that, despite being the largest and fastest growing immigrant group, it is also the group of immigrant students or language learners who show the lowest school performance, but also have become the least educated immigrant population with only 11 % of adults completing a bachelor degree (Schneider, Martinez, & Owens, 2006).

The case of the United States is interesting in illustrating this type of research. Parks (2009) presents a critical reading of how research has contributed to producing an achievement gap and how such a production has a power/knowledge effect on the forms of thinking and writing about the students who do not perform well. She concludes that:

> For minority students, the act of taking a standardized test may work to invoke stereotypes about performance even without an explicit reminder because phenomena like the achievement gap are so widely accepted as real. Thus, the achievement gap works in two ways [...] First, its acceptance as a real phenomenon impacts student performance on tests, and second, the tests then go on to produce evidence, in the form of test scores, that the phenomenon is, in fact, real. That's power/knowledge. (p. 16)

In our analysis we chose to concentrate on one document that builds on the dominant framing of many large-scale quantitative studies on achievement gaps: the OECD's report on the failure of immigrants on the grounds of the PISA 2003 dataset, "Where immigrant students succeed: A comparative review of performance and engagement in PISA 2003" (OECD, 2006). Discursively, PISA epitomizes many of the statements that nowadays constitute a kind of "regime" about what counts as mathematics education (Kanes, Morgan, & Tsatsaroni, 2014). It claims to be a wide international comparison including a diverse set of countries, and it has been quite influential for policy. Besides, it claims to assess skills for real-world application that are acquired by 15-year-olds, an age which marks the end of compulsory education in most countries (Neidorf, Binkley, Gattis, & Nohara, 2006). In accordance with our analytical strategy, our analysis of this report will illustrate three aspects of the research rationality: how research portrays students, mathematics education, and language. Whenever necessary, we complement our analysis with other documents to highlight recurrence in enunciations and statements concerning these three points.

A very first important element in these studies is the representation of a person into an identifiable record to which an aggregate of variables are attached. Each single individual student participating in the PISA test can be identified and singled out if one wants to. There are background information variables on the record, which are associated with another aggregated measurement, which is the overall achievement in the test and its subparts. Once dependent and independent variables are defined, individual students with histories and bodies disappear, metamorphosing into measurements. A dominant measurement that is highlighted in the discourse is *achievement*, in our case achievement in mathematics. But the achievement is not only taken to be the particular performance of one person in a particular test; the measurements of achievement are taken to be "students' learning" and "students' ability to apply their knowledge and experience to real-life situations" (OECD, 2006, p. 25).

Achievement—in other words, learning—is the dependent variable connected to other associated measurements of each individual student. Place of birth generates the categories of *immigrant*, who can be *first generation* student, or *second generation* student, and *native*. Connected to this there is the distinction between *home language* which may differ from the national language and *language of assessment*, and other official languages. Here, the focus is on the proportion of immigrants who speak a different home language from the language of assessment: few natives speak a different home language than the language of assessment.

Other variables called background information are students' *parents' educational level* and their *social, economic, and cultural status*. School characteristics are also possible explanatory independent variables. Furthermore, students' dispositions to education measured in their *motivation*, *beliefs about themselves*, and *perception of school* are also considered. Different analyses are used to produce results about how these variables relate and which seem to have explanatory force concerning achievement.

The results are clear: "immigrant students are motivated learners and have positive attitudes towards school. Despite these strong learning dispositions immigrant students often perform at significantly lower levels than their native peers in key school subjects, such as mathematics" (OECD, 2006, p. 3). Those who perform at the lowest point are first generation immigrants who do not speak the language of assessment at home (pp. 45–46) they tend to "perform at substantially lower levels than their low-performing native peers" (p. 41). In general, boys perform better than girls in mathematics for all the types of students (p. 49). The general performance difference among types of students is such that "in the majority of countries at least one in four immigrant students do not demonstrate basic mathematics skills as defined in the PISA 2003 assessment" (p. 3).

There is a "relationship between the relative mathematics performance of immigrant students and their relative educational and socio-economic background" (p. 9). The differences in parents' socioeconomic background can largely account for the lower performance in relation to the native peers (p. 58). In general, parents of immigrant students tend to have a lower socioeconomic situation or level of education than parents of native students (p. 58). However, parents' socioeconomic background and level of education are not the only contributing factors to explain the differences. School factors such as the tracking system or the number of immigrant students in school also have an influence. There are more differences in students' performance in relation to between-school variance than within-school variance (pp. 78–80).

The logic of this type of research is a comparative logic that is built on measurements of difference in relation to a neutrally determined norm. In particular, achievement results establish a comparison between each individual and a specified standard that is stratified in terms of proficiency levels. Low achievement on a particular standard also indicates low proficiency in relation to that standard. The measurement in itself builds on the hierarchical organization of attributes. A low measurement inevitably positions an individual in a deficit zone. This type of deficit positioning is at the very heart of the types of comparative reasoning of this kind of research. Since the measurements are taken at one point in time, they represent static pictures of positions of students, their learning and the causes for their low performance. Nevertheless, the discursive effect is a reification of a particular picture into population generalizations that are attributed to groups of population. Even though the report insists on the differences between the cases analyzed in the study, the general formulation of results still overshadows the small, and probably significant, differences.

The portrait of language (and multilingualism) is also limited to two dimensions: the correspondence between home and assessment language; and the level of proficiency in language, again measured by the average score on language proficiency tests. The category "immigrant" conflates with the category "language learner". The diversity of languages that may make part of an actual teaching and learning situation in mathematics is only slightly acknowledged; however, the analysis requires a clear-cut category that positions immigrants against the dominant cultural norm represented by the national language and the language of assessment. The result of this type of logic is a differentiation between home languages, being ranked in a secondary, less powerful position, against the desired cultural norm for communication, thinking, and learning: the national language and the corresponding language of instruction and assessment.

Drawing from the above, it becomes clearer that methodological choices in large-scale, quantitative research, educational or otherwise, are not value neutral. They are built on assumptions about what is valued as knowledge and cultural norms, and therefore produce valorized approaches and frameworks for ranking students, learning and language.

While the dominant methodological frameworks such as those undertaken in large-scale comparisons of mathematics achievement are governed by concerns of equity and excellence, they have created their own methodological imperatives and reified particular "truths" about "language learners" and their school results. Excellence, for instance, has translated into the need to have measurable standards to ascertain mathematical proficiency, to ensure that no student is left behind in the economic competition for social and economic welfare. Achievement is measured in relation to standards, and the proficiency levels of different ethnic, linguistic, or socioeconomic groups, or of different countries internationally, are seen in terms of achievement gaps. Individual students, schools, language minority groups, other disadvantaged groups, states, and countries are seen as ahead or falling behind in terms of scores on testing instruments that are developed and analyzed from a perspective of "reducing the gaps" (Schütte & Kaiser, 2011). It is important to note that the "gap-gazing effects", as Gutiérrez and Dixon-Román (2011) call it, embed in themselves important discursive effects that go beyond the limits of the good intentions of evidencing the problems of low achievers and tend to confine the students, their families, their languages, and their learning to a locked, unmovable position. Or as Parks (2009) phrases it, the whole discourse of achievement gaps and the research rationality it builds keeps the attention away from the effects that "definitions of mathematics, inequity, cultural or language hegemony" (p. 19) have for those on the wrong side of the gap.

Unpacking the research rationality of these studies is important because it brings us back to the complex, unpredictable and inequitable nature of social realities that confront educational research; all of them epitomized in the intersection of multilingualism, social class, gender, ethnicity, etc. The discourse of this type of research obliterates the messiness and uncertainties through the reduction of students and learning to well-defined numerical variables for statistical manipulation. From a wider social and political perspective it becomes evident that "the problem of failure in mathematics cannot be resolved within the boundaries of mathematics education alone [...] it demonstrates the inconsistency of a system that on the one hand demands mathematics for all but on the other hand uses it as privileged mechanism for selection and credit" (Valero & Pais, 2012, p. 173).

15.4 The Logic of Classroom Research

Besides the large-scale studies noted above, the last three decades or so have seen the emergence of qualitative, often small-scale, studies in mathematics education situated in the context of schools and classrooms, with long-term engagement in the field as compared to the more traditional notion of data collection as a one-time event. Many of these studies have looked at issues of mathematics teaching and learning and teacher education within multilingual settings.

For example, the role of code-switching, translation, translanguaging, or the use of other linguistic devices in facilitating or hindering learning mathematics has been well documented in mathematics classrooms in post-colonial countries where the medium of instruction is often the language of the colonizers, or in classrooms with immigrant learners where the medium of instruction is the language of the host country (Barwell, 2009; Gorgorió & Planas, 2001; Halai, 2009; Moschkovich, 2007; Norén, 2008, 2011; Setati, 1998; Setati & Adler, 2000; Then & Ting, 2009). It is established in these studies that, in many countries, multilingualism in mathematics is a norm rather than an exception. In most cases use of multiple languages is perceived by many participants in educational processes (politicians, administrators, leaders, teachers, and the public in general) in mainly two ways: as a "deficit" that needs to be addressed so that learners become proficient in the use of the language of instruction; or as a "scaffold" or a "resource" to support the process of learning mathematics by drawing from multiple languages.

Dilemmas and tensions have been identified in multilingual mathematics classrooms. For example, Adler (2001) identifies three interrelated "dilemmas" that lie at the heart of teaching mathematics in multilingual classrooms (p. 15). Barwell (2009) highlights at least three 'tensions" that arise in the context of multilingual mathematics classrooms (p. 60). Prediger, Clarkson, and Bose (this volume) propose a framework for bringing together three hitherto disparate processes: "code-switching [...], transitions between informal and academic (mathematical) forms of language within a given language, and transitions between different mathematical representations. Exploring the overlap between these three ideas and in particular by articulating their interconnections, new insights and implications are gained." These theoretical concepts show quite categorically the mutually constitutive relationship of language and development of mathematics knowledge.

Studies have also looked at the politics of language dynamics as they play out in classrooms illustrating the higher prestige of language of instruction as compared to the learners' home/dominant language with a consequence that official teaching and learning processes are conducted in the language of instruction while learners' language is relegated to a secondary status (Halai, 2009; Setati, 2005; Valero & Pais, 2012).

Collectively, the studies noted above have made a significant contribution in drawing attention to specific issues of multilingualism in teaching and learning mathematics. However, these in-depth qualitative mainly small-scale studies do not necessarily transcend the research rationalities of the large-scale quantitative studies. Indeed, the double effect of power is reinforced in these small-scale qualitative studies as well, because the underpinning philosophical and epistemological assumptions also take as normative the framework of language of instruction. The implication is that the multilingual learner who does not necessarily share the language of instruction as a dominant language is positioned as deficient. Moreover, languages are often seen from a structuralist paradigm as disparate bounded systems with operating distinctions on linguistic categories.

For example, in a detailed analysis of how research in mathematics education discriminates on the basis of language both within the community of research and practice, Barwell (2003) examined the pioneering work of Adler (1997, 2001) which studied in-depth a short teaching episode drawn from a larger study on teachers' knowledge of their practices in a multilingual secondary school mathematics classroom in South Africa. The teaching episode involved "Sue", a white, well-qualified, English-speaking, teacher who used a "participatory-inquiry approach" to teaching mathematics. In its conceptualization of the problem and process of research, this study has problematized the exclusive use of the language of instruction in teaching and learning and opened up spaces for use of additional languages in the classroom. It provides evidence of the issues related to "communicative competence" of learners and its implications for their learning. However, according to Barwell (2003), what remains largely uninterrogated is what constitutes mathematics and what constitutes communicative competence in mathematics from a multiplicity of perspectives. The study assumes a universal notion of mathematics built on the western/northern conceptions of mathematical thinking (e.g., explanation, reasoning, and justification), expressed in the English language. In doing so, the learners whose mother tongue is not English are positioned inadvertently as deficient. Barwell (2003) goes on to maintain that "[t]he privileging of English linguistic practices or meanings may therefore perpetuate a prevailing attitude in Sue's classroom, with the potential for teaching and research to devalue other languages and linguistic practices" (p. 41).

Drawing on a larger research project, Setati (2005) closely examined the complex relationship between language and mathematics education through an in-depth study of a lesson situated in a primary school classroom in South Africa. The study highlighted the political dynamics of language use in the classroom and found that the language that got official recognition in the classroom was invariably the language of the powerful elite or, in the case of post-colonial contexts, the language of the colonizers. Through a close scrutiny of classroom processes, Setati documents that multiple language use is a resource that the learners draw upon in the course of learning mathematics. However, invariably in the dynamics of classrooms, learners proficient in English are positioned as advantaged as compared with the learners who are learning mathematics and English simultaneously. Because of the theoretical and methodological positioning of the paper, there is a dichotomy between the home language of the learners and the language of teaching and learning. For example, the analytic framework employs a variety of discourses including mathematical discourse, nonmathematical discourse, procedural discourse, and conceptual discourse. However, it is the home language (Setswana) that is positioned as the language of conceptual discourse and English as the language in which procedural discourse took place, thereby creating a dichotomy. An implication of such a dichotomy is that learners who were learning mathematics alongside learning English were actually engaged in a procedural discourse in learning mathematics (p. 461).

On the basis of a detailed review of studies in multilingual mathematics in Africa, Setati, Chitera, and Essien (2009, p. 75) validate that research in mathematics education in multilingual settings mainly in South Africa has created a dichotomy between learning in English and learning in the home languages, giving an impression that the use of the learners' home languages for teaching and learning must necessarily exclude or be in opposition to English. Elsewhere, Setati, Molefe, and Langa (2008) expose:

> [T]hree prevalent dichotomies in research on teaching and learning mathematics in multilingual classrooms. First, is the dichotomy between using English as LoLT (language of learning and teaching) as opposed to using the learners' home language(s) as LoLT. Second, is the dichotomy about drawing on socio-political perspectives when analysing interactions in multilingual mathematics classrooms as opposed to drawing on cognitive perspectives. The third dichotomy is about gaining access to mathematical knowledge as opposed to access to English. (p. 10)

The dichotomies highlighted above have implications for understanding the process of teaching and learning from a situated sociocultural perspective. Pakistan is a case that illustrates well the dichotomies noted above. As noted in Halai (2009), it is a linguistically diverse country with Urdu as the country's national language. Urdu is the primary language of less than 10 % of the population and English is the medium of instruction in many private schools. Halai (2007, 2009) undertook observation of two small groups of students learning mathematics in an English-medium classroom and showed that language differences among the learners created new opportunities and challenges through the use of linguistic devices such as code-switching, code-mixing, and translation. Issues are noted for the learners when inappropriate translation of key mathematical terms does not enable the learners to appropriate the mathematical intent of the tasks:

> [T]he practices that students engaged in as they undertook mathematics showed movement between the language of instruction and their own language. This movement across languages involved a demonstrated need on the part of the learners to understand the language structures, grammar and vocabulary of the language of instruction. It also involved translation, which is a nuanced and complex process. Due to these complexities, questions arise about the role of code-switching in aiding the process of learning mathematics. (Halai, 2009, p. 48)

An implication of these dichotomies is that they privilege one perspective of the learner, mathematics education and language, and invariably it is the perspective of the dominant culture.

An issue is that imposition of a language and mathematics of one culture on learners who do not necessarily own it, positions the learner always as disadvantaged and "having to catch up". Language is not just a conduit of communication; embedded within it are the ways of thinking and values of people. For example, Schütte and Kaiser (2011) note that almost one-third of the students in German schools have a "migration background" and are provided support to *improve their linguistic abilities* in the German language. Looking closely at a selected classroom episode on teaching of LCM (Lowest Common Multiple), these authors illustrate the challenges for the learners who are expected to negotiate the implicit rules of the formal linguistic register or that of school mathematics concepts. Elsewhere, employing video observation of a lesson on complex fractions in a complementary school where Farsi and English are used regularly, Farsani (in press) focused on the experience of learning mathematics of bilingual, Farsi/English-speaking students of Persian heritage in the United Kingdom. Significantly it showed that complex fractions were simplified and solved in Farsi by drawing on the idiom "door dar door, nazdik dar nazdik" [far by far, near by near] (Farsani, in press). This idiom refers to the process of simplifying a complex fraction. It means the product of the two numbers furthest apart over the product of the two numbers closest to the main division line in the centre. Farsani notes that the use of this idiom opened up possibilities of learning for students of Persian heritage as they could associate with it. He raises the question of what counts as mathematics in different languages, and whether having access to more than one linguistic resource creates one or more perspectives in thinking and speaking mathematically.

The studies noted above document the challenges and issues for the learner in classrooms where they do not own the language. However, the studies do not question the almost omnipotent nature of school mathematics. The research rationalities thus honor the position of the current mathematical thought conveyed through the dominant language so that inadvertently other mathematics cultures are subjugated.

Barton (2009), through an extensive investigation, challenges the idea that mathematics is the same for everyone and goes on to argue that mathematical thought embedded in minority languages should be explored. This type of argument, however, is seldom taken in the studies noted above, which assume, through different means, that different pupils at the end will have to conform to what is legitimately prescribed as school mathematics by the curriculum.

From the discussion so far we infer that research in mathematics teaching, learning, and teacher education has been constituted by, and in turn, has constituted a dual meta-narrative: one of language as a well-defined and intact system, and the other of one school mathematics from a singular universal perspective. Thus the multilingual and multi-cultural learner remains invariably positioned as "deficient" against the power effects of this dual meta-narrative.

15.5 Concluding Remarks

In this chapter, we have focused on the research rationalities made visible in published research papers studying mathematics teaching and learning in multilingual classrooms. We have interrogated the discourses around the learner, the teaching and learning of mathematics, and language. Two types of studies were examined: large-scale, international comparative studies and small-scale, qualitative studies. The former constituted the logic of achievement gaps, and the latter the logic of classroom research. Both the theories and methods deployed by researchers contribute to the creation of the differentiation of multilingual/multicultural children. Research contributes to the construction of the idea of the "deficient multilingual child". It also provides the knowledge-based technologies of teaching and learning to engineer how multilingual/multicultural children need to be helped in order diminish the gap between their position and that of dominant cultural groups. Finally, in research, language is often portrayed as a static possession that learners have or do not have, and that allows classifying them in relation to the language use of the mainstream groups of society presented as proficient. It could be concluded that the research rationalities, as evident in the emerging body of knowledge in the field of mathematics education, have contributed to the construction of a deficient multilingual learner. This is through several unquestioned or unchallenged assumptions, implicit in meta-narratives of mathematics, mathematics teaching and learning, language and the learner him/herself. Positioned thus, educational research becomes a tool for perpetuating the values, worldviews, and ways of knowing and being of the dominant culture.

However, from a cross-disciplinary and poststructuralist perspective epistemological spaces are emerging for studying mathematics teaching and learning in multilingual and multicultural settings. For example, linguists such as Street (2003) propose literacy (and by implication language) as a "situated social practice" embedded in cultural and social contexts with concomitant power and authority relationships. Elsewhere, Makoni and Pennycook (2006) challenge the current discourse of bi/multilingual education with associated notions such as code-switching, as reproducing the historical logic of classification on which mainstream linguistic thought was built, so that "multilingualism may therefore become pluralization of monolingualism" (p. 22). They argue strongly for a renewed view that focuses on linguistic practices and people's transitions of language for the purpose of communication.

A strong implication of this theoretical positioning is to problematize the assumption that teachers and learners bring clearly defined systems of language into classrooms, because language in practice is fluid, moves across boundaries and takes meaning in context. Taking such a poststructuralist view would provide an approach to rethink the concepts with which we as researchers gaze on mathematics classrooms with multiple languages, as well as to democratize the educational process and contribute to greater equality and opportunity.

To conclude, the discussion in this chapter suggests that understanding and improving the quality of mathematics education, especially the classroom processes of teaching and learning necessarily goes beyond the search for measureable and numerical quantitative data. A more complex and nuanced approach is recommended to study the quality of mathematics learning as part of a broader reform initiative in education and development. Research rationality cannot be conceptualized without a deeper questioning of philosophical, epistemological, and ontological assumptions that underpin the traditional norms of what constitutes mathematics and by implication mathematics education research.

References

Adler, J. (1997). A participatory-inquiry approach and the mediation of mathematical knowledge in a multilingual classroom. *Educational Studies in Mathematics, 33*(3), 235–258.

Adler, J. (2001). *Teaching mathematics in multilingual classrooms*. Dordrecht, The Netherlands: Kluwer.

Arribas-Ayllon, M., & Walkerdine, V. (2008). Foucauldian discourse analysis. In C. Willig & W. Stainton-Rogers (Eds.), *The SAGE handbook of qualitative research in psychology* (pp. 91–108). London: Sage.

Barton, B. (2009). *The language of mathematics: Telling mathematical tales*. New York: Springer.

Barwell, R. (2003). Linguistic discrimination: An issue for research in mathematics education. *For the Learning of Mathematics, 23*(2), 37–43.

Barwell, R. (2009). Multilingualism in mathematics classroom: An introductory discussion. In R. Barwell (Ed.), *Multilingualism in mathematics classrooms: Global perspectives* (pp. 1–11). Bristol, England: Multilingual Matters.

Biesta, G. (2005). Against learning. Reclaiming a language for education in an age of learning. *Nordisk Pædagogik, 25*(1), 54–55.

Brown, C. L., Cady, J. A., & Lubinski, C. A. (2011). The effects of poverty and language on mathematics achievement for English language learners. In B. Atweh, M. Graven, W. Secada, & P. Valero (Eds.), *Mapping equity and quality in mathematics education* (pp. 393–406). New York: Springer.

Coleman, J. S. (1972). Coleman on the Coleman report. *Educational Researcher, 1*(3), 13–14.

Farsani, D. (in press). Complementary functions of learning mathematics in complementary schools. In A. Halai & P. Clarkson (Eds.), *Teaching and learning mathematics in multilingual classrooms: Issues for policy, practice and teacher education*. Rotterdam, The Netherlands: Sense.

Gamoran, A., & Long, D. A. (2006). *Equality of educational opportunity: A 40-year retrospective* (WCER Working Paper: No. 2006-9). Madison, WI: Wisconsin Center for Educational Research.

Gorgorió, N., & Planas, N. (2001). Teaching mathematics in multilingual classrooms. *Educational Studies in Mathematics, 47*(10), 7–33.

Gutiérrez, R., & Dixon-Román, E. (2011). Beyond gap gazing: How can thinking about education comprehensively help us (re)envision mathematics education? In B. Atweh, M. Graven, W. Secada, & P. Valero (Eds.), *Mapping equity and quality in mathematics education* (pp. 21–34). New York: Springer.

Gutstein, E. (2007). Multiple language use and mathematics: Politicizing the discussion. *Educational Studies in Mathematics, 64*(2), 243–246.

Hacking, I. (1999). *The social construction of what?* Cambridge, MA: Harvard University Press.

Halai, A. (2007). Learning mathematics in English medium classrooms in Pakistan: Implications for policy and practice. *Bulletin of Educational Research, 29*(1), 1–15.

Halai, A. (2009). Politics and practice of learning mathematics in multilingual mathematics classrooms: Lessons from Pakistan. In R. Barwell (Ed.), *Multilingualism in mathematics classrooms: Global perspectives* (pp. 47–62). Bristol, England: Multilingual Matters.

Kanes, C., Morgan, C., & Tsatsaroni, A. (2014). The PISA mathematics regime: Knowledge structures and practices of the self. *Educational Studies in Mathematics, 87*(2), 145–165.

Lubienski, S. T. (2003). Celebrating diversity and denying disparities: A critical assessment. *Educational Researcher, 32*(8), 30–38.

Makoni, S., & Pennycook, A. (2006). *Disinventing and reconstituting languages*. Clevedon, England: Multilingual Matters.

Moschkovich, J. (2007). Using two languages when learning mathematics. *Educational Studies in Mathematics, 64*(2), 121–144.

Neidorf, T. S., Binkley, M., Gattis, K., & Nohara, D. (2006). *Comparing mathematics content in the National Assessment of Educational Progress (NAEP), Trends in International Mathematics and Science Study (TIMSS), and Program for International Student Assessment (PISA), 2003 Assessments (NCES 2003–2009)*. Washington, DC: US Department of Education.

Norén, E. (2008). Bilingual students' mother tongue: A resource for teaching and learning mathematics. *Nordic Studies in Mathematics Education, 13*(4), 29–50.

Norén, E. (2011). Students' mathematical identity formations in a Swedish multilingual mathematics classroom. *Nordic Studies in Mathematics Education, 16*(1–2), 95–113.

OECD. (2006). *Where immigrant students succeed: A comparative review of performance and engagement in PISA 2003*. Paris: OECD.

OECD. (2012). *Education at a glance 2012: Highlights*. Paris: OECD.

Pais, A., & Valero, P. (2011). Beyond disavowing the politics of equity and quality in mathematics education. In B. Atweh, M. Graven, W. Secada, & P. Valero (Eds.), *Mapping equity and quality in mathematics education* (pp. 35–48). New York: Springer.

Pais, A., & Valero, P. (2012). Researching research: Mathematics education in the political. *Educational Studies in Mathematics, 80*(1), 9–24.

Parker, J. H. D. (2000). Exploring the relationship between subject knowledge and pedagogic content knowledge in primary teachers' learning about forces. *International Journal of Mathematical Education in Science and Technology, 22*(1), 89–111.

Parks, A. N. (2009). Doomsday device: Rethinking the deployment of the 'achievement gap' in equity arguments. *For the Learning of Mathematics, 29*(1), 14–19.

Popkewitz, T. S., & Brennan, M. (1997). Restructuring of social and political theory in education: Foucault and a social epistemology of school practices. *Educational Theory, 47*(3), 287–313.

Schneider, B., Martinez, S., & Owens, A. (2006). Barriers to educational opportunities of Hispanics in the United States. In M. Tienda & F. Mitchel (Eds.), *Hispanics and the future of America: Panel on Hispanics in the United States* (pp. 179–227). Washington, DC: The National Academies Press.

Schütte, M., & Kaiser, G. (2011). Equity and the quality of the language used in mathematics education. In B. Atweh, M. Graven, W. Secada, & P. Valero (Eds.), *Mapping equity and quality in mathematics education* (pp. 237–251). New York: Springer.

Setati, M. (1998). Code-switching in a senior primary class of second language learners. *For the Learning of Mathematics, 18*(2), 114–160.

Setati, M. (2005). Teaching mathematics in primary multilingual classroom. *Journal for Research in Mathematics Education, 36*(5), 447–466.

Setati, M., & Adler, J. (2000). Between languages and discourses: Language practices in primary multilingual mathematics classrooms in South Africa. *Educational Studies in Mathematics, 43*, 243–269.

Setati, M., Chitera, N., & Essien, A. (2009). Research on multilingualism in mathematics education in South Africa, 2000–2007. *African Journal of Research in Mathematics Science and Technology Education, 13*(1), 64–79.

Setati, M., Molefe, T., & Langa, M. (2008). Using language as a transparent resource in the teaching and learning of mathematics in a grade 11 multilingual classroom. *Pythagoras, 67*, 14–25.

Sims, V. M. (1930). The need for objective measurement of the socio-economic status of the child. *The Phi Delta Kappan, 13*(1), 10–15.

Street, B. (2003). What's 'new' in new literacy studies? Critical approaches to literacy in theory and practice. *Current Issues in Comparative Education, 5*(2), 77–91.

Then, D. C., & Ting, S. H. (2009). A preliminary study of teacher code switching in secondary English and Science in Malaysia. *Teaching English as a Second Language Electronic Journal, 13*(1), 1–17.

Valero, P. (2008). Discourses of power in mathematics education research: Concepts and possibilities. *PNA, 2*(2), 43–60.

Valero, P. (2013). Mathematics for all and the promise of a bright future. In B. Ubuz, Ç. Haser, & M. A. Mariotti (Eds.), *Proceedings of the 8th Congress of European Research in Mathematics Education (CERME 8)* (pp. 1804–1814). Antalya, Turkey: Middle East Technical University.

Valero, P., & Pais, A. (2012). Mathematics education between utopia and reality: Examining research in contexts of conflict, poverty and violence. In A. Halai & D. Wiliam (Eds.), *Research methodologies from the 'South'* (pp. 159–177). Karachi, Pakistan: Oxford University Press.

Correction to: Mathematics Education and Language Diversity

Correction to:
R. Barwell et al. (eds.), *Mathematics Education and Language Diversity*, New ICMI Study Series,
https://doi.org/10.1007/978-3-319-14511-2

The original version of the book was published in 2015 with exclusive rights reserved by the Publisher. As of March 2021 it has been changed to an open access publication: © The Editor(s) (if applicable) and The Author(s) 2021.
All chapters in the book are licensed under the terms of the Creative Commons Attribution-NonCommercial-NoDerivatives 4.0 International License.
Any third party material is under the same Creative Commons license as the book unless specified otherwise below.

The updated version of this book can be found at https://doi.org/10.1007/978-3-319-14511-2

R. Barwell et al. (eds.), *Mathematics Education and Language Diversity*,
New ICMI Study Series, https://doi.org/10.1007/978-3-319-14511-2_16

ICMI Study 21 Discussion Document: Mathematics Education and Language Diversity

Prepared by the International Programme Committee Released: December 2009

Executive Summary

Around the world, mathematics is learned and taught in situations of language diversity. Whether through historical multilingualism, migration, colonization, globalisation or other factors, mathematics classrooms frequently involve multiple language use, learning through second or additional languages, learning through minority or oppressed languages, or through majority or dominant languages. Increasing recognition and awareness of this long-standing reality have led to a growing body of research that seeks to understand the relationship between different facets of language diversity and mathematics learning and teaching. It is time to critically review this work, consider implications for mathematics classroom practices, and set an agenda for future research.

The principal aims of ICMI Study 21 "Mathematics education and language diversity" are:

- To gather together a community of researchers who are currently addressing issues of language diversity as they relate to mathematics education;
- To reflect on the current state of research on these issues and propose a research agenda for the future;
- To disseminate findings from research to date and issues for future work to the wider mathematics education research community and to practitioners.

In order to fulfil these aims and following the tradition and example of past ICMI Studies, the International Programme Committee is planning and conducting the following activities:

- The publication of this Discussion Document lays out the background, theoretical foundations, focus, rationale, and orientation for the Study, along with sample research questions;
- The organization of an invitational "working conference" to gather researchers who work on language diversity issues in mathematics education;

R. Barwell et al. (eds.), *Mathematics Education and Language Diversity*, New ICMI Study Series, https://doi.org/10.1007/978-3-319-14511-2

- The preparation and publication of a Study Volume to present the 'state of the art' and set out an agenda for future research;
- The preparation and publication of materials for practitioners to share key findings, concepts and issues and thereby inform mathematics classroom practice.

Note that, although this work is referred to as a 'study', neither ICMI nor the IPC will be conducting an empirical research project.

Discussion Document

Focus of the Study

Across the world, the teaching and learning of mathematics occurs in contexts of linguistic and cultural diversity. How do we work with, and work within, this diversity to enhance the learning and teaching of mathematics? In particular, how can the range and complexity of learners' language backgrounds be most effectively used to promote their mathematical learning? These questions are central to this Study.

Research within the field of mathematics education uses a variety of words to characterise language diversity, including, but not limited to, 'multilingual', 'bilingual', 'second language', 'additional language' or 'minority language'. In this document, for convenience the term 'multilingualism' is used.

All of the words mentioned above involve, however, assumptions about the nature of language diversity. These assumptions are often problematic. The Study will, therefore, also seek to examine how constructs about language, discourse and multilingualism that have their origins in language studies are pertinent to mathematics education. Equally, the Study will consider how mathematics educators can conceptualize language diversity in ways that are consistent with the major objectives of mathematics education research and practice.

Given the growth of research in this area and the increasing importance and significance attached to language diversity and multilingualism, this ICMI Study is timely: now is an appropriate time to build on the collective wisdom that exists internationally. ICMI is well placed to support and offer expert guidance on the teaching and learning of mathematics in the context of language diversity in our classrooms.

Rationale for the Study

One of the most important goals of mathematics education is to facilitate successful mathematical learning. This learning occurs in complex linguistic environments; complexity derives from the multiple language backgrounds, mathematical languages and semiotic systems that are always present. In some cases, learners and teachers feel at home in this environment, perhaps because all participants share similar language backgrounds. In other cases, learners and teachers may feel that they face ‘language barriers’. These barriers may vary from a lack of basic vocabulary to other, less obvious social, cultural or political issues. Whether learners and teachers feel at home in a linguistically diverse environment or not, they routinely find ways to discuss and learn mathematics.

Multilingualism is steadily rising up the political and educational agenda internationally, driven by a combination of economic and political integration. For example, both the European Union and the African Union have policies for the development and management of multilingualism. Economic change leads to migration of people and hence of languages, as well as new demands for foreign language skills, all of which have an impact on education in general, and mathematics education in particular.

In many countries, a global language, such as English, is linked to dominant discourses through high-status jobs or access to the dominant class. As a result, education systems come under pressure to use the global language for learning and teaching. Language in education policies can lead to a class divide in schooling generally, and mathematics education in particular, because students who have learnt mathematics in their local languages are expected to “switch” to the global language at some point in their education. Many students may be ill-prepared to engage in such change.

The three previous paragraphs highlight different dimensions of language diversity in different contexts: that of classroom practices, that of international policies, and that of national language policies. There are inconsistencies and contradictions between these dimensions. These inconsistencies emerge in relation to the somewhat simplistic, often incompatible, views of language evident across these dimensions. A better theoretical understanding of language as it relates to mathematics education is needed, a perspective that is congruent with these and other dimensions.

Framing Multilingualism

Multilingualism is a contested concept. Furthermore, different theoretical lenses offer particular ways of seeing how multiple language resources interact in any context of practice. Quite frequently multilingualism is defined in terms of **plural monolingualism** – the idea that multiple languages are used discretely and distinctly, even where individuals may be speakers of several languages. Hence boundaries between languages are sharper. Languages may be associated with particular activities, institutions or contexts on a language-by-language basis, so that one language may be seen as ‘for school’, while another language is seen as ‘for home’. An alternative approach is to think in terms of **hybridized multilingualism** – the idea that languages are *interwoven* within the fabric of social life, to the extent that the boundaries between them can be hard to clearly identify (and, indeed, conceptually problematic). Of course, the nature of the hybridity, the emphasis given to different component languages, the values associated with 'purity' or qualifications in one or another language vary from region to region, as well as social context. The hybrid language mix used and valued in mathematics classrooms may be different from the mix used and valued in bars or tea-houses. Of course, the above terms are somewhat idealised, in that language use is more fluid, complex and subtle than a simple two-sided distinction. Besides these two views, there are still other understandings that exist and that could be productive for mathematics education.

In the same way that multilingualism has different meanings, many other related terms such as bilingual, indigenous, native speaker or language learner are also problematic and there is no consensus about their use, meaning or value. We cannot, however, avoid using such terms and they have helped researchers identify the diversity of settings in which multilingualism is apparent. Some examples of these settings are:

- Multilingual societies where people are used to using several languages (e.g. South Africa);
- Multilingual societies with more than one official language, one of them with a higher status than others (e.g. Catalonia; Wales);
- Bilingual societies in which two languages are sometimes used in support of minority language(s) (e.g. Peru) or in which the use of minority languages in the classroom are restricted by law (e.g. some states in the USA);
- Societies previously seen as monolingual where immigration has made language diversity more salient (e.g. Europe, Australia);
- Societies in which a foreign language is taught through subjects like mathematics (e.g. Czech Republic);
- Societies in which mathematics education for indigenous language speakers is conducted in a majority colonising language (e.g. Africa, South America);
- Societies in where languages of instruction change across primary, secondary, and tertiary levels and/or have changed during specific historical moments (e.g. Pakistan, Algeria).

Mathematics Education and Multilingualism

Until the 1980s multilingualism has received little attention from many researchers in mathematics education. However, students and teachers regularly make use of different languages available to them in their mathematical work. In recent years, researchers have examined several aspects of this situation, including the role of multilingualism in students' learning and understanding of mathematics, the relationship between multilingualism and mathematical attainment and the ways in which mathematics teachers deal with multilingualism in their classrooms.

The teaching and learning of mathematics in multilingual contexts is thus a substantive issue in mathematics education, and one around which there has been recent but considerable growth in research and practical knowledge. Participants in this field span a wide range of countries and contexts on every continent, as indicated above.

Research to date on multilingualism and the teaching and learning of mathematics has served to highlight some of the challenges faced by teachers and learners, as well as to identify some successful teaching strategies and approaches to learning. Researchers have consistently argued that multilingualism is potentially an advantage for learning mathematics and have sought to refute the long-standing assumption that multilingualism is a *problem* in mathematics classrooms. Most studies have been located in single classrooms or in a small number of classrooms and have generally focused on single age groups, although there are some exceptions.

Not enough is known about the specific benefits, challenges and interventions appropriate for different levels of education (pre-primary, primary, secondary and tertiary). And little is known about whether and how the demands of multilingualism and mathematics change with different domains of mathematics (e.g. number, algebra, probability). Tackling these and other issues would lead to valuable additional knowledge with implications for practitioners and policy makers.

Given the widespread occurrence of multilingualism, it is surprising that it is rarely mentioned in much published research in mathematics education, mathematics education policy or mathematics curricula. Where multilingualism is mentioned, it is often as an after-thought, a challenge, or a problem to be overcome, but mostly just silence. Multilingualism may therefore be an unmentioned feature, an unmarked factor in much of the work of researching and organising mathematics education. Moreover, this work may also have contribution to make to research and practice in what are considered "monolingual" settings. Thus, we hope this Study will be of interest and value to all those working in mathematics education in general, not just those with an interest in multilingualism or working in multilingual classrooms.

Orientation

As the preceding sections show, thinking about language diversity in the learning and teaching of mathematics involves a good deal of complexity. In this section, we highlight some of the different perspectives from which this learning and teaching can be examined. Each perspective reveals slightly different issues, features and questions; each highlights a different part of the complexity. We see these different perspectives as relevant and important and we expect to see them informing the Study as it progresses.

The mathematical aspects of learning and teaching mathematics in multilingual classrooms: As noted above, the existing literature focuses on multilingualism and the learning and teaching of mathematics in general. But there has been little systematic focus on whether and how the demands of multilingualism and mathematics change with different domains of mathematics. In the mathematics classroom, we are not teaching and learning undefined and vague objects and processes, but *mathematical* objects and processes, with their own differing and specific natures and structures. Is research in this area sufficiently focused on the specificity of the mathematics in our on-going research discourse? Is it the same to have multilingualism in a class where the teacher is introducing objects such as the idea of a "vertical asymptote," in comparison to a class where the teacher is introducing non-mathematical objects, mathematical objects from other domains, or mathematical objects being presented with non-symbolic language? How do we include consideration of the intersection of the language of mathematics with the mathematics itself in the multilingual classroom? And finally, of course, we must be clear about the mathematical practices that are intervening in our empirical studies. Various analytic tools may contribute to this aspect, including, for example, work on semiotics or functional linguistics.

The psychological aspects of learning and teaching mathematics in multilingual classrooms: Any research concerning issues like learning, understanding, thinking, feeling or knowing necessarily involves some kind of psychological perspective. Many mathematics educators have been informed to a greater or lesser extent by socio-cultural psychology, derived from the work of Vygotsky, although this is by no means the only perspective. Equally, research on the psychology of language learning, processing and use may offer useful insights, such as research in the psycholinguistics of bilingualism or multilingualism.

The sociolinguistic aspects of learning and teaching mathematics in multilingual classrooms: Multilingual mathematics classrooms do not exist in a vacuum; they are influenced by the society in which they are situated. Sociolinguistics offers ideas on and insights into the nature of language use in multilingual settings. Such work includes, for example, research on code-switching, as well as research into the distribution of different languages and language varieties across different groups, professions, locations, etc. Such work is relevant to multilingual mathematics classrooms, where the use of multiple/mixed languages is potentially a relevant feature of interaction, and which, therefore, needs to be carefully conceptualised.

The political aspects of learning and teaching mathematics in multilingual classrooms: Languages are implicated in issues of power and access, whether to education, employment or positions of authority. Language use is often a marker of social class, with English often (though not always) having a particular cachet. The politics of language influence multilingual mathematics classrooms in many ways, including, for example, the choice of 'official' language, the desire of students or their parents to be taught in a particular language, or the challenges of accommodating competing languages. In some cases, language policy or practice in multilingual mathematics classrooms (including the mathematics curriculum) may be implicated in the maintenance or suppression of threatened languages or languages of marginalised groups. These languages may include 'non-standard' languages, such as creoles, or various kinds of 'street' language.

The cultural aspects of learning and teaching mathematics in multilingual classrooms: Language and culture are intertwined. In any culture, different quantitative and spatial relationships (i.e. mathematical) relationships and practices can be generated, organized and transmitted informally to solve immediate needs. In this way mathematics and language are incorporated into the core of the learning-by-doing processes of a community and are therefore part of what we call culture. From this standpoint, language and mathematics are generated within multiple cultural communities, which include, but are not limited to, academic mathematics. The transition from mathematical ideas embedded in learning-by-doing processes of any culture to more abstract mathematical objects is of particular interest.

The discursive aspects of learning and teaching mathematics in multilingual classrooms: The business of learning and teaching mathematics is conducted largely through interaction, including talk, as well as the use of written symbols, diagrams, charts and other texts. In multilingual classrooms, these various texts may involve the use of several languages. Discursive perspectives on this interaction can provide fine-grained analysis of how the joint construction of mathematical thinking is mediated by these various texts and languages. These aspects also relate mathematical language and discourse. Ways of communicating in a mathematics classroom to construct knowledge meanings and ideas are different from those in other domains or classrooms. While mathematical language and discourse may be expressed in the language of instruction, communication processes such as reading, writing and talking specifically about mathematics have their own features.

Of course, the above perspectives all overlap —they are ways of understanding the same situation. Indeed, at a meta-level, we also need to account for the epistemological orientation of this Study. As the range of perspectives summarised above shows, our work potentially draws on many areas of expertise, notably various branches of applied linguistics. Applied linguists do not often visit mathematics classrooms, and even when they do, they rarely have any mathematical training. We cannot, therefore, expect applied linguistics to provide 'off-the-shelf' concepts that we can straightforwardly apply. On the other hand, applied linguists can provide important expertise concerning the nature of language diversity in society. We need to make use of this expertise to enhance our own work. Ultimately, we hope that this work will lead to the development of tools and ideas that offer mathematics researchers, teachers and educators a way forward in developing teaching and promoting learning in multilingual mathematics classrooms. Of course, it could mean that ideas, tools or categories from linguistics are integrated with mathematics education and developed into new or modified ideas that help us see the way forward for mathematics teaching and learning in multilingual classrooms.

Themes for the Study

Provided below is a series of themes proposed by the International Programme Committee. Around each theme is a selection of questions that can serve as an initial focus for potential papers for the Study. These themes and questions are a starting point and could be adapted, modified or developed as the Study progresses and is finalised.

1. *Teaching mathematics in diverse language contexts*
   This theme focuses on language issues in the teaching of mathematics in different language contexts. The assumption here is that language issues that emerge in different contexts are not only shaped by the complexities of the language of mathematics, but are also shaped by the linguistic contexts in which mathematics is taught and learned:

   a) What strategies and resources can educators in multilingual classrooms use to teach mathematics effectively to learners who are in the process of learning the language of instruction?
   b) How do assessment and curriculum systems relate to mathematics teaching and language policies in diverse language contexts?
   c) Which current teaching practices are sensitive to the relationship between multilingualism and mathematics learning?
   d) What are the relationships between teaching language and teaching mathematics?

2. *Teacher education for diverse language contexts*
   The focus here is on issues in and for teacher education in diverse language contexts. An assumption is that teacher education principles and practices are rooted in the real world of the classrooms, and therefore must take into account the different language contexts in which mathematics is taught:

   a) Which current practices in teacher education are sensitive to the relationships between multilingualism and mathematics teaching and learning?
   b) What can be done to prepare teachers to teach mathematics effectively in multilingual classrooms?
   c) What kind of data from multilingual classrooms would be useful in designing teacher education programs?
   d) What knowledge and skills do teachers need to teach in multilingual classrooms and what are teachers' perspectives on this question?

3. *Researching mathematics teaching and learning in multilingual contexts*
   This theme focuses on the theories and methods that relate to doing research in multilingual contexts:

   a) What types of theories and methods enable the development of research in this area?
   b) What ethical issues arise in pursuing this kind of research and how can researchers address them?
   c) On what basis can researchers interpret the mathematical worlds of students who come from linguistic backgrounds with which they are not familiar?
   d) To what extent is mathematics education research sensitive to linguistic diversity?

4. *Mathematics, multilingualism and society*
   Diverse language settings reflect the broader social, cultural and political issues. When considering classroom dynamics, an assumption is that they are shaped by and go on to shape the broader social and political settings:
   a) To what extent and for what purposes does research in multilingual contexts need to address multicultural issues?
   b) What role does teaching mathematics in diverse language settings play in reproducing or challenging prevailing social patterns?
   c) How can researchers engage productively with policy makers involved in mathematics education to address language diversity?
   d) What is the relationship between the teaching and learning of mathematics in multilingual settings, and wider social discourses?
5. *Student mathematics learning and experiences in multilingual classrooms*
   This theme brings forth issues regarding students' learning and students' experiences of learning mathematics in different language contexts. An underlying assumption is that it is important to focus on learners in order to support them to learn mathematics effectively:
   a) What are the characteristics of students' mathematical discussions and explanations in different languages, in multiple classroom contexts, and in multiple mathematical domains?
   b) What are the demands on multilingual students learning mathematics in different mathematical domains (i.e. algebra, geometry, etc.) and at different ages?
   c) How do students themselves see and describe their experiences in multilingual mathematics classrooms?
   d) What are students' strengths and resources, what can we learn from successful students, and how can instruction build on these resources, strengths, and successes in linguistically diverse settings?

Participation in the Study

Design of the Study - The ICMI Study 21 on Mathematics education and language diversity is designed to enable researchers and practitioners around the world to share research, theoretical work, projects descriptions, experiences and analyses. It will consist of two components: the *Study Conference* and the *Study Volume*.

1. The ***Study Conference*** will be held in São Paolo, Brazil, on 16 - 20 September, 2011, the number of participants to be invited being limited to at most 120. It is hoped that the Conference will attract not only established researchers but also some "newcomers" to the field and mathematics teachers with interesting and refreshing ideas or promising work in progress, as well as participants from countries usually under-represented in mathematics education research meetings.

   Participation in the Study Conference is only by invitation, based on a submitted contribution. Proposed contributions will be reviewed and selections made according to the quality of the work, the potential to contribute to the advancement of the Study, with explicit links to the themes and related questions outlined in this Discussion Document, and the need to ensure diversity among the perspectives. Accepted papers will appear in the Conference Proceedings that will be published by ICMI as a CDROM, and will form the basis of the Study's scientific work.

An invitation to the Conference does not imply that an oral presentation of the submitted contribution will be made during the Conference, as the International Program Committee (IPC) may decide to organize it in other ways that facilitate the Study's effectiveness and productivity. This will be a working conference, every participant being expected to make a scientific contribution. We therefore hope that the participants will represent a variety of backgrounds, expertise, experience and nationalities that will lead to a suitable coverage of the Study theme, its different topics and the related questions.

It is the IPC's hope that the Conference will attract mathematics education researchers, mathematics teacher educators, policy makers and linguists who are interested and do work in the area of mathematics education and language diversity as well as mathematics teachers in multilingual classrooms.

Unfortunately an invitation to participate in the Conference does not imply financial support from the organisers, and participants should finance their attendance at the Conference. It is hoped that this invitation will help participants to get appropriate support from their own countries. Funds are being sought to provide partial support for participants from non-affluent countries, but the number of such grants will be limited.

2. The ***Study Volume,*** a post-conference publication, will appear in the New ICMI Study Series (NISS), published by Springer. Acceptance of a paper for the Conference does not ensure automatic inclusion in this book. The Study Volume will be based on selected contributions as well as on the outcome of the Conference. The exact format of the Study Volume has not yet been decided but it is expected to be an edited coherent book that can hopefully serve as a standard reference in the field for a foreseeable future. A report on the Study will be presented during the 12th International Congress on Mathematical Education (ICME-12), to be held in Seoul, Korea, on July 8-15, 2012.

Call for Contributions - The International Programme Committee hereby invites individuals or groups to submit original contributions on specific questions, problems or issues related to the topic of the Study for consideration by the Committee. A submission should represent a significant contribution to knowledge about the Study topic and may address questions from one or more of the Study themes (see section F above), or further issues relating to these, but it should clearly identify its primary focus. The IPC welcomes high-quality proposals from researchers and practitioners who can make solid practical and scientific contributions to the Study. New researchers in the field, teachers and participants from countries under-represented in mathematics education research meetings are especially encouraged to submit contributions. To ensure a rich and varied scope of resources for the Study, participation from countries with different economic levels or with different cultural heritage and practices is encouraged.

Those who would like to participate should prepare a paper addressing matters raised in this document or other issues related to the topic of the Study. Papers concerning work that is ongoing or yet to be carried out are also welcome. Research questions should be carefully stated and expected results should be formulated, if possible with reference to earlier and related work. These papers should be submitted **no later than 30 November 2010**, to both co-chairs of the Study by e-mail. All such documents will be regarded as input to the planning of the Study Conference and will assist the IPC in making decisions on the invitations, to be issued no later than 28 February, 2011.

General Guidelines on Submission of Papers
The papers must be written in English.

The format of papers must be as follows:

- A maximum of 8 pages, including references and figures.
- Times New Roman 14-point font, 16-point line space, and 6 points between paragraphs; occupying a frame of 170 by 247 mm.
- The title (in 16-point bold capitals), author(s) name(s) (in 14-point bold), and affiliation(s) of author(s) (in 14-point italics) should appear in this order centered, all in Times New Roman.
- The paper must begin with an abstract of up to 10 lines, single-spaced, in italics.
- Video clips may be referred to in the paper and a link should be provided.

Further technical details about the format of submissions will be available on the Study website as soon as it is launched on 28 February 2010. This website will be progressively updated with all study and travel information.

Study Timeline

DATE	*ACTIVITY*
28 February 2010	Launch of study website
30 November 2010	Deadline for submission of conference papers
28 February 2011	Letters of acceptance of papers and invitation to authors
15 April 2011	Closing date for registration and submission of final versions of papers
16 - 20 September 2011	Study Conference *(in São Paolo, Brazil)*

Inquiries

Inquiries on all aspects of the Study and suggestions concerning the content of the Study Conference should be sent to both co-chairs:

Mamokgethi Setati setatrm@unisa.ac.za or funkymaths@yahoo.co.uk
Maria do Carmo Santos Domite mcdomite@usp.br **or** mcdomite@gmail.com

Members of the International Programme Committee

Mamokgethi Setati (South Africa), Co-chair
College of Science, Engineering and Technology, University of South Africa
Maria do Carmo Santos Domite (Brazil), Co-chair
Faculdade de Educação, Universidade de São Paulo
Bernard R. Hodgson (Canada)
Département de mathématiques et de statistique,
Université Laval (Canada), *ex officio, Secretary-General of ICMI.*
Richard Barwell (Canada)
Mathematics Education Research Unit, Faculty of Education University of Ottawa
Phil Clarkson (Australia)
School of Education (Victoria), Australian Catholic University
Anjum Halai (Tanzania)
Institute for Educational Development, Eastern Africa Aga Khan University Dar es Salaam
Mercy Kazima (Malawi)
Department of Curriculum and Teaching Studies, Chancellor College, University of Malawi
Sinfree Makoni (USA)
African and African American Studies, Pennsylvania State University
Judit Moschkovich (USA)
Education Department, University of California, Santa Cruz
Núria Planas (Spain)
Faculty of Education, Universitat Autònoma de Barcelona
Paola Valero (Denmark)
Department of Education, Learning and Philosophy, Aalborg University
Martha Villavicencio (Peru)
Unit Graduate School of Education at the Universidad Nacional Mayor de San Marcos, Lima

Index

R. Barwell et al. (eds.), *Mathematics Education and Language Diversity*, New ICMI Study Series, https://doi.org/10.1007/978-3-319-14511-2